Smart Agriculture

Volume 16

Series Editors

The book series Smart Agriculture presents progress of smart agricultural technologies, which includes, but not limited to, specialty crop harvest robotics, UAV technologies for row crops, innovative IoT applications in plant factories, and big data for optimizing production process. It includes both theoretical study and practical applications, with emphasis on systematic studies. AI technologies in agricultural productions will be emphasized, consisting of innovative algorithms and new application domains. Additionally, new crops are emerging, such as hemp in U.S., and covered as well. This book series would cover regions worldwide, such as U.S., Canada, China, Japan, Korea, and Brazil.

The book series Smart Agriculture aims to provide an academic platform for interdisciplinary researchers to provide their state-of-the-art technologies related to smart agriculture. Researchers of different academic backgrounds are encouraged to contribute to the book, such as agriculture engineers, breeders, horticulturist, agronomist, and plant pathologists. The series would target a very broad audience – all having a professional related to agriculture production. It also could be used as textbooks for graduate students.

Pankaj Kumar • Ashish R. Warghat
Editors

Transforming Medicinal Plant Agriculture

Implementing Soil-less Systems and AI Innovations for Smart Farming

Springer

Editors
Pankaj Kumar
Department of Biotechnology
Dr. Yashwant Singh Parmar University of Horticulture and Forestry
Solan, Himachal Pradesh, India

Ashish R. Warghat
Fermentation and Phytofarming Technology Division
CSIR - Institute of Himalayan Bioresource Technology
Palampur, Himachal Pradesh, India

ISSN 2731-3476 ISSN 2731-3484 (electronic)
Smart Agriculture
ISBN 978-981-95-3962-8 ISBN 978-981-95-3963-5 (eBook)
https://doi.org/10.1007/978-981-95-3963-5

First Edition

This Springer imprint is published by the registered company Springer Nature Singapore Pte Ltd.
The registered company address is: 152 Beach Road, #21-01/04 Gateway East, Singapore 189721, Singapore

Preface

In recent years, the landscape of agricultural practices has undergone a transformative shift, driven by the convergence of advanced technologies and the urgent need for sustainable and efficient crop production systems. This book, *Transforming Medicinal Plant Agriculture: Implementing Soil-Less Systems and AI Innovations for Smart Farming*, presents a comprehensive account of how modern innovations are redefining the cultivation of medicinal plants. As global health increasingly relies on plant-derived therapeutics, the demand for high-quality medicinal plants has continued to rise. However, traditional soil-based agriculture often faces challenges such as overexploitation, illegal collection, soil degradation, labor intensiveness, and inconsistent yields, making it insufficient to meet this growing demand. In response, this book explores how soil-less systems such as hydroponics and vertical farming, when integrated with AI, robotics, sensor technologies, and other smart solutions, are revolutionizing the production of medicinal plants.

This volume brings together contributions from experts across biotechnology, agronomy, artificial intelligence, and smart farming. It begins with foundational insights into soil-less cultivation techniques, including vertical farming and controlled environment agriculture, and moves into specialized topics, such as interventions to combat salinity stress and improve physiological performance in medicinal plant production. The discussion expands to smart urban agriculture and elicitation strategies in hydroponics, followed by the integration of robotics in cultivation systems. Innovations such as nanotechnology have also been covered for their potential to boost plant growth and enhance the biosynthesis of secondary metabolites critical for medicinal efficacy. A significant portion of this book addresses the digital transformation of agriculture. Topics include smart monitoring systems using the Internet of Things (IoT), the application of AI in precision farming, and the integration of IoT with bioinformatics tools for nutrient analysis and climate-responsive management. These technologies collectively support data-driven decision-making and automation in soil-less farming environments.

Importantly, this book also considers the broader ecosystem that supports smart agriculture. Discussions on intellectual property rights highlight the legal and regulatory frameworks required to protect innovations in medicinal plant farming.

Additionally, we have explored entrepreneurship opportunities, business challenges, and effective marketing strategies for medicinal plants in the digital era. This book has aimed to provide a well-rounded synthesis that balances scientific depth with applied relevance. It will serve as a valuable resource for academic researchers, students, agronomists, technology developers, and agricultural entrepreneurs. By bridging scientific advancements with practical applications, we hope to support the global shift toward smart, sustainable, and scalable medicinal plant agriculture. We express our deep appreciation to all contributors for their expertise and dedication. It is our sincere belief that this work will foster innovation, inform practice, and inspire future research in the evolving field of smart farming.

Solan, Himachal Pradesh, India — Pankaj Kumar
Palampur, Himachal Pradesh, India — Ashish R. Warghat

Contents

About the Editors and Contributors

About the Editors

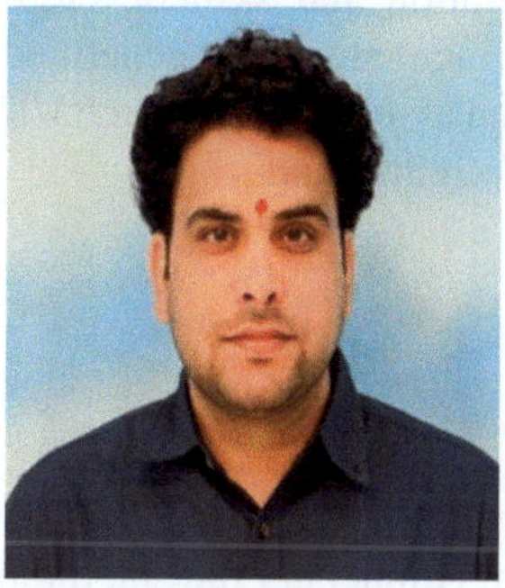

Pankaj Kumar is presently working as Assistant Professor (Biotechnology), Department of Biotechnology, Dr. Yashwant Singh Parmar University of Horticulture and Forestry, Solan, Himachal Pradesh, India. With more than nine years of specialized expertise in Medicinal plant biotechnology, Dr. Kumar has significantly contributed to understanding high altitude adaptation of medicinal plants, and secondary metabolite production through advanced cell and tissue engineering techniques. His research delves into the molecular mechanisms underlying metabolite production, utilizing cutting-edge comparative transcriptomics and metabolomics approaches. Additionally, Dr. Kumar has pioneered work in the hydro-aero cultivation of medicinal plants, focusing on their industrial applications. Dr. Kumar's academic credentials are further bolstered by his post-doctoral research at prestigious institutions such as the National Institute of Plant Genome Research and the Institute of Himalayan Bioresource Technology. He is a prolific author with over 70 research and review papers, 30 book chapters, and seven notable books, including "Plant Stress Memory: Methods and Protocols (Springer), "*Agricultural Biotechnology- Latest Research and Trends*" (Springer) and "*Ornamental Horticulture: Latest Cultivation Practices and Breeding*" (Springer). His innovative contributions have led to the publishing of five Indian patents, and he has been honored with numerous awards, such as the Young Scientists Fellowship Award-2021 by ICMR and the Scientist's Pool Scheme Award-2020 by CSIR.

Ashish R. Warghat Warghat is currently working as Principal Scientist at CSIR-Institute of Himalayan Bioresource Technology, Palampur. He did his Ph.D. in Biotechnology from the Defence Institute of High-Altitude Research (DIHAR), Defence Research and Development Organization, Leh-Ladakh. His research interests are particularly on High Altitude Himalayan Medicinal Plants with the intervention in the field of Aeroponic and Hydroponic farming for metabolite-enriched plant biomass production and understanding of the mechanism of nutrients towards growth and development. He has published more than 45 research papers in high reputed international journals. He has developed 10 most significant technologies in the field of smart agriculture and transferred it to the industries. He has also contributed significantly in the area of start-ups and entrepreneurs' generation and skill development activities and total of 12 incubatee were benefitted significantly. Currently, he is an associate editor in *BMC Biology* and *BMC Biotechnology* Journals.

Contributors

Aditi Sharma Fermentation and Phytofarming Technology Division, CSIR-Institute of Himalayan Bioresource Technology, Palampur, Himachal Pradesh, India

Academy of Scientific and Innovative Research (AcSIR), Ghaziabad, India

Akhil Rana Business Development and Marketing Unit, CSIR-Institute of Himalayan Bioresource Technology, Palampur, Himachal Pradesh, India

Ali Baba Eshawu School of Biotechnology, Faculty of Applied Sciences and Biotechnology, Shoolini University, Solan, Himachal Pradesh, India

Amit Kumar Fermentation and Phytofarming Technology Division, CSIR-Institute of Himalayan Bioresource Technology, Palampur, Himachal Pradesh, India

Academy of Scientific and Innovative Research (AcSIR), Ghaziabad, India

Amitabha Acharya Nanobiology Lab, Fermentation and Phytofarming Technology, CSIR-Institute of Himalayan Bioresource Technology, Palampur, Himachal Pradesh, India

Academy of Scientific and Innovative Research (AcSIR), Ghaziabad, India

Anchal Thakur Nanobiology Lab, Fermentation and Phytofarming Technology, CSIR-Institute of Himalayan Bioresource Technology, Palampur, Himachal Pradesh, India

Academy of Scientific and Innovative Research (AcSIR), Ghaziabad, India

Ashish R. Warghat Fermentation and Phytofarming Technology Division, CSIR-Institute of Himalayan Bioresource Technology, Palampur, Himachal Pradesh, India

Academy of Scientific and Innovative Research (AcSIR), Ghaziabad, India

Avika Subba Regional Cum Facilitation Centre Northern Region-I, NMPB, Ministry of Ayush, GOI, New Delhi, India

Research Institute in Indian Systems of Medicine (RIISM), Mandi, Himachal Pradesh, India

National Medicinal Plants Board, New Delhi, India

Chandresh Kumari School of Biotechnology, Faculty of Applied Sciences and Biotechnology, Shoolini University, Solan, Himachal Pradesh, India

Deepika Sharma Business Development and Marketing Unit, CSIR-Institute of Himalayan Bioresource Technology, Palampur, Himachal Pradesh, India

Harpreet Kaur Nanobiology Lab, Fermentation and Phytofarming Technology, CSIR-Institute of Himalayan Bioresource Technology, Palampur, Himachal Pradesh, India

Academy of Scientific and Innovative Research (AcSIR), Ghaziabad, India

Hemant Sood Department of Biotechnology and Bioinformatics, Jaypee Institute of Information Technology, Waknaghat, Himachal Pradesh, India

Hena Dhar School of Biotechnology, Faculty of Applied Sciences and Biotechnology, Shoolini University, Solan, Himachal Pradesh, India

Jaspreet Singh Orchid Laboratory, Department of Botany, Panjab University, Chandigarh, India

Karnika Thakur School of Biotechnology, Faculty of Applied Sciences and Biotechnology, Shoolini University, Solan, Himachal Pradesh, India

Kritika Sharma Artificial Intelligence for Computational Biology Lab, CSIR-Institute of Himalayan Bioresource Technology, Palampur, Himachal Pradesh, India

Academy of Scientific and Innovative Research (AcSIR), Ghaziabad, India

Mahinder Partap Agriculture Research Organization (ARO), Volcani Institute, Rishon LeZion, Israel

Manvi Sharma School of Biotechnology, Faculty of Applied Sciences and Biotechnology, Shoolini University, Solan, Himachal Pradesh, India

Marbi Ete School of Biotechnology, Faculty of Applied Sciences and Biotechnology, Shoolini University, Solan, Himachal Pradesh, India

Meenakshi Thakur Regional Cum Facilitation Centre Northern Region-I, NMPB, Ministry of Ayush, GOI, New Delhi, India

Research Institute in Indian Systems of Medicine (RIISM), Mandi, Himachal Pradesh, India

National Medicinal Plants Board, New Delhi, India

Mohammad Moghaddam Department of Horticultural Science, Faculty of Agriculture, Ferdowsi University of Mashhad, Mashhad, Iran

Nasrin Ferhadi Greenhouse and Controlled Environments Research Center (GCER), Horticultural Science Research Institute (HSRI), AREEO, Lahijan, Iran

Nedhi Jasrotia School of Biotechnology, Faculty of Applied Sciences and Biotechnology, Shoolini University, Solan, Himachal Pradesh, India

Neha Sharma Division of Crop Improvement and Seed Technology, ICAR-Central Potato Research Institute, Shimla, Himachal Pradesh, India

Novin Chamling Sikkim Regional Centre, G.B. Pant National Institute of Himalayan Environment, Gangtok, Sikkim, India

Pankaj Kumar Department of Biotechnology, College of Horticulture, Dr. Yashwant Singh Parmar University of Horticulture and Forestry, Solan, Himachal Pradesh, India

Phate Pratik Vinayak Fermentation and Phytofarming Technology Division, CSIR-Institute of Himalayan Bioresource Technology, Palampur, Himachal Pradesh, India

Academy of Scientific and Innovative Research (AcSIR), Ghaziabad, India

Pratibha Jaryal Orchid Laboratory, Department of Botany, Panjab University, Chandigarh, India

Premika Rai Sikkim Regional Centre, G.B. Pant National Institute of Himalayan Environment, Gangtok, Sikkim, India

Promila Pathak Orchid Laboratory, Department of Botany, Panjab University, Chandigarh, India

Puja Acharya Sikkim Regional Centre, G.B. Pant National Institute of Himalayan Environment, Gangtok, Sikkim, India

Reza Azadi Gonbad Tea Research Center, Horticultural Science Research Institute, AREEO, Lahijan, Iran

Rolika Gupta Division of Microbiology, Career Point University, Hamirpur, Himachal Pradesh, India

Sandeep Rawat Sikkim Regional Centre, G.B. Pant National Institute of Himalayan Environment, Gangtok, Sikkim, India

Seema Devi Fermentation and Phytofarming Technology Division, CSIR-Institute of Himalayan Bioresource Technology, Palampur, Himachal Pradesh, India

Academy of Scientific and Innovative Research (AcSIR), Ghaziabad, India

Shashi Rani Fermentation and Phytofarming Technology Division, CSIR-Institute of Himalayan Bioresource Technology, Palampur, Himachal Pradesh, India

Academy of Scientific and Innovative Research (AcSIR), Ghaziabad, India

Shiv Rattan Robert H. Smith Faculty of Agriculture, Food and Environment, The Hebrew University of Jerusalem, Jerusalem, Israel

Shrutika Surada School of Biotechnology, Faculty of Applied Sciences and Biotechnology, Shoolini University, Solan, Himachal Pradesh, India

Shriyanshu Thakur Division of Microbiology, Career Point University, Hamirpur, Himachal Pradesh, India

Sitanshu Das Sikkim Regional Centre, G.B. Pant National Institute of Himalayan Environment, Gangtok, Sikkim, India

Sudesh Kumar Yadav CSIR-Institute of Himalayan Bioresource Technology, Palampur, Himachal Pradesh, India

Sukhjinder Singh Business Development and Marketing Unit, CSIR-Institute of Himalayan Bioresource Technology, Palampur, Himachal Pradesh, India

Sultan Mahamud Sikkim Regional Centre, G.B. Pant National Institute of Himalayan Environment, Gangtok, Sikkim, India

Susai Arul Selvan School of Biotechnology, Faculty of Applied Sciences and Biotechnology, Shoolini University, Solan, Himachal Pradesh, India

Tarun Pal School of Bioengineering and Food Technology, Faculty of Applied Sciences and Biotechnology, Shoolini University, Solan, Himachal Pradesh, India

Tatenda Justice Gunda School of Biotechnology, Faculty of Applied Sciences and Biotechnology, Shoolini University, Solan, Himachal Pradesh, India

Vikram Patial Pharmacology and Toxicology Laboratory, Dietetics & Nutrition Technology, CSIR-Institute of Himalayan Bioresource Technology, Palampur, Himachal Pradesh, India

Academy of Scientific and Innovative Research (AcSIR), Ghaziabad, India

Vinayak Sirkeck School of Biotechnology, Faculty of Applied Sciences and Biotechnology, Shoolini University, Solan, Himachal Pradesh, India

Vishal Acharya Artificial Intelligence for Computational Biology Lab, CSIR-Institute of Himalayan Bioresource Technology, Palampur, Himachal Pradesh, India

Academy of Scientific and Innovative Research (AcSIR), Ghaziabad, India

Chapter 1
Introduction of Soilless Farming

Seema Devi, Phate Pratik Vinayak, Amit Kumar, Pankaj Kumar, and Ashish R. Warghat

Abstract Worldwide, traditional, and herbal medicines have been essential to health and utilized to cure numerous acute and chronic diseases with negligible or very less adverse effects. Medicinal plants (MPs) abundant in bioactive molecules known as secondary metabolites, including alkaloids, flavonoids, and terpenoids, have been applied to treat diseases due to their diverse therapeutic potential. For the commercial level production of herbal, medicinal, cosmetic, and food companies rely entirely on naturally grown plants or standard agricultural procedures. An increasing need for MPs, along with concerns about their quality, purity, consistency, and sustainability, has necessitated a re-evaluation of traditional production methods. Soilless agriculture, including hydroponics, aeroponics, aquaponics, and vertical farming, provides revolutionary approaches for MPs and biomass production. Automated conditions inside these high-tech advanced cultivation techniques can enhance the uniformity, concentration, and yield of secondary metabolites in MPs by facilitating precise regulation of environmental parameters, including lighting, temperature, humidity, pH, and electrical conductivity. Integration of soilless agriculture with computer technology and artificial intelligence (AI) can enhance automation, yield estimation, nutrient delivery, and disease prediction and diagnosis. To meet the standard grade biomass requirements by industrial guidelines, these advanced techniques represent the most effective farming approaches for rapidly increasing plant and biomass production quality. This book chapter aims to elucidate recent advancements in soilless agriculture and the integration of AI and com-

S. Devi · P. P. Vinayak · A. Kumar · A. R. Warghat (✉)
Fermentation and Phytofarming Technology Division, CSIR-Institute of Himalayan Bioresource Technology, Palampur, Himachal Pradesh, India
Academy of Scientific and Innovative Research (AcSIR), Ghaziabad, India
e-mail: ashish.warghat@csir.res.in

P. Kumar
Department of Biotechnology, College of Horticulture, Dr. Yashwant Singh Parmar University of Horticulture and Forestry, Solan, Himachal Pradesh, India
e-mail: pksharmabiotech@gmail.com

P. Kumar, A. R. Warghat (eds.), *Transforming Medicinal Plant Agriculture*, Smart Agriculture 16, https://doi.org/10.1007/978-981-95-3963-5_1

puter technology to produce MPs used in the medicinal, cosmetic, and food industries.

Keywords Hydroponics · Vertical farming · Artificial intelligence · Medicinal plants · Secondary metabolites

1.1 Introduction

Medicinal plants (MPs) are essential to human growth and have been utilized since ancient times. The World Health Organization reports that 80% of the global population depends primarily on plant-derived drug products for essential healthcare requirements. Plants serve as living chemical factories for generating an immense number of secondary metabolites, which become the foundation for numerous commercial medicine and herbal therapies. Secondary metabolites do not directly contribute to plant growth and development but are crucial for plant adaptation in adverse environmental conditions or defense mechanisms [1]. In the past few decades, consumer need for MPs and their products has increased significantly, primarily due to the rising interest in alternative treatments and the pursuit of innovative therapies. Due to overgrazing, excessive harvesting, and poor agricultural practices, many MPs are now threatened or endangered. These potential threats to plant populations, effectiveness, and viability resulting from traditional cultivation and inadequate harvesting procedures present significant hurdles to progress in innovative plant-derived pharmaceutical products [2]. The current imperative is a shift toward innovative agricultural methods that integrate food and plant production with conserving the environment [3–5]. Soilless agriculture of many MPs, fruit crops, and vegetables has gained popularity recently due to their capacity to generate substantial biomass, high yields, and a diverse array of pharmacologically active metabolites.

Soilless agriculture, including hydroponics, aeroponics, aquaponics, and vertical farming (VF), provides a revolutionary approach to medicinal and food crop production, facilitating a more uniform and high-quality plant biomass generation. VF is a contemporary agricultural method that involves cultivating crops in vertically stacked layers within a controlled indoor environment, primarily employing hydroponic or aeroponic systems. Hydroponics is a technique in which plant roots are directly submerged in a nutrient-rich solution, supported by inert substrates such as pebbles, perlite, and vermiculites [6]. Aeroponics is a cultivation technique that utilizes a nutrient-rich spray to grow plants without soil or other solid media [7, 8]. Aeroponic farming differs from hydroponic methods as plant roots are not submerged in liquid media; instead, they directly absorb nutrition in the form of mist. The aquaponics farming system combines hydroponics and recirculating aquaculture to synchronously cultivate crops and fishes [4]. These high-tech advanced agriculture practices meticulously regulate environmental factors critical to plant growth, including light quantity and wavelength, temperature, moisture, nutrient compositions, and aeration. By ensuring these variables, such systems facilitate throughout-the-year cultivation, superior plant performance, minimize reliance on synthetic pesticides, and maximize the efficacy of nutrient utilization and output in less space and time [9, 10]. These cultivation approaches provide a sustainable and dependable framework to produce several commercially significant crops, including MPs, spices, aromatics, flowers, and vegetables [6, 11, 12]. Soilless agriculture has overtaken traditional approaches, which have not integrated sophisticated technology as well as other sectors [13]. The advent of computational technology and artificial technology (AI) has transformed agriculture with Internet of Things (IoT) technologies and innovative precision farming techniques. The implementation of IoT technology in agriculture can facilitate enhanced farm management, diminish waste, and augment agricultural output with low environmental repercussions [14, 15].

In soilless agriculture, agriculturalists fine-tune the nutrient and media composition in the irrigated solution and adjust the synthesis of specific secondary metabolites in plants [6]. Multiple studies have shown that plants produced in artificial growing mediums have enhanced growth and phytochemical content relative to those cultivated in soil, promoting rapid growth and improved nutrient composition compared to traditional farming methods [12, 16, 17]. These advantages highlight the importance of soilless agriculture in overcoming the limitations of traditional practices and enhancing crop production.

The book chapter conducted a comprehensive examination of different types of soilless agriculture techniques and the application of AI and IoT in MPs production.

1.2 Different Classes of Soilless Agriculture

Growing plants without soil, in solid or liquid media, where nutrients are artificially supplied to the plants for growth and development is known as soilless farming. The purpose of soilless farming is to promote plant growth while regulating the amounts of water, mineral salts, and above all dissolved oxygen [8, 12]. Roots quickly take up nutrients and oxygen when they are submerged in flowing water. Plant growth will be delayed if there is not enough oxygen present. On the other hand, plant development is accelerated if the water is oxygen-rich. To maximize both yield and quality, the grower must thus balance the combination of water, nutrients, and oxygen with the requirements of the plan [18, 19]. This strategy has several socioeconomic benefits in addition to the ability to address the growing global food crisis, malnutrition, and the effective use and management of natural resources, all of which contribute to ecological sustainability by ensuring a steady supply of wholesome food throughout the year [20, 21]. For all countries with limited land area, a constantly shifting climate, and developing food issues with indigenous populations, it is a genuinely effective crop-growing method. Soilless farming technology primarily focuses on solid media cultures, hydroponics, aeroponics, and aquaponics [3, 22]. Cereals, vegetables, fruits, condiments, flowers, medicinal crops, and fodder crops are among the many crops cultivated in soilless cultures [12, 23]. Crops grown in soilless media have higher yields and nutritional benefits, which may be a way to combat the problems of malnutrition and global food security.

1.2.1 Hydroponics

The process of growing plants without soil by using mineral fertilizer solutions in a liquid medium is known as hydroponics [24]. Various inert media, including peat moss, charcoal, gravel, rock wool, perlite, coco peat, and coconut coir, are used as supporting agents in hydroponics [8]. Numerous crops, such as lettuce, cucumbers, tomatoes, medicinal plants, and ornamental flowers, have been grown successfully in hydroponics [12, 25]. Compared to soil-based cultivating systems, it offers many advantages, such as quick growth, high yields, simplicity of handling, efficient water use, and less use of fertilizer. To identify signs of nutrient deficiencies or toxicity in the plant system, the nutrient content of the solution is controlled and monitored. In a regulated, boosted environment, hydroponics produces superior yield, biomass, and bioactive molecule synthesis compared to traditional farming.

1.2.2 Aeroponics

Aeroponics is an innovative farming method that does not require soil because plant roots are suspended in the air. By keeping the right balance of moisture, nutrients, and oxygen in the roots, the basic concept is to apply a nutrient-rich solution with a fine mist, allowing the plants to grow as best they can [7]. To facilitate efficient nutrient uptake, this system typically uses pumps, timers, and spray nozzles to deliver a well-oxygenated mist containing water and nutrients to the roots at predetermined intervals. Aeroponics ensures accurate nutrient delivery and accelerates root oxygenation by enclosing plant roots in a shaded chamber and applying nutrient-dense solutions regularly. This precise and efficient nutrient delivery system enhances overall plant development, reduces the risk of root diseases, and conserves resources [26]. The nutrient-mist system creates a great growing environment for plants while using the least amount of water possible. Using this method, plants are placed horizontally at the top of the growing container and supported by panels made of plastic or polystyrene. To suspend the root system, these panels are usually made of inert materials like steel, plastic, and plastic film [27]. The controlled parameters are equal nutrient levels, electrical conductivity (EC), pH, temperature, humidity, light intensity, atomization spray duration and interval, and availability of oxygen [7]. Compared to crops grown in soil, crops grown in an aeroponic system yield more and have comparable amounts of phenolics, flavonoids, and antioxidants.

1.2.3 Aquaponics

Aquaponics is a highly innovative technique that integrates hydroponics and aquaculture, two separate technologies, into a single, cohesive system. "Ponics" refers to the process of growing plants without soil using water as the main medium, whereas "aqua" refers to the practice of raising fish in very crowded tanks [28]. Fish, which are typically kept in tanks, produce ammonium as waste. This ammonia is later transformed by the system's beneficial bacteria into nitrites and ultimately nitrates, which are essential nutrients for plant growth. After that, the nutrient-rich water is pumped to the hydroponic beds where herbs and vegetables are being cultivated. By absorbing the nutrients, these plants purify the water and eliminate toxins for the fish. The closed-loop cycle is subsequently completed by returning the clean water to the fish tanks [29]. The bacterial community is essential to the effective cycling of nutrients in an aquaponic system. Fish waste produces hazardous ammonia, which is converted to nitrite by the Nitrosomonas and Nitrobacter bacteria. Additionally, these bacteria complete the crucial cycle of nutrient conversion and uptake by converting nitrite into nitrate, an essential nutrient for the plants cultivated in the system [28]. Regardless of the weather, aquaponics is a very successful farming technique that makes it possible to grow fish and vegetables all year round [30]. Furthermore, a wide variety of foods can be grown in the regulated environment of aquaponic systems, increasing community nutritional security and dietary diversity [31].

1.2.4 Vertical Farming

Vertical farming (VF) is a modern agricultural technique that entails the cultivation of crops in vertically arranged tiers within a regulated indoor setting [32]. It allows for planned production and rapid growth by controlling the environment and nutrient solutions for hydroponic crops. With minimal nutrient runoff, water loss, and chemical treatments, precision control guarantees the year-round production of high-quality products. The automatic system and sensor technologies constitute another critical element of VF. Numerous growing spaces and indoor farms around are presently utilizing this technology to ensure reliable and consistent growth conditions year-round. Moreover, automated technologies are employed to diminish operational costs [33, 34]. As the name suggests, these systems often grow plants vertically, usually in layers that are stacked horizontally but can also be placed on walls that are oriented vertically, to enhance space-use efficiency [35, 36].

1.3 Artificial Intelligence in Soilless Agriculture

Technological and scientific advancement in soilless agriculture has significantly enhanced precision irrigation through the collection, dissemination, and application of very accurate control methods [15]. Modern computer technology including artificial intelligence (AI), machine learning (ML), deep learning (DL), Internet of Things (IoT), and digital sensor technology monitor essential parameters including humidity, temperature, disease prediction, stress evaluation, and yield estimation to monitor the plant life cycle and characteristics to achieve increased production [37, 38].

The IoT and sensor technology can be employed to monitor and automate the regulation of pH, water level, temperature, and light intensity. During winter, water often freezes in certain areas, in hydroponics and VF, potentially obstructing farming. The hydroponics farm's water temperature sensors can identify temperature decreases and alert the grower accordingly. Similarly, EC and pH sensors can identify mineral deficiencies, enabling the appropriate circulation of minerals through a pump in sufficient quantities. ML in hydroponics and VF have empowered farmers to automate plant growth and monitor and control pH and EC values in the nutrient reservoir [39]. Utilizing ML with the IoT will enhance agricultural productivity and quality [40, 41].

1.3.1 Application of Computer Technology in Soilless Agriculture

1.3.1.1 Yield Production and Estimation

ML and DL algorithms have emerged as powerful, robust, and reliable models for estimating crop yield and production. These computation models require an extensive dataset of agricultural yields to generate yield predictions. This dataset may encompass variety and cultivar types, climate data, geographical location, nutrient details, pesticide quantity, and crop management practices. Yanes et al. [42] built an automated aquaponics system that uses DL algorithms to monitor the growth rate and weight of lettuce. The ML model is built using features related to crop size and fresh weight. The model determined the crop quantity and natural weight with an error margin of 0.5 g (8.3%) and 30 mm (18.7%), respectively. Four machine learning approaches support vector regressor (SVR), extreme gradient boosting (XGB), random forest (RF), and deep neural network (DNN) were examined to forecast the production of lettuce cultivated in a hydroponic system. The cultivation of lettuce was conducted using three types of hydroponic systems, with samples collected after 50 days of transplanting for each system over 2 years. Crop yield was forecasted based on characteristics like leaf count, water utilization, and dry weight.

Among all machine learning methods, XGB exhibited superior performance, achieving a maximum R2 of 0.94 and a minimal RMSE of 8 [43].

1.3.1.2 Precise Nutrient Delivery and Irrigation

The integration of AI with IoT-enabled sensors and devices has led to a higher rate of automation in smart agriculture techniques. A link between agriculture and AI is the utilization of recommendation systems to enhance productivity by identifying nutritional deficiencies in plants, optimizing resource allocation, reducing environmental impact, and preventing economic damage. Upon detecting any lack of nutrients and water the sensors linked to the VF system promptly transmit the relevant data to the cloud via a micro-controller. The cloud ML algorithm functions and forecasts the appropriate course of action [44]., examined four CNN models: DenseNet 201, ResNet101V2, MobileNet, and VGG16, to identify deficiencies of nitrogen, phosphorus, and potassium in hydroponically grown basil. The author generated a custom dataset for the ML model using a camera, capturing a total of 1757 photos. Through data improvement and preliminary analysis, the total number of images was increased to 10,000. Of all CNN models, VGG16 attained the best reliability at 93.82% [45]. developed a limited dataset utilized to forecast the most effective nutrient concentrations in aquaponics for the ideal growth of fish and plants. Three ML classifiers, including linear SVM, Adaboost, and Gradient Boosting, were utilized to predict nutritional levels. Linear SVM attained 75% accuracy in predicting suitable nutrition.

1.3.1.3 Disease Prediction and Detection

To enhance a farmer's revenue from crop growth and output, it is essential to identify and diagnose plant diseases. Tao et al. [46] proposed an automated leaf sorting method utilizing DL for hydroponic systems. The hydroponics system for lettuce cultivation has aberrant leaves that must be removed before final packaging. Aberrant lettuce leaves in the system were categorized as yellow, wilted, and decayed, with decayed leaves being more detrimental due to the presence of infectious agents. Leaves are categorized as healthy or aberrant based on their color and texture. The DeepLabV3+ model was employed in this system, utilizing key approaches such as ResNet-101, ResNet-50, Xception-65, and Xception-71. ResNet-101 achieves the highest pixel precision of 99.24% in segmentation, while ResNet-50 offers a rapid segmentation rate. Guo et al. [47] proved the advantages of backpropagation neural networks and image purification for detecting mildew disease in the root area of mulberry cuttings cultivated aeroponically. This technique attained a reliability of 80% in detecting mildew disease on the root area of mulberry cuttings.

AI methodologies coupled with digital technology in soilless agriculture techniques enhance the environmental sustainability and climate resilience of agricultural practices, markedly decreasing crop losses and field growth while also minimizing the application of fertilizers, and pesticides [48]. Considering the conclusions outlined in the previous paragraph, AI, ML, DL algorithms, and IoT smart technologies can enhance the overall productivity of a soilless agriculture production system [15, 49]. Standardizing processes is essential to fully leverage technological advances for soilless agriculture applications, especially regarding device standardized procedures. Innovative study opportunities remain to explore and examine smart agriculture systems for food and MPs production through smart and precision farming.

1.4 Conclusion and Remarks

Soilless agriculture techniques are gaining global recognition, providing benefits for both farmers and consumers in contemporary agriculture. The present book chapter elucidated the functions of hydroponic, aeroponic, aquaponic, and VF systems in augmenting MPs and quality biomass production. Additionally, the book chapter also examines the efficacy of remotely monitoring factors such as EC, pH, humidity, temperature, and irrigation through IoT devices and sensor-based technology [50]. Much research has demonstrated how ML and IoT tools have enabled farmers to anticipate crop diseases and yields across various agricultural systems [51, 52]. Facing potential challenges of global warming, population increase, and resource constraints, the utilization of IoT technology in conventional and soilless agriculture is becoming progressively essential. By leveraging the potential offered by soilless agriculture techniques and AI, agriculture can reduce risks, enhance processes, and create a more resilient food supply chain. The IoT and soilless agriculture techniques can resolve challenges within the agricultural sector while satisfying the demands of an expanding worldwide community. Capturing the full capabilities of IoT in agriculture and realizing its advantages for farmers, consumers, and the environment necessitates additional study, growth, and joint effort among stakeholders. Further soilless agriculture techniques can be scaled up by applying cutting-edge technologies such as precision farming and robotics to fill the needs of people in urban areas for food and medicinally important bioactive molecules production.

Acknowledgments The authors acknowledge the Council of Scientific and Industrial Research (CSIR), Government of India, under the project "Development of industrially sustainable products and processes based on Himalayan Bioresources using advanced fermentation and phytofarming technologies" (MLP-0207) for providing financial support. The authors are thankful to the Director, CSIR-IHBT, Palampur for providing necessary facilities.

Author Contribution Seema devi: Literature reviewed, Writing the manuscript, Phate Pratik Vinayak: Literature reviewed, Writing the manuscript, Amit Kumar: Literature reviewed, Writing the manuscript, Pankaj Kumar: Literature reviewed, Writing the manuscript, Ashish R. Warghat: Conceptualisation, supervision, Writing and editing the manuscript. All the authors have read and approved the final manuscript.

References

1. Kumar A, Partap M, Warghat AR (2024) Jasmonic acid: a versatile phytohormone regulating growth, physiology, and biochemical responses. J Plant Growth Regul:1–24
2. Atanasov AG, Waltenberger B, Pferschy Wenzig EM, Linder T, Wawrosch C, Uhrin P, Temml V, Wang L, Schwaiger S, Heiss EH, Rollinger JM (2015) Discovery and resupply of pharmacologically active plant-derived natural products: a review. Biotechnol Adv 33(8):1582–1614
3. Erekath S, Seidlitz H, Schreiner M, Dreyer C (2024) Food for future: exploring cutting-edge technology and practices in vertical farm. Sustain Cities Soc:105357
4. Okomoda VT, Oladimeji SA, Solomon SG, Olufeagba SO, Ogah SI, Ikhwanuddin M (2023) Aquaponics production system: a review of historical perspective, opportunities, and challenges of its adoption. Food Sci Nutr 11(3):1157–1165
5. Jaeger SR et al (2024) Vertical farming (plant factory with artificial lighting) and its produce: consumer insights. Curr Opin Food Sci:101145
6. Partap M, Kumar P, Kumar A, Joshi R, Kumar D, Warghat AR (2020) Effect of elicitors on morpho-physiological performance and metabolites enrichment in Valeriana jatamansi cultivated under aeroponic conditions. Front Plant Sci 11:01263
7. Lakhiar IA, Gao J, Syed TN, Chandio FA, Buttar NA (2018) Modern plant cultivation technologies in agriculture under controlled environment: a review on aeroponics. J Plant Interact 13(1):338–352
8. Partap M, Rattan S, Kanika, Ashrita, Sood A, Kumar P, Warghat AR (2021) Hydroponic and Aeroponic cultivation of economically important crops for production of quality biomass. In: Agric Biotechnol, pp 573–585
9. Eldridge BM, Manzoni LR, Graham CA, Rodgers B, Farmer JR, Dodd AN (2020) Getting to the roots of aeroponic indoor farming. New Phytol 228(4):1183–1192
10. Shamshiri R, Kalantari F, Ting KC, Thorp KR, Hameed IA, Weltzien C, Ahmad D, Shad ZM (2018) Advances in greenhouse automation and controlled environment agriculture: a transition to plant factories and urban agriculture. Int J Agric Biol Eng 11(1):1–22
11. Thakur K, Partap M, Kumar D, Warghat AR (2019) Enhancement of picrosides content in Picrorhiza kurroa Royle ex Benth. mediated through nutrient feeding approach under aeroponic and hydroponic system. Ind Crop Prod 133:160–167
12. Sharma D, Partap M, Warghat AR, Bhargava B (2024) Hydroponic cultivation enhances the morpho-physiological traits and quality flower production in three cultivars (*marigold scarlet* red, marigold orange, and marigold yellow) of French marigold (*Tagetes patula* L.). Sci Hortic 327:112803
13. Pathan M, Patel N, Yagnik H, Shah M (2020) Artificial cognition for applications in smart agriculture: a comprehensive review. Artif Intell Agric 4:81–95
14. Goel RK, Yadav CS, Vishnoi S, Rastogi R (2021) Smart agriculture–urgent need of the day in developing countries. SUSCOM 30:100512
15. Rathor AS, Choudhury S, Sharma A, Nautiyal P, Shah G (2024) Empowering vertical farming through IoT and AI-driven technologies: a comprehensive review. Heliyon
16. Fatima RT, Dantas MV, Lima GS, Oliveira VKN, Soares LAA, Silva AAR, Gheyi HR, Guedes MA, Nóbrega JS, Fernandes PD (2023) Salicylic acid does not relieve salt stress on gas exchange, chlorophyll fluorescence, and hydroponic melon growth. Braz J Biol 83:e274595
17. Chowdhury M, Samarakoon UC, Altland JE (2024) Evaluation of hydroponic systems for organic lettuce production in controlled environment. Front Plant Sci 15:1401089
18. Woznicki T, Møllerhagen PJ, Heltoft P, Kusnierek K (2021) Growing potatoes (*Solanum tuberosum* l.) hydroponically in wood fiber—a preliminary case-study report. Agronomy 11(7):1369
19. Kusnierek K, Heltoft P, Møllerhagen PJ, Woznicki T (2023) Hydroponic potato production in wood fiber for food security. npj Sci Food 7(1):24

20. Gonnella M, Renna M (2021) The evolution of soilless systems towards ecological sustainability in the perspective of a circular economy. Is it really the opposite of organic agriculture? Agronomy 11(5):950
21. Mir YH, Mir S, Ganie MA, Shah AM, Majeed U, Chesti MH, Mansoor M, Irshad I, Javed A, Sadiq S, Wani FJ (2022) Soilless farming: an innovative sustainable approach in agriculture. PIJ 11(6):2663–2675
22. Garzón J, Montes L, Garzón J, Lampropoulos G (2023) Systematic review of technology in aeroponics: introducing the technology adoption and integration in sustainable agriculture model. Agronomy 13(10):2517
23. Rattan S, Partap M, Kumar S, Warghat AR (2022) Nutrient feeding approach enhances the vegetative growth biomass, volatile oil composition, and myristicin content in hydroponically cultivated *Petroselinum crispum* (mill.) Nyman. J Appl Res Med Aromat Plants 26:100359
24. Roy K, Kajgrova L, Mraz J (2022) TILA feed: a bio-based inventory for circular nutrients management and achieving bioeconomy in future aquaponics. New Biotechnol 70:9–18
25. Mubeen B, Hasnain A, Jie, W, Zheng H, Peijnenburg WJ, Rozali SE, Rasool R, Naqvi SAH, Rao MJ, Sohail MA, Moustafa M (2023) Enhanced production of active photosynthetic and biochemical molecules in *Silybum marianum* L. using biotic and abiotic elicitors in hydroponic culture. Molecules 28(4):1716
26. Faisal A, Rajashekar V, Biswal P, Mukherjee A, Bhowmick GD (2024) Advancing agriculture with Aeroponics: a critical review of methods, benefits, and limitations. In: Hydroponics: The Future of Sustainable Farming, pp 311–322
27. Maucieri C, Nicoletto C, Van Os E, Anseeuw D, Van Havermaet R, Junge R (2019) Hydroponic technologies. In: Aquaponics food production systems, vol 10. Springer, Cham, pp 978–973
28. Masabni J, Niu G (2022) Aquaponics. In: Plant factory basics, applications and advances. Academic Press, pp 167–180
29. Dayawansha IHDRS et al (2024) Exploring the potential of aquaponics in addressing food security and future food demand in Sweden
30. Yahya F, El Samrani A, Khalil M, Abdin AED, El-Kholy R, Embaby M, Negm M, De Ketelaere D, Spiteri A, Pana E, Takavakoglou V (2023) Decentralized wetland-aquaponics addressing environmental degradation and food security challenges in disadvantaged rural areas: a nature-based solution driven by Mediterranean living labs. Sustainability 15(20):15024
31. Sunny AR, Islam MM, Rahman M, Miah MY, Mostafiz M, Islam N, Hossain MZ, Chowdhury MA, Islam MA, Keus HJ (2019) Cost effective aquaponics for food security and income of farming households in coastal Bangladesh. Egypt J Aquat Res 45(1):89–97
32. Oh S, Lu C (2023) Vertical farming-smart urban agriculture for enhancing resilience and sustainability in food security. J Hortic Sci Biotechnol 98(2):133–140
33. Sengodan P et al (2022) An overview of vertical farming: highlighting the potential in Malaysian high-rise buildings. Pertanika J Sci Technol 30(2)
34. Van Gerrewey T, Boon N, Geelen D (2021) Vertical farming: the only way is up? Agronomy 12(1):2
35. Ji Y, Kusuma P, Marcelis LF (2023) Vertical farming. Curr Biol 33(11):R471–R473
36. Kumar S (2020) Automation in hydroponics farming
37. Liakos KG, Busato P, Moshou D, Pearson S, Bochtis D (2018) Machine learning in agriculture: a review. Sensors 18(8):2674
38. Subeesh A, Mehta CR (2021) Automation and digitization of agriculture using artificial intelligence and internet of things. Artif Intell Agric 5:278–291
39. Alipio MI, Cruz AEMD, Doria JDA, Fruto RMS (2017) A smart hydroponics farming system using exact inference in Bayesian network. In: In 2017 IEEE 6th global conference on consumer electronics (GCCE). IEEE, pp 1–5
40. Musa A, Hassan M, Hamada M, Aliyu F (2022) Low-power deep learning model for plant disease detection for smart-hydroponics using knowledge distillation techniques. Journal of Low Power Electronics and Applications 12(2):p.24

41. Elbasi E, Zaki C, Topcu AE, Abdelbaki W, Zreikat AI, Cina E, Shdefat A, Saker L (2023) Crop prediction model using machine learning algorithms. Appl Sci 13(16):9288
42. Yanes AR, Martinez P, Ahmad R (2020) Towards automated aquaponics: a review on monitoring, IoT, and smart systems. J Clean Prod 263:121571
43. Rafael GJ, Pierre M, Frank E, Heidi W, Sibylle D, Daniel C, Matthew R, Gemma M, Daniel M, Guillermo G, Gustavo S (2022) Evidence for increasing global wheat yield potential. Environ Res Lett 17(12)
44. Gul Z, Bora S (2023) Exploiting pre-trained convolutional neural networks for the detection of nutrient deficiencies in hydroponic basil. Sensors 23(12):5407
45. Dhal P, Azad C (2022) A comprehensive survey on feature selection in the various fields of machine learning. Appl Intell 52(4):4543–4581
46. Tao W, Zhao L, Wang G, Liang R (2021) Review of the internet of things communication technologies in smart agriculture and challenges. Comput Electron Agric 189:106352
47. Guo Y, Gao J, Tunio MH, Wang L (2022) Study on the identification of mildew disease of cuttings at the base of mulberry cuttings by Aeroponics rapid propagation based on a BP neural network. Agronomy 13(1):106
48. Krishnan RS, Julie EG, Robinson YH, Raja S, Kumar R, Thong PH (2020) Fuzzy logic based smart irrigation system using internet of things. J Clean Prod 252:119902
49. Fasciolo B, Awouda A, Bruno G, Lombardi F (2023) A smart aeroponic system for sustainable indoor farming. Proc CIRP 116:636–641
50. Lakshmanan R, Djama M, Selvaperumal SK, Abdulla R (2020) Automated smart hydroponics system using internet of things. Int J Electr Comput Eng Syst (IJECE) 10(6):6389–6398
51. Bhargava YV, Chittoor PK, Bharatiraja C, Verma R, Sathiyasekar K (2022) Sensor fusion based intelligent hydroponic farming and nursing system. IEEE Sensors J 22(14):14584–14591
52. Musa P, Sugeru H, Wibowo EP (2023) Wireless sensor networks for precision agriculture: a review of NPK sensor implementations. Sensors 24(1):51

Chapter 2
Vertical Farming for Medicinal Plants: A Smart Approach

Aditi Sharma, Shashi Rani, Amit Kumar, Pankaj Kumar, and Ashish R. Warghat

Abstract Medicinal plants have been utilized to treat human health issues for hundreds of years. Presently, much of the pharmaceutical industry depends on traditional agriculture practices and growing techniques to produce medicinal plants. This strategy is primarily constrained by its significant reliance on seasonal variations and chemical inclusions resulting in incompatible biochemical composition. The obstacles indicate that traditional agriculture practices may be inadequate to meet the future demands for medicinal plants globally. Consequently, new, and sustainable production methods are requisite. Vertical farming (VF) technology is now recognized as a viable option, facilitating enhanced production of medicinally significant plants at both local and economic levels. In VF growers can precisely regulate environmental factors important to plant growth, including lighting quality and wavelength, temperature, humidity, and nutrient media. By maximizing these attributes, VF systems facilitate year-round production, enhance plant quality, eradicate the necessity for chemical pesticides, and boost the efficiency of water and nutrient utilization. Recent breakthroughs in artificial intelligence (AI), machine learning (ML), the Internet of Things (IoT), and computer vision technologies inside VF systems improve precision, automation, and early disease and yield forecasting, hence offering greater control in this high-tech technique. These strengths highlight the importance of VF in overcoming the limitations of traditional approaches and enhancing

A. Sharma
Fermentation and Phytofarming Technology Division, CSIR-Institute of Himalayan Bioresource Technology, Palampur, Himachal Pradesh, India

S. Rani · A. Kumar · A. R. Warghat (✉)
Fermentation and Phytofarming Technology Division, CSIR-Institute of Himalayan Bioresource Technology, Palampur, Himachal Pradesh, India

Academy of Scientific and Innovative Research (AcSIR), Ghaziabad-201002, India
e-mail: ashish.warghat@csir.res.in

P. Kumar
Department of Biotechnology, College of Horticulture, Dr. Yashwant Singh Parmar University of Horticulture and Forestry, Solan, Himachal Pradesh, India
e-mail: pksharmabiotech@gmail.com

P. Kumar, A. R. Warghat (eds.), *Transforming Medicinal Plant Agriculture*, Smart Agriculture 16, https://doi.org/10.1007/978-981-95-3963-5_2

medicinal plant productivity. This chapter explores the potential of VF systems to enhance plant growth and evaluates the performance of different medicinal plant production. Furthermore, the book chapter emphasizes contemporary advancements in VF through computer technology, which enhances control and automation, hence accelerating plant development compared to traditional growing methods.

Keywords Vertical farming · Medicinal plants · Internet of Things · Sustainable agriculture · Secondary metabolites

2.1 Introduction

Since the emergence of human civilization, plants have been a primary source for curing, healing, and alleviating various diseases. Medicinal and aromatic plants hold significant value due to their therapeutic properties. These plants find applications across various industries, including food, beverages, supplements, perfumery, cosmetics, and medicine [1, 2]. Plants possess a diverse array of low molecular weight chemical molecules, categorized into primary and secondary metabolites. Primary metabolites play a crucial role in supporting plant growth and development. while secondary metabolites are crucial for facilitating communication between plants and their surroundings, thus enabling adaptation and defense against stress [3–5]. These secondary metabolites include alkaloids, tannins, flavonoids, and phenolics, which contribute significantly to modern and traditional medicine. While some primary metabolites exhibit medicinal properties, secondary metabolites are primarily contributors to the therapeutic potential of medicinal plants [6]. In developing countries, 65–80% of the population seeks herbal medicines for primary healthcare. [7]. However, the escalating demand for medicinal plants and environmental challenges such as climate stress, soil degradation, deforestation, and habitat loss threaten their sustainable production [2]. Traditional wild collection remains a cost-effective method, but it poses risks to biodiversity and the conservation of plant species. Additionally, over-harvesting in the wild can deplete important medicinal plant populations, by reducing their availability for future generations. As urbanization and industrialization limits available arable land, alternative cultivation methods are necessary to ensure a uninterrupted supply of high-quality medicinal plants [8].

Given the growing global demand for herbal medicines and the ecological concerns surrounding wild harvesting, vertical farming (VF) presents a viable, innovative, and sustainable alternative for the large-scale cultivation of medicinal plants. VF involves multilayered, precisely regulated indoor plant production systems designed to optimize environmental parameters, including light intensity, temperature, humidity, carbon dioxide concentration, water availability, and nutrient supply, to enhance plant growth [9]. By extending cultivation vertically rather than horizontally, VF enhances land-use efficiency, reduces dependency on soil-based agriculture, and ensures year-round, high-yield production [10]. Unlike traditional farming, which is vulnerable to climate variations, VF provides a controlled environment where medicinal plants can grow with minimal exposure to external stressors. These advanced cultivation systems employ LED lighting, hydroponic or aeroponic nutrient delivery systems, and precise climate control mechanisms to maximize plant growth and metabolite production. The controlled environment of VF offers several advantages for medicinal plant production. These include consistent phytochemical composition, protection from environmental stressors, reduced microbial and pollutant contamination, and sustainable land use [11]. Unlike traditional farming methods, which may require chemical pesticides and herbicides, VF minimizes the need for agrochemicals, leading to safer and purer plant-based medicines. Additionally, the ability to fine-tune environmental factors in VF allows for the enhancement of specialized bioactive compounds, ensuring medicinal plants have optimal therapeutic properties (Fig. 2.1). By integrating advanced technologies such as automated nutrient delivery systems, artificial intelligence (AI), Internet of Things (IoT), machine learning (ML) for monitoring plant health, and energy-efficient LED lighting, VF can revolutionize the medicinal plant industry [28]. As research continues to refine these methods, VF is poised to play a pivotal role in addressing future healthcare needs while promoting environmental sustainability [29]. Tables 2.1 and 2.2 represents the medicinal plants with their therapeutical potential grown under vertical farming.

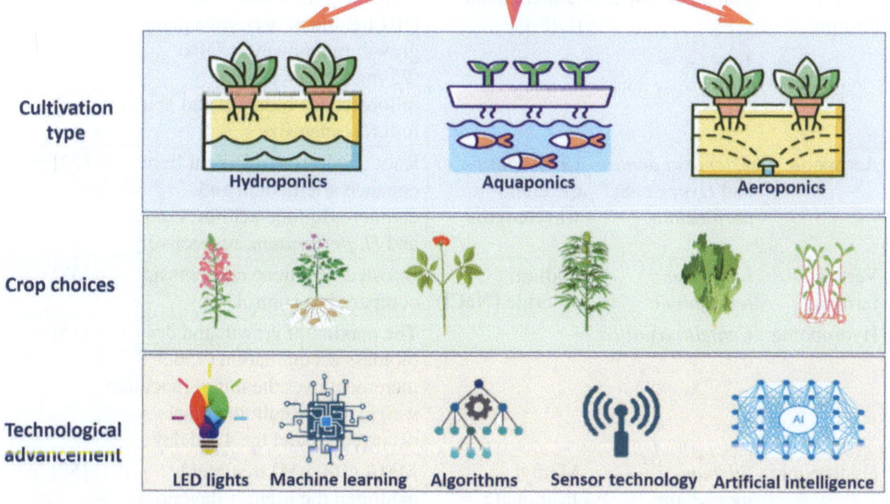

Fig. 2.1 A schematic representation of the prevailing challenges associated with conventional agriculture and vertical farming

Table 2.1 Medicinal plants cultivated in vertical farming and their therapeutic potential

System type	Plant species	Treatments	Plant response	Reference
Hydroponic	Sweet basil (*Ocimum basilicum*), Korean mint (*Agastache rugosa*), and sweet wormwood (*Artemisia annua*)	Ethanol (CH_3CH_2OH)	A lower concentration of ethanol treatments increased the shoot dry weight of sweet basil and sweet wormwood and higher concentration of ethanol treatment reduces shoot fresh weight and leaf area in the sweet basil and Korean mint	[52]
Vertical farming	*Achyranthes japonica* Nakai microgreens	UV-B stress	12 h of UV-B treatment before harvesting, at a specific intensity and wavelength, induce UV-B stress and enhanced bioactive compounds in microgreens	[53]
Vertical farming	*Artemisia annua*	Copper or zinc fortified nutrient solution	No effect on plant height and yield, decreased artemisinin content, and increased amounts of phenolic acids and flavonoids	[54]
Vertical farming	*Cannabis sativa*	Methyl Jasmonate (MeJA)	MeJA (100 μM) treatment is suitable for enhancing cannabinoid production while controlling plant height	[55]
Vertical farming	*Brassica oleracea* var. acephala variety of kales	Green tea water extract (GTE)	The total glucosinolate content increased from 9.56 μmol/g in the control group to 16.81 μmol/g in the GTE-treated group (treated with 0.5 g/L GTE)	[56]
Vertical farming	*Cannabis sativa*	Diniconazole (DIN)	Increasing the concentration of DIN treatment decreases plant growth rate however, DIN (25 mg L^{-1}) increased inflorescence biomass and apical inflorescence size	[57]
Aeroponic	*Artemisia annua* and *Hypericum perforatum*	LED lighting of different wavelengths	Root exposed to different light enhance artemisinin and coumaroylquinic acid in *A.annua* and *H. perforatum*, respectively	[22]
Vertical farming	*Limonium tetragonum*	Sodium chloride (NaCl)	Boosted flavanone contents as compared to control	[58]
Hydroponic	*Centella asiatica*	–	The maximum growth and dry biomass accumulation (156.3% increment over the initial inoculum weight) in the cultured plants occurred around the 42nd day	[15]
Hydroponic	*Silybum marianum*	Methyl Jasmonate (MeJA)	MeJA (100 μM) treatment exhibited the highest flavonoid (304 μg g^{-1}), phenolic (372 μg g^{-1}), and SOD (16.2 U g^{-1}) contents	[59]

Table 2.1 (continued)

System type	Plant species	Treatments	Plant response	Reference
Hydroponic	*Taraxacum officinale* and *Reichardia picroides*	pH	A lower pH value (pH = 4.0) seemed to be beneficial to nutritional and dietary value in both species	[21]
Hydroponic	*Salvia miltiorrhiza*	Sodium chloride (NaCl)	No effect on plant height and root length of *salvia miltiorrhiza*	[60]
Aquaponic	Basil	Different irrigation ratios of hydroponic: aquaponic solutions	Leaf area increased with increasing the ratio of hydroponic: aquaponic solution to 2:1	[61]
Aquaponic	*Cannabis sativa*	Micronutrient fertilizer	Adding K^+ fertilizer (75 and 113 mg L^{-1}) with micronutrients to aquaponic solution increased the harvest index by 16% and 22% compared with the control, respectively	[62]
Hydroponic	*Bacopa monnieri*	Silver NPs (AgNPs)	Plants treated with AgNPs exhibited higher levels of protein and carbohydrates, but comparably reduced levels of total phenolic content (TPC), catalase (CAT), and peroxidase (POX) enzyme activity	[63]
Hydroponic	*Carum copticum*	Titanium dioxide (TiO_2)	The exposure of TiO_2 NPs (300 ppm) and light spectrum showed a substantial impact on the anthocyanin and total phenolic content (TPC) in the shoot part	[64]
Hydroponic	*Calotropis procera*	Magnetic NPs and salicylic acid	Enhanced phenolics, flavonoids, and aromatic metabolites	[65]
Aeroponic	*Withania sominifera*	Salicylic acid (SA)	SA (150 mg L^{-1}) treatment increases chlorophyll and carotenoid content, plant height, root length, root volume, number of leaves per plant, root fresh weight, root dry weight, shoot fresh weight, and shoot dry weight	[66]
Aeroponic	*Scutellaria baicalensis*	Electro-stimulation	The concentration of aglycones was drastically increased (approx. 2.5-fold) after treatment	[67]
Aeroponic	*Valeriana jatamansi*	Methyl Jasmonate (MeJA)	Maximum content of valerenic acid and hydroxy valerenic acid was detected in leaf (2.47 and 8.37 mg/g) and root (1.78 and 7.89 mg/g) at treatment levels of 100 µM and 150 µM methyl jasmonate, respectively	[17]

Table 2.2 Medicinal plants successfully grown in vertical farming

Medicinal plant	System type	Trade name	Family	Medicinal use	Part used	Estimated Annual Trade (MT)	Reference
Silybum marianum	Hydroponic	Milk thistle	Asteraceae	Anticancerous, neurotoxicity, depression, lung, and prostate diseases	Seed	<10	[12]
Picrorhiza kurroa	Aeroponic and hydroponic system	Katuka	Plantaginaceae	Hepatoprotective, antiallergic, antiasthmatic, antioxidant, anticancerous, and immune modulator	Root (rhizome)	1000–2000	[13]
Euphorbia peplus	Hydroponic	Tithymalus peplus; Esula peplus.	Euphorbiaceae	Treatment of digestive and respiratory complaints, inflammation, and injuries, antibacterial, anti-tumor, and cell manifestations	Stems, leaves, roots, and latex	–	[14]
Centella asiatica	Hydroponic	Brahmi booti	Apiaceae	Alzheimer's, leprosy, varicose veins, ulcer, lupus	Leaf, whole plant	500–1000	[15]
Plectranthus amboinicus	Hydroponic	Patharcur	Lamiaceae	Anti-inflammatory, analgesic, anti-diuretic and anti-hyperglycemic potentials	Leaf and root	10–50	[16]
Valeriana jatamansi	Aeroponic	Tagar	Valerianaceae	Cytotoxicity, neuronal problems, insomnia, antioxidant, antiviral activities	Root (rhizome)	1000–2000	[17]
Calendula officinalis	Hydroponic	Gulasharfi	Asteraceae	Anti-inflammatory agent, anti-HIV activity, antitumoral activity, antimutagenic activity	Flower	100–200	[18]
Echinacea purpurea	Hydroponic	Eastern purple coneflower	Asteraceae	Antianxiety, antidepression, cytotoxicity, and antimutagenicity activity	Root	<10	[19]
Withania somnifera	Aeroponic and hydroponic	Ashwagandha	Solanaceae	Increase energy, improve overall health and longevity, and prevent diseases	Root, leaf, seed	2000–5000	[20]

Taraxacum officinale	Hydroponic	Kanphul	Asteraceae	Hepatoprotective, anticolitis, immunoprotective, antiviral, antifungal, antibacterial, antiarthritic, antidiabetic, antiobesity, antioxidant and anticancer effects	Root (rhizome)	<10	[21]
Artemisia annua	Aeroponic	Seeme davana	Asteraceae	Anti-malarial, antimicrobial and antidiabetic activity	Whole plant	1000–2000	[22]
Acalypha indica	Hydroponic	Khokali	Euphorbiaceae	Wounds healing activity, cardiac disorders, biliousness, rheumatism, hemorrhages, ulcers, amenorrhoea,	Whole plant	100–200	[23]
Achillea millefolium	Hydroponic	Brinjasif	Asteraceae	Anti-inflammatory, analgesic, antioxidant, and antiseptic activity	Whole plant	100–200	[24]
Moringa oleifera	Hydroponic	Sahenjana	Moringaceae	Anticancer, anti-hypertension, anti-diabetes, and anti-epilepsy activity	Leaf, seed, bark (stem)	500–1000	[25]
Ocimum basilicum	Hydroponic	Sweet basil	Lamiaceae	Anti-aging, anti-depressant, detoxifier, and immune modulator	Leaf, root, whole plant, fruit (seed)	200–500	[26]
Bacopa monnieri	Hydroponic	Jal Brahmi	Scrophulariaceae	Anti-inflammatory, analgesic, antipyretic, sedative, free radical scavenging, and anti-lipid and peroxidative properties	Whole plant	1000–2000	[27]

This chapter presents an in-depth review of advancements in VF technology. Furthermore, it explores the integration of VF with AI, and IoT in stimulating the biosynthesis of secondary metabolites in therapeutically significant plant species.

2.2 Types of Vertical Farming

Vertical farming (VF) is an advanced agricultural practice that maximizes productivity while minimizing resource use. It can be categorized into three primary types: hydroponics, aquaponics, and aeroponics (Fig. 2.2).

2.2.1 Hydroponics

Hydroponics is a method of cultivation that eliminates the need for soil, where plant roots are immersed in nutrient-rich aqueous solutions containing essential macronutrients and micronutrients. Inert substrates like gravel, sand, or sawdust provide mechanical support for root structures [30]. Hydroponic farming significantly

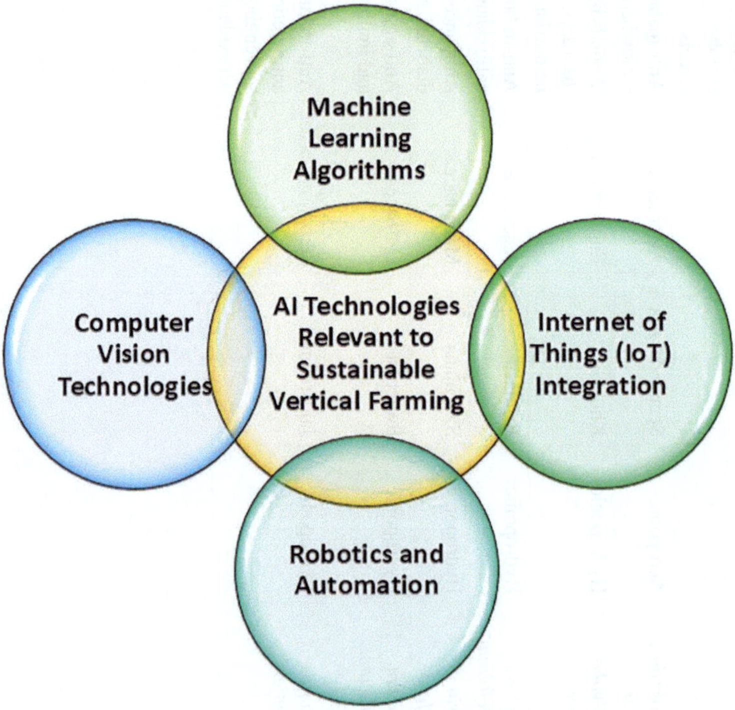

Fig. 2.2 Classification of vertical farming systems and associated technological advancements

enhances sustainability by operating within controlled environments, ensuring optimal nutrient delivery, and minimizing the need for chemical inputs. This system has been integrated into modern sustainable agriculture to address global food demand while using 60–70% less water compared to conventional soil-based agriculture [31]. Hydroponics is extensively employed in both commercial and residential environments to grow a range of items, including leafy greens, tuber crops, tomatoes, and herbs. Its aim to enhance plant production and sustainability, reduce water usage, and elevate agricultural yields has recently gained prominence.

2.2.1.1 Types of Hydroponics

A range of techniques, such as water cultures or container systems with artificial substrates, are utilized to maximize yield. Hydroponic techniques are further subdivided into the following techniques:

Water Culture

Water culture represents the simplest version of a dynamic hydroponic system, known for its affordability and suitability for fast-growing, plants such as lettuce. In this deepwater culture system, plants are supported on a floating platform, typically made of styrofoam or similar material, which rests on the nutrient solution. An external oxygen supply is provided to the plant roots using aquarium air pumps, ensuring optimal aeration and nutrient uptake [32].

Water Stream Hydroponics

This system employs a polymeric film as a growing surface, allowing for efficient nutrient delivery and root support. This system is highly automated, reducing manual labor while optimizing resource utilization. It operates on a pre-programmed setup, ensuring precise control over nutrient supply, water flow, and environmental conditions [33, 34]. Additionally, its cost-effectiveness and minimal substrate requirements make it a sustainable choice for large-scale hydroponic farming.

Wick System

This is a simple and passive hydroponic method that operates without the need for electricity. Plant roots are placed in pots filled with a growing medium like perlite, connected to a nutrient solution reservoir through a nylon wick. Water and nutrients are reached to the roots through capillary action, and the supply can be regulated by using larger or multiple wicks [35, 36]. This system is best suited for small plants, herbs, and spices but is less effective for water-intensive crops.

Drip System

In a drip system, plant roots are typically exposed to a nutrient solution delivered via a drip mechanism. An electric pump ensures the continuous and precise supply of nutrients from a central reservoir. This method is recognized for its cost-effectiveness and simplicity compared to other hydroponic systems. It enables the systematic cultivation of a wide range of plant species while promoting efficient water conservation [37]. A pumping mechanism distributes the nutrient solution to each plant root in optimal proportions [38]. Plants are generally supported in a mildly absorbent growing medium, which facilitates the gradual release of nutrients, ensuring sustained uptake [39].

Nutrient Film Technique [NFT] Systems

With this technique, there is no need for a growing medium, and maintenance expenses are decreased because plant roots are suspended in a nutritional solution inside plastic baskets. A submerged pump ensures a regular flow of the nutrient solution, which moves through a growing tube or tray, bathing the roots before draining back into the reservoir [33]. The NFT utilizes a thin, recirculating film of nutrient solution that flows through channels, directly nourishing the bare roots [40]. A dense root mat develops at the channel base, enhancing nutrient absorption. The excess solution collects at the lower end and returns to the reservoir, making NFT an efficient system for growing various vegetables.

Deep Flow Technique System

In the deep flow technique [DFT], plants grow in drainage channels filled with a continuously flowing nutrient solution, typically 4–6 cm deep, which keeps the roots submerged. Nutrient solutions are recirculated: when the water level in the culture bed drops below a preset value, the solution is replenished and supplied to the plant roots at regular intervals [41]. This hydroponic system eliminates the need for soil by suspending the roots of the plants in a nutrient-rich medium. Oxygen for root respiration is provided by dissolved oxygen in the water. However, since the dissolved oxygen concentration in aerated hydroponic water is much lower than in soil, plant roots in DFT are thought to endure continuous hypoxic stress, regardless of aeration levels [42].

Ebb and Flow [Flood and Drain] System

In the flood and drain system, the grow tray is periodically flooded with nutrient solution, which is then drained back into the reservoir using an underwater pump controlled by a timer. This system is versatile and can accommodate various growing media such as gravel, peat moss, vermiculite, perlite, or granular rock wool [34]. The nutrient solution floods the growing medium for a short period, after which it drains back, drawing air into the root zone and providing oxygen. Plants absorb water and nutrients from the moist medium. While the fluctuating root environment may not be ideal for optimal growth, plant performance is generally satisfactory [43]. The system uses gravity to return the excess solution to the reservoir for reuse, making it a sustainable option for hydroponic cultivation.

Floating Raft System [FRS]

FRS is an inexpensive hydroponic method typically used for dense, short-cycle greenhouse cultivation of leafy vegetables, especially those grown for the fresh-cut market [44]. In this system, plants are supported by a floating raft constructed from expanded plastic, allowing their roots to remain submerged in the nutrient solution.

2.2.2 *Aquaponics*

Aquaponics is a recirculating bio-integrated system that synergizes hydroponic plant cultivation with aquaculture. In this closed-loop system, nutrient-rich effluent from fish tanks undergoes biological filtration, wherein nitrifying bacteria such as Nitrosomonas convert toxic ammonia into bioavailable nitrate, which is then assimilated by plants for growth [45]. Simultaneously, plants function as biofilters, removing excess nutrients and purifying the water before it is recirculated into the fish tanks. Additionally, plant uptake of CO_2 from fish respiration and the heat-absorbing capacity of aquaculture water contribute to microclimate stabilization within greenhouse environments, reducing energy costs [46]. This integrated system exemplifies sustainable intensification by enhancing nutrient recycling, optimizing water usage, and improving productivity [47].

2.2.3 *Aeroponics*

Aeroponics represents the most water-efficient vertical farming technique, using up to 90% less water compared to conventional hydroponic systems [45]. Unlike hydroponics and aquaponics, aeroponics eliminates the need for any solid or liquid growing medium. Instead, plant roots are suspended in a controlled environment of air and intermittently misted with a fine spray of nutrient-rich solution, ensuring maximum oxygen availability, and promoting optimal nutrient absorption. The aeroponic systems further enhance sustainability by leveraging gravity for automatic drainage, thereby minimizing energy expenditure on water recirculation [48]. This system has been shown to overcome several physiological constraints associated with hydroponic cultivation, including improved root oxygenation, enhanced nutrient-use efficiency, and increased biomass accumulation [49]. Aeroponics is a promising technique for high-yield, resource-efficient plant production in controlled environment agriculture. Aeroponics eliminates the need for containers or grow trays to hold water, as it uses a nutrient-rich mist instead. Additionally, the misting system supplies enhanced oxygen to the roots, promoting accelerated growth, while being free from the use of fertilizers or pesticides [50, 51].

2.3 Medicinal Plant Cultivation in Vertical Farming

Medicinal plant cultivation in VF offers a sustainable solution to meet the growing demand for herbal remedies while addressing land and environmental challenges. Various medicinal species, including sweet basil [*Ocimum basilicum],* Maskbala [*Valeriana jatamansi*], Mint [*Mentha piperita],* and Ginseng [Panax ginseng] are successfully grown in VF systems, benefiting from the controlled environment that enhances their medicinal properties (Table 2.2). Below are some studies that demonstrate the successful cultivation of medicinal plants in VF systems.

2.3.1 Ocimum basilicum *[Sweet basil]*

Sweet basil [*Ocimum basilicum*] is an aromatic culinary herb belonging to the Lamiaceae family. The medicinal value of basil is extensively recognized in modern as well as traditional medical systems. Currently, *O. basilicum* is valued for its characteristic fragrance and is widely used in the food processing, cosmetic, and pharmaceutical industries [68]. In a study, seedlings of three Genovese basil cultivars "Eleonora" [Enza Zaden, Enkhuizen, Netherlands], "Aroma 2" [Fenix, Belpasso, Italy], and "Italiano Classico" [La Semiorto, Sarno, Italy] were grown 15 days after sowing under soil-less cultivation. The research conducted a thorough and multifaceted evaluation of the useful and exploratory response of basil, emphasizing the advantageous outcomes of managing pre-harvest variables in soil-less cultivation. These factors contributed to improvements in biomass production and the accumulation of specialized metabolites [69].

Ocimum basilicum [sweet basil], *Agastache rugosa* [Korean mint], and *Artemisia annua* [sweet wormwood] are annual herbaceous plants known for producing organic bioactive molecules that serve as defense mechanisms in response to stressors. Humans have utilized these bioactive compounds for medical and nutritional reasons for generations [70]. This investigation involved cultivating these plants under VF while continually exposing them to external ethanol treatment at doses of 0, 0.5, 1, 2, 4, and 8 mM. The results showed that low-concentration ethanol treatments [0.5, 1, and 2 mM] enhanced the shoot dry biomass of *O. basilicum* and

A. annua relative to the untreated plants. In contrast, high-concentration ethanol treatments [4 and 8 mM] induced significant oxidative stress in the plants [52].

2.3.2 Salvia officinalis *[Sage]*

Salvia officinalis commonly known as sage is a renowned evergreen aromatic and medicinal plant belonging to the Labiatae family. A study was performed to assess the impact of water spraying durations and Indole-3-butyric acid [IBA] exposure on the adventitious root formation and quality of cuttings from sage cultivars. The study emphasized the following primary variables: [a] spraying durations of 5 min [I_5] and 10 min [I_{10}], [b] IBA exposure at 0 g L^{-1} [C0, control] and 1 g L^{-1} [C1], and [c] two sage cultivars, "Farina Silver Blue" and "La Siesta." The findings indicated that sage cuttings cultivated in aeroponic environments exhibited superior performance when subjected to pre-treatment with external auxin [IBA]. Overall, cuttings cultivated with a 10-min spraying duration [I_{10}] demonstrated superior quality characteristics relative to those produced with a 5-min interval [I_5]. The results provided underscore the possibility of aeroponics for long-term cutting production [71].

2.3.3 Picrorhiza kurroa *[Kutki]*

Picrorhiza kurroa Royle ex Benth. also known as kutki, belongs to the family Scrophulariaceae. *P. kurroa* cultivated in hydroponic and aeroponic systems under controlled conditions to assess growth and biomass production. The aeroponic system outperformed the hydroponic system in enhancing biomass production and plant quality, while the hydroponic system excelled in promoting rootlet proliferation and elongation. These results suggest that aeroponics is a superior method for the large-scale production of high-quality *P. kurroa* biomass for pharmaceutical applications [13].

2.3.4 Mentha piperita *[Mint]*

Mentha piperita, *Mentha spicata*, and *Mentha suaveolens* can be cultivated in aeroponic system. In aeroponic systems, increased aeration of the nutrient solution delivers more oxygen to the plant roots compared to conventional hydroponic systems, promoting growth, and reducing the risk of pathogen formation. Among the species evaluated, *M. aquatica* demonstrated superior performance in terms of yield-related traits, including leaf number, leaf area, and both fresh and dry weights of the plant, shoots, and leaves. These findings highlight the high potential of all *Mentha* species for cultivation in aeroponic systems [72].

2.3.5 Valeriana officinalis *[Common Valerian]*

Valeriana officinalis var. common [family Valerianaceae] was evaluated for biomass production and essential oil content and composition under four different cultivation systems: aeroponic, floating, growing media [a perlite and vermiculite mix], and soil. The dry roots and rhizomes of valerian are utilized in contemporary phytomedicinal items, encompassing moderate sedatives, antianxiety therapies, and gastrointestinal preparations [73]. The outcomes from the valerian cultivation experiments indicate that floating cultivation can produce yields equivalent to, or superior to, those from conventional field production [74].

2.3.6 Artemisia annuaa *[Annual Wormwood]*

Artemisia annua belongs to the family Asteraceae, a medicinal herb known for its anti-malarial properties. The effect of copper [Cu] and zinc [Zn] supplementation on *A. annua* was investigated under a hydroponics system. The Zn16X treatment led to a 35-fold enhancement in zinc accumulation in the plant's aerial parts, without impacting its height or yield, indicating no toxicity. Although Cu and Zn treatments reduced the artemisinin content, they significantly enhanced the production of phenolic acids and flavonoids, especially in the Zn4X treatment, which also showed elevated levels of caffeoylquinic acid. These findings suggest that Cu and Zn enrichment can stimulate secondary metabolite production, potentially enhancing the therapeutic value of *A. annua* in hydroponic systems within vertical farming [54].

2.3.7 Panax ginseng *[Ginseng]*

Ginseng [*Panax ginseng*] is a valuable medicinal plant widely used in pharmaceuticals and cosmetics. Traditional open-field [OF] cultivation is subject to environmental fluctuations and pesticide contamination, which can affect yield and quality. This study investigates the potential of a VF platform for stable and sustainable ginseng berry production. Compared to OF cultivation, VF ensures higher yields per area, improved phytochemical composition, and minimal pesticide contamination. This study demonstrates the feasibility of VF as a sustainable alternative for year-round ginseng berry production, enhancing its potential use in the pharmaceutical and cosmetic industries [75].

2.4 Leveraging Artificial Intelligence for Advancements in Vertical Farming

Artificial Intelligence (AI) has revolutionized modern agricultural practices by integrating machine learning (ML), deep learning (DL), the Internet of Things (IoT), and cloud computing, thus enabling precision farming and automation in VF systems [76]. AI-driven smart farming technologies offer real-time monitoring, data-based decision-making, and automation, enhancing crop productivity, resource efficiency, and sustainability [77] (Fig. 2.3). The deployment of AI in VF has evolved from simple environmental control systems to adaptive, self-learning frameworks capable of optimizing plant growth, predicting harvest times, and detecting diseases at early stages [78]. The success of VF relies on numerous automated procedures, including automated processing, bed management and refilling, water supply, lighting, fertilizing, and crop monitoring via visual systems [79].

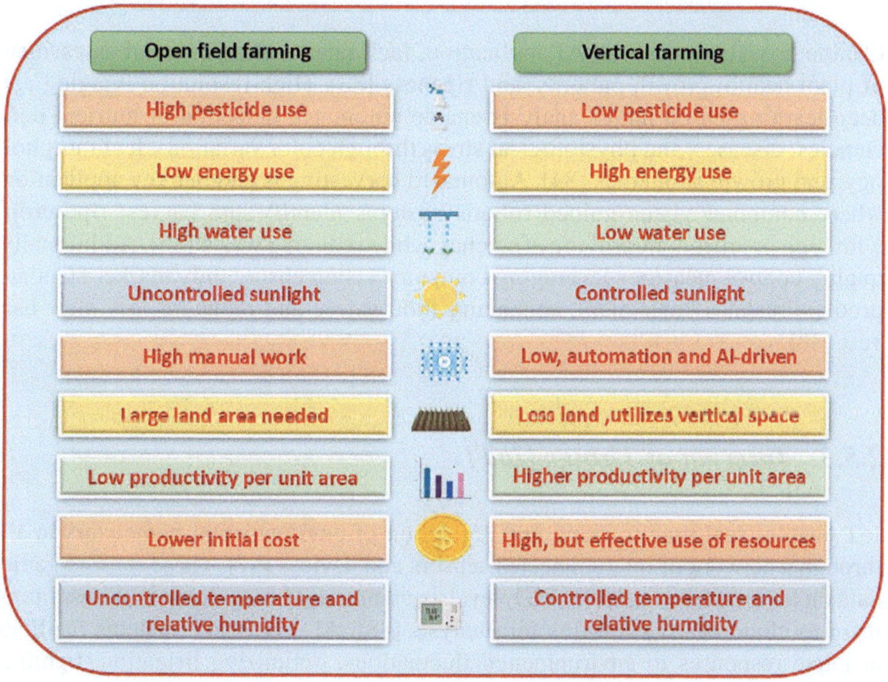

Fig. 2.3 Integration of AI Technologies in Smart and Sustainable Vertical Farming

2.5 AI-Driven Technologies in Sustainable Vertical Farming

2.5.1 Machine Learning Algorithms

Machine learning (ML), a subset of AI, enables predictive modeling for optimizing crop yields, planting schedules, and resource allocation. ML techniques process vast datasets, identifying patterns in environmental variables such as temperature, humidity, light intensity, and nutrient levels [80]. ML algorithms are categorized into supervised, unsupervised, and reinforcement learning models, each contributing to aspects of VF such as disease detection, weather prediction, yield estimation, and irrigation management [81, 82]. These predictive capabilities facilitate adaptive farming strategies, mitigating risks associated with fluctuating environmental conditions.

2.5.2 Computer Vision Technologies

Computer vision, a crucial AI application, facilitates real-time visual assessment of plant health, growth patterns, and ripeness [83]. High-resolution imaging and deep learning-based image analysis enable automated detection of nutrient deficiencies, diseases, and physiological stress through color variations, leaf morphology, and growth anomalies [84]. Automated harvesting is another key application, where computer vision-guided robotic systems identify and harvest ripe crops with high precision, improving efficiency while reducing waste [85]. Additionally, quality control measures leveraging computer vision ensure only market-standard produce reaches consumers, enhancing food safety and reducing loss after harvest [86].

2.5.3 Internet of Things [IoT]

IoT enables real-time tracking and regulation of environmental parameters in VF through a network of interconnected sensors and devices [87]. These sensors gather data on temperature, humidity, CO_2 levels, and nutrient concentrations, transmitting them to cloud-based platforms for analysis [88]. AI-driven IoT systems facilitate adaptive responses to environmental fluctuations, optimizing irrigation, lighting, and ventilation for maximum yield [89, 90]. However, challenges such as data security, interoperability, and high implementation costs remain key concerns in IoT deployment in VF [91].

2.5.4 *Robotics and Automation*

Robotics and automation play a crucial role in minimizing labor dependency and enhancing precision in VF operations [92]. Robotic systems are employed in planting, pruning, nutrient delivery, and harvesting, ensuring consistency and efficiency while reducing operational costs [93]. Agricultural robots [AgBots] integrate IoT and AI to perform complex tasks such as real-time crop monitoring, pest control, and automated sorting and packaging. The application of drones and autonomous systems further enhances resource utilization, contributing to sustainable and scalable farming practices [94].

2.6 Advantages of Vertical Farming

- Vertical farming significantly enhances productivity per unit area, yielding up to 80% more harvest compared to conventional farming. This increased productivity is attributed to three key contributors: (1) the higher yield per square meter, enabled by the reducing the land footprint through the vertical stacking of crop cultivation units, (2) an improved photosynthetic rate, facilitated by an optimized combination of environmental conditions, and (3) the ability to produce crops perpetually [95, 96].
- Another advantage of VF is the reduced or eliminated need for soil, which subsequently mitigates the risk of pest and disease infestations. It also enables continuous production over the year, unaffected by natural disruptions such as floods, heavy rainfall, snowfall, droughts, or pest and disease outbreaks [96, 97].
- VF reduces transportation costs by shortening the distance between food production and urban areas, thus minimizing fossil fuel consumption associated with transporting produce from rural regions to cities. Additionally, VF uses 70–95% less water than traditional farming practices. This efficiency is due to hydroponic cultivation, where irrigation water is provided through a closed-loop system, eliminating the impact of droughts, and stabilizing the climate [98–102].
- Nutrient loss to soil and aquatic environments is minimized due to the recirculation of nutrient solutions in a closed circuit [79]. VF also promotes organic food production, as it does not require the use of pesticides due to limited exposure to the external environment, although the potential for pest contamination cannot be entirely ruled out [97].
- Consumers benefit from fresh produce that retains its original nutrient composition. Additionally, VF contributes to urban greening, helping to mitigate rising temperatures and reduce air pollution in cities. The production cycle is notably

faster in vertical farming, as studies demonstrate that soil-less agricultural methods, such as vertical farming and hydroponics, accelerate crop growth and improve product quality [13]. Furthermore, VF systems are energy-efficient, promote early plant growth, and allow for easier harvesting [41].

2.7 Disadvantages of Vertical Farming

- One of the main challenges of vertical farming is the significant upfront cost required to set up the system. This includes the expense of remote monitoring systems, software, automated shelving systems, and climate control infrastructure [103, 104].
- The absence of insects in VF systems presents a potential issue for crop pollination, which may require artificial intervention to ensure effective fertilization.
- VF systems also incur high energy costs, as plant growth is entirely dependent on artificial lighting, which can significantly increase electricity consumption [105].
- Furthermore, improperly managed excess nutrients in vertical farming can pose a risk of contaminating urban water systems.
- Another challenge is the generation of waste, including plant residues and other byproducts, which can accumulate around VF facilities and contribute to environmental pollution.
- Lastly, there may be a shortage of skilled labor required to operate and manage VF systems initially. Consequently, a well-trained workforce will need to be developed through specialized training programs [106].

2.8 Conclusion and Future Directions

The present chapter provides a thorough examination of IoT and ML methodologies employed in VF for the cultivation of medicinal plants. The chapter also examines the efficacy of remotely monitoring VF factors such as EC, pH, humidity, root zone temperature, and irrigation period through IoT devices. VF methods, including hydroponics, aeroponics, and aquaponics, use less land and resources while producing more biomass and plants continuously and sustainably. Numerous related issues, including soil-borne diseases, chemical pesticides and fertilizers, and the wasting of water and nutrients, can be reduced through the implementation of VF or other soil-less agricultural methods. VF can improve the output and quality of medicinal plants by refining growth conditions. Nonetheless, its extensive application encounters obstacles, requiring additional research to enhance the processes for optimizing cultivation and elevating the standard of medicinal plant production. In conclusion, VF has come out as an intriguing approach for improving the global production of medicinal plants. VF and AI have an opportunity to make significant improvements to the meet requirements of medicinal plant production for plant-based products and herbal drug formulation by implementing strategy development, nurturing partnerships, and promoting environmental sustainability.

Acknowledgement The authors acknowledge the Council of Scientific and Industrial Research (CSIR), Government of India, under the project "Development of industrially sustainable products and processes based on Himalayan Bioresources using advanced fermentation and phytofarming technologies" (MLP-0207) for providing financial support. The authors are thankful to the Director, CSIR-IHBT, Palampur for providing necessary facilities.

Author Contribution Aditi Sharma: Literature reviewed, Table and figure preparation, Writing the manuscript, Shashi Rani: Literature reviewed, Table and figure preparation, Writing the manuscript, Amit Kumar: Literature reviewed, Table and figure preparation, Writing the manuscript, Pankaj Kumar: Literature reviewed, Writing and editing the manuscript, Ashish R. Warghat: Conceptualisation, Literature reviewed, supervision, Writing and editing the manuscript. All the authors have read and approved the final manuscript.

References

1. Novak J, Blüthner WD (2020) Medicinal, aromatic and stimulant plants. Springer. ISBN 13978-3030387921
2. Hakimi Y, Fatahi R, Naghavi MR, Zamani Z, Shokrpour M (2021) Medicinal plants and boosting the immune system. In: Fourth international conference on agricultural sciences, medicinal plants and traditional medicine, pp 1–33
3. Erb M, Kliebenstein DJ (2020) Plant secondary metabolites as defenses, regulators, and primary metabolites: the blurred functional trichotomy. Plant Physiol 184(1):39–52
4. Mahajan M, Kuiry R, Pal PK (2020) Understanding the consequence of environmental stress for accumulation of secondary metabolites in medicinal and aromatic plants. J Appl Res Med Aromat Plants 18:100255
5. Sun Y, Fernie AR (2024) Plant secondary metabolism in a fluctuating world: climate change perspectives. Trends Plant Sci 29(5):560–571
6. Mipeshwaree Devi A, Khedashwori Devi K, Premi Devi P, Lakshmipriyari Devi M, Das S (2023) Metabolic engineering of plant secondary metabolites: prospects and its technological challenges. Front Plant Sci 14:1171154
7. Johnson OO, Ayoola GA (2015) Antioxidant activity among selected medicinal plants combinations (multi-component herbal preparation). Int J Res Health Sci 3(1):526–532
8. Asigbaase M, Adusu D, Anaba L, Abugre S, Kang-Milung S, Acheamfour SA, Adamu I, Ackah DK (2023) Conservation and economic benefits of medicinal plants: insights from forest-fringe communities of Southwestern Ghana. Trees Forests People 14:100462
9. SharathKumar M, Heuvelink E, Marcelis LFM (2020) Vertical farming: moving from genetic to environmental modification. Trends Plant Sci 25:724–727
10. Eigenbrod C, Gruda N (2014) Urban vegetable for food security in cities. A review. Agron Sustain Dev 35:483–498
11. Dsouza A, Dixon M, Shukla M, Graham T (2025) Harnessing controlled-environment systems for enhanced production of medicinal plants. J Exp Bot 76(1):76–93
12. Karimi G, Vahabzadeh M, Lari P, Rashedinia M, Moshiri M (2011) "Silymarin", a promising pharmacological agent for treatment of diseases. Iran J Basic Med Sci 14(4):308

13. Thakur K, Partap M, Kumar D, Warghat AR (2019) Enhancement of picrosides content in *Picrorhiza kurroa* Royle ex Benth. mediated through nutrient feeding approach under aeroponic and hydroponic system. Ind Crops Prod 133:160–167
14. Bafort F, Bouhadada A, Ancion N, Crutzen N, Jijakli MH (2022) Optimization of hydroponic production of the medicinal plants *Euphorbia peplus* and *Artemisia annua* in container. In XXXI international horticultural congress (IHC2022): international symposium on advances in vertical farming. 1369:171–178
15. Prasad A, Pragadheesh VS, Mathur A, Srivastava NK, Singh M, Mathur AK (2012) Growth and centelloside production in hydroponically established medicinal plant-*Centella asiatica* (L.). Ind Crops Prod 35(1):309–312
16. Dass SM, Chai TT, Wong FC (2022) Investigation on the effects of hydroponic-planting versus soil-planting using *Plectranthus amboinicus* as a plant model. J Trop Life Sci 12(1):53–61
17. Partap M, Kumar P, Kumar A, Joshi R, Kumar D, Warghat AR (2020) Effect of elicitors on morpho-physiological performance and metabolites enrichment in *Valeriana jatamansi* cultivated under aeroponic conditions. Front Plant Sci 11:01263
18. Stewart CL, Lovett-Doust L (2003) Effect of phosphorus treatment on growth and yield in the medicinal herb *Calendula officinalis* L. (Standard Pacific) under hydroponic cultivation. Can J Plant Sci 83(3):611–617
19. Ahmadi F, Samadi A, Sepehr E, Rahimi A, Shabala S (2021) Improving essential oil compositions of purple coneflower (*Echinacea Purpurea* L.) medicinal plant using novel growing media and nutrition pattern in hydroponics
20. Von Bieberstein P, Xu YM, Gunatilaka AL, Gruener R (2014) Biomass production and withaferin a synthesis by *Withania somnifera* grown in aeroponics and hydroponics. HortScience 49(12):1506–1509
21. Alexopoulos AA, Marandos E, Assimakopoulou A, Vidalis N, Petropoulos SA, Karapanos IC (2021) Effect of nutrient solution pH on the growth, yield and quality of *Taraxacum officinale* and *Reichardia picroides* in a floating hydroponic system. Agronomy 11(6):1118
22. Paponov M, Flate J, Ziegler J, Lillo C, Paponov IA (2023) Heterogeneous nutrient supply modulates root exudation and accumulation of medicinally valuable compounds in *Artemisia annua* and *Hypericum perforatum* front. Plant Sci 14:1174151
23. Venkatachalam P, Jayalakshmi N, Geetha N, Sahi SV, Sharma NC, Rene ER, Sarkar SK, Favas PJ (2017) Accumulation efficiency, genotoxicity and antioxidant defense mechanisms in medicinal plant *Acalypha indica* L. under lead stress. Chemosphere 171:544–553
24. Pedneault K, Dorais M, Léonhart S, Angers P, Gosselin A (2014) Time-course accumulation of flavonoids in hydroponically grown *Achillea millefolium* L. Can J Plant Sci 94(2):383–395
25. Managa LR, du Toit ES, Prinsloo G (2021) Variations in the leaf metabolite profile between hydroponic and field grown *Moringa oleifera* Lam. genotypes. Biochem Syst Ecol 97:104302
26. Indira D, Sabitha Rani A (2024) Comparative analysis of growth parameters in hydroponic and soil-grown systems of *Ocimum basilicum* L. (Basil). Plant Sci Arch 9(2):26–32
27. Datta S, Raturi PP, Singh I, Nath A, Singh RC, Sawant L (2024) Comparative analysis of phytochemical contents in hydroponically grown *Bacopa monnieri* (L.) Pennel with the soil grown plants
28. Rathore S, Rai S (2024) Interplay of intellectual property rights, technological innovation, and sustainable food security: a comprehensive analysis. In: E-financial strategies for advancing sustainable development: fostering financial inclusion and alleviating poverty. Springer Nature, Cham, pp 351–361
29. Zobayed S et al (2020) Transplant production in closed systems, blueberry. In: Plant factory: an indoor vertical farming system for efficient quality food production, 2nd edn. Academic Press, San Diego, CA, pp 318–323
30. Partap M, Rattan S, Kanika, Ashrita, Sood A, Kumar P, Warghat AR (2021) Hydroponic and Aeroponic cultivation of economically important crops for production of quality biomass. In: Agricultural biotechnology: latest research and trends, pp 573–585
31. Royston RM, Pavithra MP (2018) Vertical farming a concept. Int J Eng Tech 4:500–506

32. Swain A, Chatterjee S, Vishwanath M (2021) Hydroponics in vegetable crops: a review. Pharma Innov 10(6):629–634
33. Koriesh EM, Abo El-Soud IH (2020) Medicinal plants in hydroponic system under water-defcit conditions—a way to save water. In: Technological and modern irrigation environment in Egypt. Springer, Cham, pp 131–153
34. Mihrete TB (2025) Crop substrates for sustainable hydroponic farming
35. Eldridge BM, Manzoni LR, Graham CA, Rodgers B, Farmer JR, Dodd AN (2020) Getting to the roots of aeroponic indoor farming. New Phytol 228(4):1183–1192
36. Kannan M, Elavarasan G, Balamurugan A, Dhanusiya B, Freedon D (2022) Hydroponic farming–a state of art for the future agriculture. In: Materials today: proceedings
37. Gentry M (2019) Local heat, local food: integrating vertical hydroponic farming with district heating in Sweden. Energy 174:191–197
38. Rouphael Y, Colla G (2009) The influence of drip irrigation or sub irrigation on zucchini squash grown in closed-loop substrate culture with high and low nutrient solution concentration
39. Gaikwad DJ, Maitra S (2020) Hydroponics cultivation of crops. In: Protected cultivation and smart agriculture, pp 279–287
40. Domingues DS, Takahashi HW, Camara CA, Nixdorf SL (2012) Automated system developed to control pH and concentration of nutrient solution evaluated in hydroponic lettuce production. Comput Electron Agric 84:53–61
41. Son JE, Kim HJ, Ahn TI (2020) Hydroponic systems. In: Plant factory. Academic, Cambridge, pp 273–283
42. Pramono S, Nuruddin A, Ibrahim MH (2020) Design of a hydroponic monitoring system with deep flow technique (DFT). In: AIP conference proceedings, vol 2217. AIP Publishing
43. Jones JB Jr (2014) Complete guide for growing plants hydroponically. CRC Press
44. Maggini R, Kiferle C, Guidi L, Pardossi A, Raffaelli A (2011) Growing medicinal plants in hydroponic culture. In: International symposium on advanced technologies and management towards sustainable greenhouse ecosystems, vol 952. Greensys, pp 697–704
45. Barui P, Ghosh P, Debangshi U (2022) Vertical farming-an overview. Plant Arch 22(2):09725210
46. Lennard W, Goddek S (2019) Aquaponics: the basics. In: Aquaponics food production systems, p 113
47. Samikannu R, Koshariya AK, Poornima E, RameshS, KumarA, BoopathiS (2023) Sustainable development in modern aquaponics cultivation systems using IoT technologies. In: Human agro-energy optimization for business and industry. IGI Global, pp 105–127
48. Garzón J, Montes L, Garzón J, Lampropoulos G (2023) Systematic review of technology in aeroponics: introducing the technology adoption and integration in sustainable agriculture model. Agronomy 13(10):2517
49. Fasciolo B, Awouda A, Bruno G, Lombardi F (2023) A smart aeroponic system for sustainable indoor farming. Proc CIRP 116:636–641
50. Engler N, Krarti M (2021) Review of energy efficiency in controlled environment agriculture. Renew Sust Energ Rev 141:110786
51. Khan E, Ahmed T, Anis MI, Rehman SA (2023) Pest recognition and cloud monitoring in aeroponic system
52. Shin J, Lee Y, Hahm S, Lee K, Park J (2023) Effects of exogenous ethanol treatment in nutrient solution on growth and secondary metabolite contents of three herb species in an indoor vertical farming system. Plants 12(22):3842
53. Kim YL, Yeom MS, Sim HS, Lee GO, Kang IJ, Yang GS, Yun JG, Son KH (2024) Effect of pre-harvest intermittent UV-B exposure on growth and secondary metabolites in *Achyranthes japonica* Nakai microgreens in a vertical farm. Horticulturae 10(10):1040
54. Son YJ, Park JE, Lee N, Ju YW, Pyo SH, Oh C, Yoo G, Nho CW (2024) Copper-or zinc-fortified nutrient solution in vertical farming system enriches copper or zinc and elevates phenolic acid and flavonoid contents in Artemisia annua L. Agronomy 14(1):135

55. Hahm S, Lee Y, Lee K, Park J (2024) Optimization of cannabinoid production in hemp through methyl Jasmonate application in a vertical farming system. Horticulturae 10(11):1165
56. Ju YW, Pyo SH, Park SW, Moon CR, Lee S, Benashvili M, Park JE, Nho CW, Son YJ (2024) Treatment of water extract of green tea during kale cultivation using a home vertical farming appliance conveyed catechins into kale and elevated glucosinolate contents. Curr Res Food Sci 9:100852
57. Hahm S, Lee B, Bok G, Kim S, Park J (2023) Diniconazole promotes the yield of female hemp (*Cannabis sativa*) inflorescence and cannabinoids in a vertical farming system. Agronomy 13(6):1497
58. Jang SN, Kang MJ, Kim YN, Jeong EJ, Cho KM, Yun JG, Son KH (2023) Physiological and biochemical responses of *Limonium tetragonum* to NaCl concentrations in hydroponic solution. Front Plant Sci 14:1159625
59. Mubeen B, Hasnain A, Mehboob R, Rasool R, Riaz A, Elaskary SA, Shah MM, Faridi TA, Ullah I (2022) Hydroponics and elicitation, a combined approach to enhance the production of designer secondary medicinal metabolites in Silybum *marianum*. Front Plant Sci 13:897795
60. Gengmao Z, Quanmei S, Yu H, Shihui L, Changhai W (2014) The physiological and biochemical responses of a medicinal plant (*Salvia miltiorrhiza* L.) to stress caused by various concentrations of NaCl. PLoS One 9(2):e89624
61. Roosta HR (2014) Comparison of the vegetative growth, eco-physiological characteristics and mineral nutrient content of basil plants in different irrigation ratios of hydroponic: aquaponic solutions. J Plant Nutr 37(11):1782–1803
62. Yep B, Zheng Y (2020) Potassium and micronutrient fertilizer addition in a mock aquaponic system for drug-type *Cannabis sativa* L. cultivation. Can J Plant Sci 101(3):341–352
63. Krishnaraj C, Jagan EG, Ramachandran R, Abirami SM, Mohan N, Kalaichelvan PT (2012) Effect of biologically synthesized silver nanoparticles on *Bacopa monnieri* (Linn.) Wettst plant growth metabolism. Process Biochem 47(4):651–658
64. Baharmoghadam M, Keramat B, Nadernejad N, Mozafari H, Pourseyedi S (2024) Improving essential oil production and antioxidant properties of *Carum copticum* L. by titanium dioxide nanoparticles and light spectrum interaction under hydroponic conditions. Plant Cell Tissue Organ Cult (PCTOC) 159(2):53
65. Adabavazeh F, Pourseyedi S, Nadernejad N, Razavizadeh R, Mozafari H (2022) Evaluation of synthesized magnetic nanoparticles and salicylic acid effects on improvement of antioxidant properties and essential oils of *Calotropis procera* hairy roots and seedlings. Plant Cell, Tissue and Organ Culture (PCTOC) 151(1):133–148
66. Mohit Rabary P, Movahedi Z, Ghabooli M, Rostami M (2020) Exogenous salicylic acid improves photosynthetic pigments and morphological traits of four medicinal plants in an aeroponic system. J Plant Mol Breed 8(1):1–9
67. Grzelka K, Matkowski A, Ślusarczyk S (2023) Electrostimulation improves plant growth and modulates the flavonoid profile in aeroponic culture of *Scutellaria baicalensis* Georgi. Front Plant Sci 14:1142624
68. Barátová S, Mezeyová I, Hegedúsová A, Andrejiová A (2015) Impact of biofortification, variety and cutting on chosen qualitative characteristic of basil (Ocimum basilicum L.). Acta Fytotechn Zootech 18(3):71–75
69. Ciriello M, Formisano L, El-Nakhel C, Corrado G, Pannico A, De Pascale S, Rouphael Y (2021) Morpho-physiological responses and secondary metabolites modulation by preharvest factors of three hydroponically grown genovese basil cultivars. Front Plant Sci 12:671026
70. Hussain MS, Fareed S, Ansari S, Rahman MA, Ahmad IZ, Saeed M (2012) Current approaches toward production of secondary plant metabolites. J Pharm Bioallied Sci 4:10
71. Scaltrito E, Cristiano G, Sdao AE, Gruda NS, Loconsole D, De Lucia B (2024) Influence of water spraying intervals and indole-3-butyric acid concentrations on salvia rooted cuttings quality in a closed aeroponics system. Sci Hortic 337:113452
72. Yaghobvand M, Mumivand H, Raji MR, Banan A (2022) Evaluation of growth, yield and morphological and physiological characteristics of five mint (Mentha spp.) species in the Aeroponic system under greenhouse. J Hortic Sci 36(3):603–618

73. Das S, Chaudhari AK, Singh VK, DwivedyAK, Dubey NK (2023) Chitosan based encapsulation of Valeriana officinalis essential oil as edible coating for inhibition of fungi and aflatoxin B1 contamination, nutritional quality improvement, and shelf life extension of Citrus sinensis fruits. Int J Biol Macromol 233:123565
74. Tabatabaei SJ (2008) Effects of cultivation systems on the growth, and essential oil content and composition of valerian. Int J Geogr Inf Syst 14(1–2):54–67
75. Park JE, Kim H, Kim J, Choi SJ, Ham J, Nho CW, Yoo G (2019) A comparative study of ginseng berry production in a vertical farm and an open field. Ind Crops Prod 140:111612
76. Sharma A (2020) AI and robotics for reducing waste in the food supply chain: systematic literature review, theoretical framework and research agenda
77. TalaviyaT, Shah D, Patel N, Yagnik H, Shah M (2020) Implementation of artificial intelligence in agriculture for optimisation of irrigation and application of pesticides and herbicides. Artif Intell Agric 4:58–73
78. Kaur G, Singh B (2022) Artificial intelligence (AI) based smart agriculture for sustainable development. Manag Account 57(6):54–57
79. Oh S, Lu C (2023) Vertical farming-smart urban agriculture for enhancing resilience and sustainability in food security. J Hortic Sci Biotechnol 98(2):133–140
80. Popkova EG (2022) Vertical farms based on hydroponics, deep learning, and AI as smart innovation in agriculture. In: Smart innovation in agriculture. Springer Nature, Singapore, pp 257–262
81. Siropyan M, Celikel O, Pinarer O (2022) Artificial intelligence driven vertical farming management system. In: Proceedings of the World congress on engineering 2022
82. Siregar RRA, Seminar KB, Wahjuni S, Santosa E (2022) Vertical farming perspectives in support of precision agriculture using artificial intelligence: a review. Computers 11(9):135
83. Shasteen K, Kacira M (2023) Predictive modeling and computer vision-based decision support to optimize resource use in vertical farms. Sustainability 15(10):7812
84. Pillutla BK (2022) Monitoring plants growth in indoor vertical farms using computer vision and AI techniques. Northern Illinois University
85. Franchetti B, Ntouskos V, Giuliani P, Herman T, Barnes L, Pirri F (2019) Vision based modeling of plants phenotyping in vertical farming under artificial lighting. Sensors 19(20):4378
86. Aquino HL, Concepcion RS, Dadios EP, Sybingco E, Bandala AA (2020) Trend forecasting of computer vision application in aquaponic cropping systems industry. In: 2020 IEEE 12th international conference on humanoid, nanotechnology, information technology, communication and control, environment, and management (HNICEM). IEEE, pp 1–6
87. Chin YS, Audah L (2017) Vertical farming monitoring system using the internet of things (IoT). In: AIP conference proceedings. AIP Publishing
88. Bhowmick S, Biswas B, Biswas M, Dey A, Roy S, Sarkar SK (2019) Application of IoT-enabled smart agriculture in vertical farming. In: Advances in communication, devices and networking: proceedings of ICCDN 2018. Springer, Singapore, pp 521–528
89. Halgamuge MN, Bojovschi A, Fisher PM, Le TC, Adeloju S, Murphy S (2021) Internet of things and autonomous control for vertical cultivation walls towards smart food growing: a review. Urban For Urban Green 61:127094
90. Nugroho ED, Putrada AG, Rakhmatsyah A (2021) Predictive control on lettuce NFT-based hydroponic IoT using deep neural network. In: 2021 international symposium on electronics and smart devices (ISESD). IEEE, pp 1–6
91. Putri RE, Wibowo M, Ardli J, Andasuryani A (2023) Monitoring and controlling of vertical farming system using internet of things (IoT). In: AIP conference proceedings, vol 2592. AIP Publishing
92. Krishnan A, Swarna S (2020) Robotics, IoT, and AI in the automation of agricultural industry: a review. In: 2020 IEEE Bangalore humanitarian technology conference (B-HTC), pp 1–6
93. Hayashi S, Yamamoto S, Tsubota S, Ochiai Y, Kobayashi K, Kamata J, Kurita M, Inazumi H, Peter R (2014) Automation technologies for strawberry harvesting and packing operations in Japan. J Berry Res 4(1):19–27
94. Petropoulou AS, van Marrewijk B, de Zwart F, Elings A, Bijlaard M, Van Daalen T, Jansen G, Hemming S (2023) Lettuce production in intelligent greenhouses—3D imaging and computer vision for plant spacing decisions. Sensors 23(6):2929

95. Khalil HI, Wahhab KA (2020) Advantage of vertical farming over horizontal farming in achieving sustainable city, Baghdad city-commercial street case study. In: IOP conference series: materials science and engineering, vol 745. IOP Publishing, p 012173
96. Mir MS, Naikoo NB, Kanth RH, Bahar FA, Bhat MA, Nazir A, Mahdi SS, Amin Z, Singh L, Raja W, Saad AA (2022) Vertical farming: the future of agriculture: a review. J Pharm Innov 11(2):1175–1195
97. Cowan N, Ferrier L, Spears B, Drewer J, Reay D, Skiba U (2022) CEA systems: the means to achieve future food security and environmental sustainability? Front Sustain Food Syst 6:891256
98. Benke K, Tomkins B (2017) Future food-production systems: vertical farming and controlled-environment agriculture. Sustain Sci Pract Policy 13(1):13–26
99. Kalantari F, Tahir OM, Joni RA, Fatemi E (2018) Opportunities and challenges in sustainability of vertical farming: a review. J Landsc Ecol 11(1):35–60. https://doi.org/10.1515/jlecol-2017-0016
100. Avgoustaki DD, Xydis G (2020) How energy innovation in indoor vertical farming can improve food security, sustainability, and food safety? In: Advances in food security and sustainability, vol 5. Elsevier, pp 1–51
101. Gurung L, Rawal JS, Puspa RC, Joshi GR, Mandal A (2024) Vertical farming in urban agriculture: opportunities, challenges, and future directions. Big Data Agric 6:89–95
102. Shi X, Shi C, Tablada A, Guan X, Cui M, Rong Y, Zhang Q, Xie X (2025) A review of research Progress in vertical farming on façades: design, technology, and benefits. Sustainability 17(3):921
103. Bafort F, Jijakli MH (2024) Digital agriculture: a solution for sustainable food and nutritional security. Vertical farming of medicinal plants. pp 129–177. ISBNs 9783031435485, 3031435486, 9783031435478
104. Son JE, Goto E, Kacira M, Orsini F (eds) (2022) Innovative technologies for vertical farming. Frontiers Media SA
105. Buscher J, Bakunowitsch J, Specht K (2023) Transformative potential of vertical farming—an urban planning investigation using multi-level perspective. Sustainability 15(22):15861
106. Moghimi F, Asiabanpour B (2023) Economics of vertical farming in the competitive market. Clean Techn Environ Policy 25(6):1837–1855

Chapter 3
Controlled Environment Agriculture for Medicinal Plant Production

Pratibha Jaryal, Promila Pathak, and Jaspreet Singh

Abstract Controlled Environment Agriculture (CEA) offers an innovative and efficient method for cultivating medicinal plants by precisely regulating environmental factors such as temperature, humidity, light, CO_2 concentration, and nutrient supply. Techniques including hydroponics, aeroponics, and vertical farming enable soil-less cultivation, resulting in faster growth rates, higher biomass yield, and enhanced accumulation of valuable secondary metabolites. Hydroponic systems have proven effective for producing pharmaceutical-grade plants with improved bioactive compound content, while aeroponics optimizes nutrient delivery and root oxygenation through fine misting techniques. Vertical farming further maximizes space utilization by stacking multiple crop layers, facilitating year-round production unaffected by seasonal and climatic limitations. Although challenges such as high initial costs and species-specific cultivation requirements remain, CEA promotes sustainable, pesticide-free, and scalable medicinal plant production. The integration of these advanced technologies can support global medicinal plant demands, reduce pressure on wild populations, and improve agricultural resource use efficiency. This review underscores the potential of CEA as a cornerstone in modern medicinal plant cultivation and sustainable agriculture.

Keywords Aeroponics · CEA system · Hydroponics · Medicinal plant · Vertical farming

P. Jaryal · P. Pathak (✉) · J. Singh
Orchid Laboratory, Department of Botany, Panjab University, Chandigarh, India

P. Kumar, A. R. Warghat (eds.), *Transforming Medicinal Plant Agriculture*, Smart Agriculture 16, https://doi.org/10.1007/978-981-95-3963-5_3

3.1 Introduction

3.1.1 Controlled Environment Agriculture (CEA)

Controlled Environment Agriculture (CEA) is a highly advanced and intensive method of hydroponic farming in which plants are cultivated in a precisely regulated environment to maximize the efficiency and effectiveness of horticultural practices. It is a closed-system novel farming method that uses a small footprint to grow crops [1]. CEA techniques demand sound knowledge of chemistry, horticulture, engineering, plant physiology, plant pathology, computers, and entomology. Crops may be grown vertically in warehouses, buildings, shipping containers, or specially designed structures that allow the temperature and humidity to be regulated [2]. The term, controlled environment agriculture, has been around since the 1960s [3, 4]. Despommier [5] was among the first to highlight the advantages of Controlled Environment Agriculture (CEA) and vertical farming systems, which were designed as enclosed systems to address the challenges posed by conventional farming in urban environments such as pest infestations, pollution, and waste management issues. According to Hashimoto [6], the CEA system operates similarly to industrial production lines, with various cultivation components working in a coordinated manner. These systems typically consist of an external structure that encloses stacked platforms or tiers, where crops are cultivated under optimized conditions for nutrient distribution [7]. Plant development is supported by artificial lighting and equipment like dehumidifiers and fans [7].

CEA requires six essential components:

1. A thermally insulated structure with opaque walls to maintain internal environmental stability.
2. A multi-level framework to accommodate crops and an integrated lighting system.
3. Pumps for dissipating heat from lighting and managing humidity levels.
4. A carbon dioxide supply system to promote efficient photosynthesis.
5. A mechanism for nutrient delivery.
6. Environmental control units to control electrical conductivity and pH levels, ensuring proper nutrient flows and uptake.

3.1.2 Importance of CEA in Medicinal Plant Production

Controlled Environment Agriculture (CEA) plays a vital role in modern medicinal plant production by offering a high level of control over environmental variables such as temperature, humidity, light intensity and quality, CO_2 concentration, and nutrient availability. This precise control allows growers to produce medicinal plants with consistent and standardized levels of bioactive compounds, which is crucial for ensuring the efficacy and safety of herbal medicines and pharmaceutical formulations. CEA systems enable year-round cultivation, overcoming the limitations of seasonal growth cycles and variable field conditions, which is especially beneficial for plants that require specific microclimates or are difficult to grow in open fields. Moreover, by manipulating environmental factors, CEA can enhance the synthesis of valuable secondary metabolites such as alkaloids, flavonoids, and terpenoids, increasing the therapeutic potency of the plant material. These systems also provide protection from pests, pathogens, and abiotic stress factors like drought and extreme weather, reducing the need for chemical inputs and ensuring a cleaner, more sustainable crop. Additionally, CEA supports the conservation and propagation of rare, slow-growing, or endangered medicinal species by providing optimal and replicable growth conditions, thereby reducing the pressure on wild populations. With the integration of technologies such as hydroponics, aeroponics, and automated monitoring, CEA also ensures efficient use of water and nutrients, minimizes environmental impact, and allows for traceable, scalable production that meets the regulatory standards required for medicinal plant use in global markets.

3.2 Key Technologies in Controlled Environment Agriculture

3.2.1 The Vertical Farming Models

Vertical farming is a multi-level indoor agriculture system that uses advanced technologies like recycled water, LED lighting, climate control, and hydroponics to optimize crop growth and reduce reliance on soil and seasonal conditions. The plants are supported in substrates like rock wool or perlite, where they receive essential nutrients, or their roots are directly exposed to flowing nutrient solution using the nutrient-film technique [8]. Air conditioning in vertical farming enables CO_2 enrichment and precise temperature control to boost plant growth. Unused water and nutrients are recycled, making the system efficient and sustainable. This method supports Clean, Green, Gourmet (CGG) food production and is suitable for growing various crops, herbs, and pharmaceuticals. Vertical farming reduces reliance on chemicals and minimizes climate-related risks by operating in controlled environments (Fig. 3.1). It supports sustainable, demand-driven production, enhances food security, and can repurpose urban buildings, offering a solution to both environmental challenges and unused infrastructure. Vertical farming enables mass production of medicinal plants with precise control over growth conditions and delivery timing. It operates independently of environmental and seasonal factors, reduces land and resource use, and ensures consistent quality. The system allows for optimized growth, faster production cycles, and manipulation of plant traits like flowering and active compounds. Vertical farming offers numerous benefits across environmental, economic, social, and political dimensions. It enables year-round, high-yield food production using fewer resources, with reduced

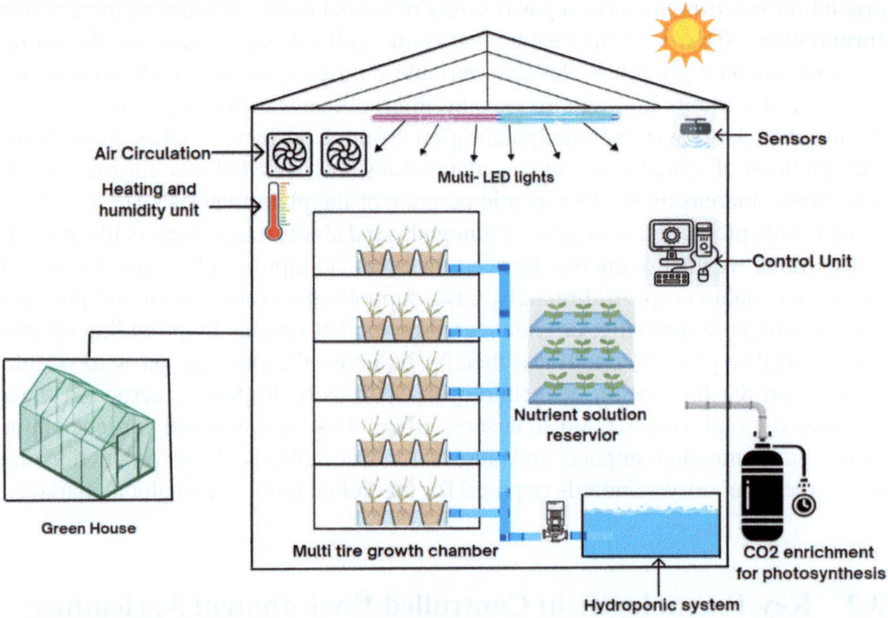

Fig. 3.1 Schematic representation of a vertical farming system within a controlled environment

emissions and minimal chemical use [9]. Economically, it cuts costs by eliminating soil, heavy machinery, and long-distance transport while creating export opportunities. Socially, it generates jobs and offers new roles for rural workers. Politically, it supports climate goals, enhances biosecurity, and improves energy resilience. However, challenges include high start-up costs, limited crop variety, smaller production volumes, and the need for skilled labor and investment. While vertical farming won't replace traditional farming, it is ideal for areas with space constraints, harsh climates, or access to cheap renewable energy. The propagation of some medicinal plants and the enhancement of the secondary metabolites using vertical farming are represented in Table 3.1.

3.2.2 Hydroponic in Medicinal Plant Production

Hydroponics is a soil-less cultivation method where plants grow in a nutrient-rich water solution, allowing precise control over nutrient supply and environmental conditions. This technique is increasingly significant in medicinal plant production due to its ability to produce high-quality plants with enhanced concentrations of valuable bioactive compounds. Hydroponic systems have proven effective for cultivating medicinal plants with enhanced secondary metabolite production and consistent quality [16, 17]. Controlled environments in hydroponic greenhouses reduce

Table 3.1 Enhanced production of secondary metabolites and clonal propagation in some medicinal plants using CEA

Medicinal Plant(s)	Cultivation Methodology	Key Findings	Reference
Artemisia annua and *Hypericum perforatum*	Aeroponics	Root exposure to white, blue, and red light enhanced the accumulation of artemisinin in shoots in shoots by 2.3-, 2.5-, and 1.9-fold, respectively. In *H. perforatum*, root exposure to white, blue, red, and green light enhanced the accumulation of coumaroylquinic acid in leaves by 89, 65, 84, and 74%,	[10]
Silybum marianum	Hydroponics with elicitors (MeJA, AgNPs, *A. niger*)	Enhanced production of secondary metabolites, including flavonoids and phenolics, through elicitor treatments in hydroponic systems.	[11]
Caralluma edulis, Leptadenia reticulata, Tylophora indica	Aeroponics	Successful clonal propagation of threatened medicinal plants with improved rooting and growth parameters.	[12]
Helichrysum cymosum	Hydroponics, aquaponics, and field cultivation	Aquaponic-cultivated H. cymosum yielded extracts with significantly lower antioxidant capacity and polyphenol contents than hydroponics and field-cultivated plants. However, there were no significant differences between field and hydroponic systems regarding antioxidant capacities and polyphenol contents	[13]
Withania somnifera	Hydroponics, Aeroponics,	Hydroponic techniques are optimal in reproducibly and efficiently generating withaferin A.	[14]
Picrorhiza kurroa	Hydroponics, Aeroponics,	*Picrorhiza kurroa* cultivation is suitable under the aeroponic system to produce quality biomass	[15]

variability caused by pests, temperature, and contaminants, resulting in pesticide-free, high-quality products such as *Calendula officinalis* [18] and *Hypericum perforatum* [19]. Hydroponics also offers a reliable platform for producing clean, pharmaceutical-grade *Centella asiatica* [20] and helps overcome germination and establishment issues common in soil cultivation [21–23]. Greenhouse hydroponics has also proven effective in overcoming germination and establishment difficulties frequently faced when cultivating medicinal plant species in soil, especially those that are hard to grow under open-field conditions [21–23].

Hydroponics generally promotes faster plant growth and higher biomass production compared to open-field cultivation, due to the direct availability of nutrients at the root zone, which are easily absorbed by the plants. Several studies have demonstrated that hydroponic systems significantly enhance both biomass yield and secondary metabolite production in medicinal plants. Léonhart et al. [16] reported that *Tanacetum parthenium*, *Achillea millefolium*, *Taraxacum officinale*, and *Calendula officinalis* adapted well to greenhouse hydroponic conditions, producing high yields and superior quality within a short time. Dorais et al. [24] found that in a floating raft system, the root and shoot dry weights of *Achillea millefolium*, *Artemisia vulgaris*, *Inula helenium*, *Stellaria media*, *Taraxacum officinale*, and *Valeriana officinalis* were significantly higher than those of field-grown plants after 50–120 days. Likewise, *Echinacea* spp., which typically require 2–4 years to mature in open fields, achieved comparable or higher biomass in hydroponic systems within a few months. Specifically, *Echinacea angustifolia* yielded similar root biomass in just 8 months from two consecutive hydroponic cycles as compared to traditional two-year field cultivation [25], while *Echinacea purpurea* showed a 2.3-fold yield increase over average soil-grown crops in North America [26]. These findings confirm that hydroponics can significantly shorten crop cycles and enhance biomass output. In addition to increased biomass, higher concentrations of secondary metabolites have also been observed. Essential oil content, in particular, was significantly higher in hydroponically grown plants such as *Pelargonium roseum*, *Cymbopogon citratus*, *Ocimum gratissimum*, *Vetiveria zizanioides*, and *Nepeta transcaucasica*, which produced 5–6 times more oil than field-grown counterparts. Moreover, the essential oil from hydroponically grown *Pelargonium roseum* had a higher geraniol content, indicating better quality [27].

While hydroponics has shown promise for many medicinal plants, some studies have reported limitations in secondary metabolite accumulation. For instance, *Echinacea angustifolia* grown hydroponically exhibited significantly lower root concentrations of caffeic acid derivatives, particularly echinacoside, compared to field-grown counterparts [28–31]. This discrepancy is likely due to ontogenetic factors, as hydroponically grown plants were harvested after only a few months, whereas field-grown plants are typically harvested after several years. These findings suggest that while many medicinal species adapt well to greenhouse hydroponic systems, others require further research to optimize cultivation protocols based on their specific biological needs. Improving the efficiency of hydroponic systems may involve careful genotype selection, as the concentration of bioactive compounds varies significantly among varieties. Greenhouse hydroponics is generally considered a profitable approach for producing medicinal plants, offering high biomass yield, clean raw material, and ease of harvesting and processing. Moreover, targeted adjustments to the nutrient solution, such as optimizing nitrogen levels, can further enhance the production of desired secondary metabolites.

3.2.3 Aeroponics in Medicinal Plant Production

The advancement of modern agriculture is increasingly driven by the demand for efficient, soil-less cultivation systems that optimize plant growth under controlled conditions. Among these, aeroponics has emerged as a high-performance crop production technology, gaining global recognition for its potential to enhance plant health, resource efficiency, and crop quality. Originally limited in use, with early applications in countries like China and Korea for producing high-quality potato seeds [32], aeroponics is now practiced across Europe, South America [33], and is being introduced in parts of Africa [34]. A pivotal NASA [35] study demonstrated that plants could grow even in the absence of soil by supplying a nutrient-rich mist directly to their roots, marking a turning point in controlled environment agriculture. In aeroponic systems, several key parameters must be optimized to ensure successful plant development. These include spray mist characteristics such as droplet size and misting frequency, which affect root oxygenation and nutrient absorption. Studies have tested various misting schedules, such as 7 s on and 10 min off for cucumber [36], and 1.5 min on and 5 min off for lettuce (*Lactuca sativa*), to identify the most effective regimes. Ultrasonic atomization foggers are often used to produce fine droplets that improve nutrient delivery and root zone aeration. Environmental parameters such as pH and electrical conductivity (EC) of the nutrient solution are crucial for nutrient uptake and must be carefully monitored using pH and EC meters. The ideal temperature range for most crops lies between 15 °C and −25 °C, while humidity and dissolved oxygen concentration also play critical roles in maintaining plant health and preventing root diseases. The nutrient reservoir must be managed meticulously to ensure continuous delivery of essential macro- and micronutrients. According to Rolot et al. [37], key elements required for vigorous plant growth include nitrogen (N), phosphorus (P), potassium (K), sulfur (S), zinc (Zn), along with carbon (C) and oxygen (O), which are supplied directly from the atmosphere. As global research continues to explore alternative agricultural technologies, aeroponics stands out for its adaptability, minimal resource use, and capacity to support plant growth in diverse and challenging environments. These innovations are reshaping the future of sustainable agriculture by enhancing yield, quality, and resource-use efficiency.

3.3 Summary

Controlled Environment Agriculture (CEA) is an advanced farming approach that optimizes the growth of medicinal plants by regulating key environmental factors such as temperature, humidity, light, CO_2, and nutrients. Techniques like hydroponics, aeroponics, and vertical farming enable soil-less cultivation, promoting faster growth, higher biomass yields, and increased production of valuable secondary metabolites. Hydroponics allows precise nutrient delivery and has demonstrated success in producing high-quality medicinal plants with enhanced bioactive compounds. Aeroponics improves nutrient uptake and root oxygenation through fine misting, while vertical farming maximizes space use and ensures year-round production. Despite challenges like high startup costs and species-specific requirements, CEA provides sustainable, pesticide-free, and scalable solutions for medicinal plant cultivation. It also supports conservation efforts by reducing reliance on wild plant populations. With ongoing advancements, CEA is poised to become a key technology in meeting global demands for medicinal plants while promoting environmental sustainability.

References

1. Harbick K, Albright LD (2016) Comparison of energy consumption: greenhouses and plant factories. Acta Hortic 1134:285–292
2. Singh AK, Yang X (2021) GREENBOX horticulture, an alternative avenue of urban food production. Agric Sci 12:1473–1489
3. Hodges CN, Groh JE, Johnson AW (1968) Controlled-environment agriculture for Coastal Desert areas. AGRIS 1968:58–68
4. Gómez C, Currey CJ, Dickson RW, Kim HJ, Hernández R, Sabeh NC, Burnett SE (2019) Controlled environment food production for urban agriculture. HortSci 54:1448–1458
5. Despommier D (2011) The vertical farm: controlled environment agriculture carried out in tall buildings would create greater food safety and security for large urban populations. J Verbr Lebensm 6:233–236
6. Hashimoto Y (1991) Computer integrated plant growth factory for agriculture and horticulture. IFAC Proc 24:105–110
7. Kozai T (2013) Resource use efficiency of closed plant production system with artificial light: concept, estimation and application to plant factory. Proc Jpn Acad 89:447–461
8. Jones J (2016) Hydroponics: a practical guide for the soilless grower. CRC Press, Boca Raton, FL
9. Giroux R, Berinstain A, Braham S, Graham T, Bamsey M, Boyd K, Dixon M (2006) Greenhouses in extreme environments: the Arthur Clarke Mars Greenhouse design and operation overview. Adv Space Res 38(6):1248–1259
10. Paponov M, Ziegler J, Paponov IA (2023) Light exposure of roots in aeroponics enhances the accumulation of phytochemicals in aboveground parts of the medicinal plants Artemisia annua and Hypericum perforatum. Front Plant Sci 14:1079656
11. Mubeen B, Hasnain A, Mehboob R, Rasool R, Riaz A, Elaskary SA, Shah MM, Faridi TA, Ullah I (2022) Hydroponics and elicitation, a combined approach to enhance the production of designer secondary medicinal metabolites in *Silybum marianum*. Front Plant Sci 13:897795
12. Mehandru P, Shekhawat NS, Rai MK, Kataria V, Gehlot HS (2014) Evaluation of aeroponics for clonal propagation of *Caralluma edulis*, *Leptadenia reticulata* and *Tylophora indica*–three threatened medicinal asclepiads. Physiol Mol Biol Plants 20:365–373
13. Nomnqophiso Z, Kambizi L, Etsassala NG, Nchu F (2024) Bioactivities of *Helichrysum cymosum* cultivated in aquaponic, hydroponic and field systems. J Med Plants Econ Dev 8(1):200
14. von Bieberstein P, Xu YM, Gunatilaka AL, Gruener R (2014) Biomass production and withaferin a synthesis by *Withania somnifera* grown in aeroponics and hydroponics. HortSci 49(12):1506–1509
15. Thakur K, Partap M, Kumar D, Warghat AR (2019) Enhancement of picrosides content in *Picrorhiza kurroa* Royle ex Benth. mediated through nutrient feeding approach under aeroponic and hydroponic system. Ind Crop Prod 133:160–167
16. Léonhart S, Pedneault K, Gosselin A, Angers P, Papadopoulos AP, Dorais M (2002) Diversification of greenhouse crop production under supplemental lighting by the use of new cultures with high economic potential. Acta Horticul 580:249–253
17. Azarmi F, Nazemieh H, Dadpour MR (2012) Growth and essential oil production of Aloisya citriodora L. grown in greenhouse using soil and floating system. Res J Biol Sci 7(4):148–151
18. Stewart CL, Lovett-Doust L (2003) Effect of phosphorus treatment on growth and yield in the medicinal herb *Calendula officinalis* L. (Standard Pacific) under hydroponic cultivation. Can J Plant Sci 83(3):611–617

19. Brechner ML, Albright LD, Weston LA (2007) Impact of a variable light intensity at a constant light integral: effects on biomass and production of secondary metabolites by *Hypericum perforatum*. In: International symposium on medicinal and nutraceutical plants, vol 756, pp 221–228
20. Prasad A, Pragadheesh VS, Mathur A, Srivastava NK, Singh M, Mathur AK (2012) Growth and centelloside production in hydroponically established medicinal plant- *Centella asiatica* (L.). Ind Crops Prod 35:309–312
21. Canter PH, Thomas H, Ernst E (2005) Bringing medicinal plants into cultivation: opportunities and challenges for biotechnology. Trends Biotechnol 23:180–185
22. Crosby GW, Cracker LE (2007) Development of soilless culture methods for production of moringa (*Moringa oleifera* lam.) root and leaf biomass. Acta Hortic 756:139–146
23. Dall'Acqua S, Aiello N, Scartezzini F, Albertin V, Innocenti G (2010) Analysis of highly secondary-metabolite producing roots and flowers of two *Echinacea angustifolia* DC var. angustifolia accessions. Ind Crops Prod 31(3):466–468
24. Dorais M, Papadopoulos AP, Luo X, Léonhart S, Gosselin A, Pedneault K, Angers P, Gaudreau L (2001) Soilless greenhouse production of medicinal plants in North Eastern Canada. Acta Horticul 554:297–303
25. Maggini R, Tozzini L, Pacifici S, Raffaelli A, Pardossi A (2012) Growth and accumulation of caffeic acid derivatives in *Echinacea angustifolia* DC. var *angustifolia* grown in hydroponic culture. Ind Crop Prod 35:269–273
26. Letchamo W, Polydeonny LV, Gladisheva NO, Arnason TJ, Livassy J, Awang DVC (2002) Factors affecting *Echinacea* quality. In: Janick J, Whipkey A (eds) Trends in new crops and new uses. ASHS Press, Alexandria, VA, pp 514–521
27. Mairapetyan SK, Tadevosyan AH, Alexanyan SS, Stepanyan BT (1999) Optimization of the N:P:K ratio in the nutrient medium of some soilless aromatic and medicinal plants. Acta Hortic 502:29–32
28. Zheng Y, Dixon M, Saxena PK (2006) Growing environment and nutrient availability affect the content of some phenolic compounds in *Echinacea purpurea* and *Echinacea angustifolia*. Planta Med 72:1407–1414
29. Maggini R, Raffaelli A, Annaheim KE, Tozzini L, Pacifici S, Guidi L, Pardossi A (2010) Effect of post-harvest handling and extraction on the content of echinacoside and cynarin in the root tissues of *Echinacea angustifolia* DC. J Food Agric Environ 8(2):266–271
30. Sabra A, Adam L, Daayf F, Renault S (2012) Salinity-induced changes in caffeic acid derivatives, alkamides and ketones in three *Echinacea* species. Environ Exp Bot 77:234–241
31. Berti M, Wilckens R, Fischer S, Hevia F (2002) Effect of harvest season, nitrogen, phosphorus and potassium on root yield, echinacoside and alkylamides in *Echinacea angustifolia* L. in Chile. Acta Hortic 576:303–310
32. Kim HS, Lee EM, Lee MA, Woo IS, Moon CS, Lee YB, Kim SY (1999) Production of high quality potato plantlets by autotrophic culture for aeroponic systems. J Kore Soc Hortic Sci 123:330–333
33. Mateus-Rodriguez JR, Haan SD, Andrade-Piedra JL, Maldonado L, Hareau G, Barker I, Chuquillanqui C, Otazú V, Frisancho R, Bastos C et al (2013) Technical and economic analysis of aeroponics and other systems for potato mini-tuber production in Latin America. Am J Potato Res 90:357–368
34. Otazú V (2014) Manual on quality seed potato production using aeroponics. International Potato Center (CIP), Lima, p 44
35. NASA Spinoff (2006) Progressive plant growing has business blooming. In: Environmental and agricultural resources. NASA Spinoff, New York, pp 64–77
36. Peterson LA, Krueger AR (1988) An intermittent aeroponics system. Crop Sci 28:712–713
37. Rolot JL, Seutin H, Michelante D (2002) Production de minitubercules de pomme de terre par hydroponie: évaluation d'un systèmecombinant les techniques "NFT" et "gravel culture" pour deux types de solutions nutritives. Biotec Agr Soc Env 6:155–161

Chapter 4
Salinity Stress Interventions for Crop Productivity in Hydroponics

Shiv Rattan

Abstract Salinity stress has a substantial impact on crop output, especially in hydroponic systems where nutrient solution stability is critical. Several strategies have been tested to reduce the negative effects of salinity on crops. These include bio-stimulants, genetic, and molecular methods, and nutrient solution optimisation. Bio-stimulants such as, amino acids, arbuscular mycorrhizal fungus, plant growth-promoting rhizobacteria (PGPR), fulvic acid, and chitosan have showed potential in improving crop salinity tolerance. Under saline conditions, these bio-stimulants enhance physiological, biochemical, and agronomical responses, resulting in higher production and improved plant health. Genetic and molecular interventions aim to harness agricultural genetic heterogeneity and create salinity-tolerant crop plants. Advances in sequence-based breeding and functional genomics have revealed novel genetic variations that control salinity tolerance, providing long-term solutions for salt stress reduction. Another successful tactic in hydroponic systems is to optimise nutrient solutions. Maintaining crop productivity in saline environments requires careful control of electrical conductivity, pH, and water quality to guarantee nutrient uptake and avoid contamination. In conclusion, crop yield in hydroponic systems under salt stress can be greatly increased by combining bio-stimulants, genetic and molecular techniques, and optimised nutrient solutions. These initiatives provide economical and long-lasting answers to the problems caused by salinity in agriculture. This book chapter will provide a complete summary of the current strategies used to alleviate salt stress in hydroponic systems. By investigating the efficacy of these tactics, proposing topics for future research, and emphasising the practical applicability of these treatments.

Keywords Salinity eustress · Hydroponics · Bio-stimulant · Population · Electrical conductivity

---

S. Rattan (✉)
Robert H. Smith Faculty of Agriculture, Food and Environment, The Hebrew University of Jerusalem, Jerusalem, Israel
e-mail: shiv.rattan@mail.huji.ac.il

P. Kumar, A. R. Warghat (eds.), *Transforming Medicinal Plant Agriculture*, Smart Agriculture 16, https://doi.org/10.1007/978-981-95-3963-5_4

4.1 Introduction

Global population growth is expected to continue throughout the century, with predictions predicting a global population of 9.7 billion in 2050 and 10.9 billion in 2100 [1]. In many regions of the world, environmental contamination from food-related activities coexists with malnutrition and overnutrition [2]. Concurrently, climate change appears as a primary risk driver, threatening food security, plant ecosystems, and dietary nutritional quality. Agriculture is the industry that is most sensitive to climate change. Climate change-induced abiotic stresses on crops, such as drought and salinity, have a negative impact on sustainable food production [3]. Agriculture's significance in this setting emphasises its critical role in determining global dynamics. It emphasises the importance of knowledgeable water resource management techniques in the agriculture sector to promote sustainability and mitigate adverse ecological consequences. Salinity in soil and water is a prevalent issue that severely limits global agricultural food production, particularly in arid and semi-arid countries. Salt stress has become a widespread problem, impacting around 20% of the world's arable land, and it is expected to climb to 50% by the end of the twenty-first century [1]. Proteins, lipids, DNA, and RNA are all molecularly harmed by high salinity levels. Salt stress causes osmotic and ionic stress, interferes with gas and nutrient exchange, and increases the synthesis of reactive oxygen species (ROS) molecules such superoxide, hydrogen peroxide, hydroxyl radical, and singlet oxygen at the cellular level [4]. Soilless cultivation, particularly hydroponic, is an effective, sustainable, resilient, and resource-efficient alternative cultivation approach that offers an opportunity to supplement traditional soil-based production [5, 6]. Soilless culture techniques provide for flexibility and intensification, ensuring high crop yields and high-quality products even in places with poor growth circumstances [7]. These systems provide effective techniques for managing salty stress on plants, preventing salinity levels over crops' tolerance thresholds, which severely impact plant growth and production. At the same time, regulated saline stress can be used to improve the overall quality of vegetable products by reducing anti-nutritional elements like nitrate and increasing secondary metabolites and sensory quality attributes [8–10].

Biostimulants are cutting-edge agronomic instruments that fall somewhere between fertilisers and plant growth regulators. They have the special capacity to increase plant productivity and growth. It improves plant nutrient utilisation efficiency, increases resistance to unfavourable environmental conditions, raises the quality of produce, and promotes the best possible uptake of nutrients in limited amounts in the soil and root system [11, 12]. Applying biostimulants to the plant or rhizosphere increases the efficiency of nutrient absorption and assimilation, increases resistance to abiotic stressors, and improves the agricultural product's overall quality [13, 14]. Plant Growth-Promoting Rhizobacteria (PGPR) play an important role in alleviating salt stress by improving water absorption, promoting the uptake of vital nutrients, and accumulating osmolytes such as proline, glutamate, glycine betaine, soluble sugars, and carbohydrates [12, 15]. Furthermore, PGPR leads to the enhancement of antioxidant enzymes [1]. Chitosan is a naturally occurring biopolymer formed from chitin found in the exoskeletons of crustaceans such as prawns, crab, and lobster. It is formed via deacetylation of chitin, which results in a positively charged polysaccharide. Chitosan is used in agriculture as a biopesticide, biofertiliser, and biostimulant to promote plant development, increase crop production, and guard against abiotic stressors and illnesses because of its biocompatibility, biodegradability, non-toxicity, and antibacterial qualities [1].

Salinity stress is a major concern in modern agriculture, especially in hydroponic systems that rely on nutrient solutions rather than soil for plant growth. Elevated salinity levels in irrigation water can have a substantial impact on plant physiological systems, resulting in stunted development, lower yields, and even mortality. This challenge is especially acute in areas where natural water sources are high in dissolved salts, necessitating novel strategies to assure crop growth. Hydroponics, an advanced and effective way of growing plants without soil, has various benefits such as greater control over nutrient delivery, less water usage, and the capacity to produce crops in non-arable regions. However, hydroponic systems' reduced buffering capacity against salt build-up due to the absence of soil also makes crops more vulnerable to salinity stress. Given the rising global need for food, resolving this issue is essential to the sustainability and effectiveness of hydroponic farming. To lessen the negative effects of salt stress on hydroponic crops, a number of tactics have been studied. These techniques can be broadly divided into three categories: biostimulants, genetic and molecular approach, and nutrient solution optimisation. High productivity and year-round, optimised output of consistent quality are supported by hydroponic systems. The hydroponic system is crucial for increasing organoleptic and bioactive quality components due to its efficiency and precision in altering nutrient solution composition, as well as its control over temperature, light, and air composition [6]. Effective eustress (positive stress), such as moderate salinity or nutritional stress, can elicit tailored plant responses that involve the activation of physiological and molecular mechanisms as well as the strategic accumulation of bioactive compounds required for adaptation to suboptimal conditions. Furthermore, the hydroponic system can be instrumental for the biofortification of vegetables with micronutrients essential or beneficial to human health, such as iodine, iron, selenium, silicon, and zinc. Accurate control of microelement concentrations and constant exposure of roots to the fortified nutrient solution without soil interaction can maximise their uptake, translocation, and accumulation in the edible plant parts.

4.2 Parameters Used for Intervention of Salinity Stress in Hydroponic Cultivation

4.2.1 Salinity Eustress

Salinity eustress refers to the positive effects that mild or moderate salinity levels can have on plants, enhancing their physiological processes and stress tolerance. Unlike salt stress, which is harmful, eustress (meaning "good stress") can stimulate adaptive mechanisms, such as improved water-use efficiency, activation of antioxidants, and enhanced production of beneficial secondary metabolites. This concept is being explored in agricultural practices to boost crop resilience and productivity in saline environments. Salinity eustress plays an intriguing function in hydroponics, where plants are cultivated in nutrient-rich water solutions without soil. There are various advantages of using modest salinity stress in controlled hydroponic systems. Salinity stress can stimulate the development of bioactive substances such as phenolics, flavonoids, and vitamins in crops like lettuce and tomatoes. These chemicals enhance the nutritional and antioxidant potential of the vegetables [16, 17]. Similarly, controlled salt levels can trigger stress-response mechanisms in plants, making them more resistant to environmental stresses. This involves improved ion regulation and antioxidant defence [16, 17]. Hydroponic systems provide perfect control over fertiliser solutions. Salinity stress can be paired with biofortification techniques to promote the uptake of important micronutrients including zinc, selenium, and iodine in edible plant portions [18]. While excessive salinity can be harmful to plants, moderate amounts can achieve a balance between sustaining output and boosting crop functionality [17]. Previous research has shown that modest salt stress in hydroponic systems might increase the phytochemical content and antioxidant activity of lettuce. For example, applying salts like KCl or NaCl in regulated amounts enhanced the resistance and nutritional value of lettuce plants [16]. Similarly, in tomato hydroponic cultivation salinity eustress has been utilised to raise levels of vitamin C, β-carotene, and lycopene, among other health-promoting substances [18]. Additionally, this lowers the fruit's nitrate levels, enhancing its overall quality. Salinity eustress in hydroponics has been used to biofortify vegetables with micronutrients such as iodine, selenium, and zinc. This is accomplished by precisely regulating the nutrient solution composition, which enhances the uptake and accumulation of these elements in edible plants [18]. Excessive levels of sodium chloride (NaCl) in irrigation water and agricultural soils disrupt physiological processes in vegetable crops, resulting in stunted development and yield decline [18, 19]. Vegetable crops can synthesise a wide spectrum of secondary metabolites to combat oxidative damage and scavenge reactive oxygen species (ROS) produced by environmental stresses [20]. These health-promoting phytochemicals, abundant in stressed plants, might increase the functional quality of fresh vegetables, benefiting the human diet [18, 21]. Numerous studies have shown that adding sodium chloride (NaCl) to the nutrient solution increases the amounts of sugars, organic acids, and amino acids in a variety of vegetable crops, including watermelon [22], pepper [23], melon [24], lettuce [25], and cauliflower [26].

4.2.2 *Biostimulant*

Biostimulants are chemicals that improve physiological processes in plants, hence increasing their growth and stress tolerance. In the context of salinity stress, biostimulants such as; plant growth-promoting rhizobacteria (PGPR), arbuscular mycorrhizal fungus, amino acids, fulvic acid, and chitosan have demonstrated great promise [1]. These biostimulants can boost root development, nutrient uptake, and stress-related metabolite synthesis, making plants more resilient to high salinity conditions. Biostimulants play a crucial role in managing salt stress in hydroponics by enhancing plant resilience and productivity under saline conditions such as improving salt tolerance, enhanced nutrient uptake, better water management, microbial support, and yield and quality boost. For example, in hydroponic lettuce farming, biostimulants like vermicompost, amino acids, PGPR, chitosan, and fulvic acid have been effective in alleviating salt-induced damage, improving growth, yield, and nutritional content [1]. Both PGPR and arbuscular mycorrhizal fungi (AMF) enhance root health and nutrient exchange, mitigating the adverse effects of salinity [1]. The study on hydroponic lettuce farming found that biostimulants increased yield, plant growth, nutrient uptake, and chlorophyll content while decreasing oxidative stress and nitrate accumulation [1]. The study evaluates how biostimulants can alleviate salt stress in hydroponically grown lettuce under 50 mM NaCl conditions. The key findings include; Vermicompost and PGPR were the most effective, increasing yields by up to 75%. Biostimulants enhanced nutrient uptake, chlorophyll content, stomatal conductance, and water status. It reduced oxidative stress markers like malondialdehyde (MDA) and nitrate content in leaves. Biostimulants activated antioxidative defence systems, showcasing their potential as a sustainable and cost-effective solution for mitigating salt-induced damage in hydroponic farming.

The recent studies suggest that nano-fertilisers are a promising technique for boosting crop resilience and productivity in saline environments [27]. They studied the use of nano-fertilisers (K_2SO_4, ZnO, and SiO_2) to combat salt stress in two alfalfa genotypes in a hydroponic system. When these fertilisers were evaluated at different salt concentrations, they discovered that Nano-K_2SO_4 markedly increased growth metrics such as shoot dry weight, plant height, and flower count. Nano-SiO_2 increased proline synthesis and antioxidant activity (SOD and catalase), especially at higher salt levels, enhancing plant stress tolerance. Meanwhile, Nano-ZnO increased microelement absorption and relative water content, particularly at moderate salinity. Silicon (Si) is abundant in soil, but its significance in plant biology is not well understood. The role of Si in soybean growth and its efficacy in salt stress reduction were examined [28]. Hydroponically grown soybeans were fed sodium metasilicate (Na2SiO3) as a source of silicon. The study found that using silicon (2.5 mM Na_2SiO_3) enhanced plant height, biomass, and chlorophyll content. When exposed to salt stress (80 mM NaCl), growth, gibberellins (GAs), and chlorophyll levels decreased, but abscisic acid (ABA) and proline levels increased. Furthermore, adding silicon to salt-stressed plants reduced the negative effects by increasing gibberellin levels, lowering ABA and proline levels, and partially recovering growth.

4.3 Nutrient Solution Optimisation

Maintaining plant health in hydroponic systems requires proper fertiliser solution management. Optimising nutritional solution composition, adjusting pH levels, and monitoring electrical conductivity can all help to reduce salinity's harmful effects. Adjusting key nutrient ratios and adding particular ions to offset sodium's effects can increase nutrient uptake and overall plant performance under saline circumstances. Optimising nutrient solutions is critical for managing salt stress in hydroponic systems. Here are several strategies: Maintaining an adequate EC level helps balance nutrient availability and osmotic pressure, so mitigating the harmful consequences of salinity. Calcium and potassium supplements can help to reduce the harmful effects of sodium ions and increase plant resistance. Incorporating silicon strengthens cell walls and increases plant salt tolerance. Chelated micronutrient versions of iron, zinc, and manganese improve nutrient absorption under saline circumstances. To avoid toxicity, the sodium and chloride levels in the solution should be balanced or monitored. Compounds such as amino acids and humic acids can improve nutrition absorption and stress tolerance. Increasing soilless nutrient solution electrical conductivity from 2.2 to 6.5 dS m^{-1} boosted tomato lycopene, total phenolic content, vitamin C, and β-carotene [29–31]. Mild to moderate salinity has been shown to improve the nutritional and bioactive value of hydroponically produced leafy greens [32–37]. Raising EC from 1.8 to 6.0 dS m^{-1} boosted both ascorbic acid and α-tocopherol levels in spiny chicory [37]. Increasing EC from 2.5 to 3.5 dS m^{-1} increased antioxidant compounds in wild rocket, and from 2.0 to 5.8 dS m^{-1} improved antioxidant activity, chlorogenic acid, cynarin, and luteolin levels in artichoke and cardoon leaves grown in a floating system [33, 35].

4.4 Genetic and Molecular Interventions

Advances in genetic engineering and molecular biology have created new opportunities for generating salt-tolerant crop cultivars [38, 39]. Marker-assisted selection, genome editing, and transgenic techniques have been used to identify and incorporate salt tolerance characteristics into crops. Understanding the genetic basis of salinity tolerance allows researchers to produce crop types that flourish in saline settings, providing consistent and high yields [40]. Reducing salt stress in hydroponic growth is a serious task, but developments in genetic and molecular biology provide promising answers. Potential interventions that could help reduce this stress include CRISPR-Cas9 and other gene-editing technologies that changes genes involved with salt tolerance, such as those that improve ion homeostasis or create osmoprotectants [41, 42]. Similarly, stress-responsive genes can be introduced using transgenic techniques, which can include genes from naturally salt-tolerant plants. For example, genes such as NHX1 (sodium/proton antiporters) can enhance the plant's ability to sequester sodium ions in vacuoles [43]. Furthermore, traditional selective breeding, together with genomic technologies, can be utilised to select and propagate salt-resistant cultivars. Similarly, molecular interventions involve priming plants with stress tolerance biochemicals. The use of chemicals such as glycine betaine, proline, or plant hormones (for example, abscisic acid) can improve salt stress tolerance by stabilising proteins and membranes [44]. RNA interference (RNAi) can be utilised to downregulate genes that cause salt stress, whereas tiny RNAs (miRNAs) can be used to boost the expression of protective pathways [45]. Plants can become more robust to environmental stressors such excessive salinity by modifying epigenetic regulators such as DNA methylation or histone modification patterns [46].

4.5 Conclusion

The growing demand for global food security, along with increasing biotic and abiotic challenges aggravated by climate change, necessitates protected cultivation of vegetable crops. Technological advancements in the management of hydroponic cultivation increase production while lowering initial infrastructure expenses. The flexibility and precision with which nutrient solution composition can be modulated, as well as the control of temperature, light, and atmospheric composition, make hydroponic systems useful in targeting organoleptic and bioactive quality components, addressing demand for improved vegetable quality. The effective use of eustress, such as mild to moderate salinity or nutritional stress, can elicit targeted plant responses by activating physiological and molecular mechanisms and strategically accumulating bioactive compounds required for adaptation to challenging environments. Vegetable crops have diverse responses to stressful situations, including physiological, biochemical, metabolic, and morphological alterations that occasionally lead to changes in the chemical composition and visual appearance of the end products. As a result, the use of eustress could be a beneficial technique in modern horticulture to meet the agro food industries and consumers' demands for high-quality, functional vegetable products. Understanding the molecular and physiological mechanisms elicited by controlled plant eustress, as well as those that facilitate micronutrient uptake in response to genotype and environmental conditions, will usher horticultural science into the era of tailoring superior sensory and functional quality vegetables. Furthermore, hydroponic cultivation allows for the effective biofortification of vegetables with micronutrients that are needed or beneficial to human health, such as iodine, iron, selenium, silicon, and zinc.

References

1. Ikiz B et al (2024) The use of biostimulants as a key to sustainable hydroponic lettuce farming under saline water stress. BMC Plant Biol 24(1):808

2. Kumar R et al (2023) Vertical farming and organic farming integration: a review. Org Farming:291–315
3. Sadak MS et al (2024) Changes in growth, photosyn thetic pigments and antioxidant system of Hordeum vulgare plant grown under salinity stress via signal molecules application. Vegetos:1–17
4. Sadak MS et al (2023) Exogenous calcium reinforces photosynthetic pigment content and osmolyte, enzymatic, and non-enzymatic antioxidants abundance and alleviates salt stress in bread wheat. Plants 12(7):1532
5. Partap M et al (2021) Hydroponic and aeroponic cultivation of economically important crops for production of quality biomass. In: Agricultural biotechnology: latest research and trends. Springer, Singapore, pp 573–585
6. Rattan S et al (2022) Nutrient feeding approach enhances the vegetative growth biomass, volatile oil composition, and myristicin content in hydroponically cultivated *Petroselinum crispum* (Mill.). J Appl Res Med 26:100359
7. Gruda NS et al (2022) Advances in soilless culture and growing media in today's horticulture-an editorial. Agronomy 12(11):2773
8. Bonasia A et al (2017) Nutritional, biophysical and physiological characteristics of wild rocket genotypes as affected by soilless cultivation system, salinity level of nutrient solution and growing period. Front Plant Sci 8:246365
9. Gruda N et al (2018) Impacts of genetic material and current technologies on product quality of selected greenhouse vegetables-a review. Eur J Hortic Sci 83(5):319–328
10. Conversa G et al (2021) Reduction of nitrate content in baby-leaf lettuce and *Cichorium endivia* through the soilless cultivation system, electrical conductivity and management of nutrient solution. Front Plant Sci 12:645671
11. Rady MM et al (2023) Exploring the role of novel biostimulators in suppressing oxidative stress and reinforcing the antioxidant defense systems in *Cucurbita pepo* plants exposed to cadmium and lead toxicity. Agronomy 13(7):1916
12. Dasgan HY et al (2023) Bio-fertilizers reduced the need for mineral fertilizers in soilless-grown capia pepper. Horticulturae 9:188
13. Bulgari R et al (2019) Biostimulants application in horticultural crops under abiotic stress conditions. Agronomy 9(6):306
14. Rouphael, Colla (2020) Biostimulants in agriculture. Front Plant Sci 11:511937
15. Moncada A et al (2020) Alleviation of salt stress by plant growth-promoting bacteria in hydroponic leaf lettuce. Agronomy 10(10):1523
16. Carillo P et al (2021) Regulated salinity eustress in a floating hydroponic module of sequentially harvested lettuce modulates phytochemical constitution, plant resilience, and post-harvest nutraceutical quality. Agronomy 11:1040
17. Santander C et al (2022) Salinity eustress increases the biosynthesis and accumulation of phenolic compounds that improve the functional and antioxidant quality of red lettuce. Agronomy 12(3):598
18. Rouphael, Kyriacou (2018a) Enhancing quality of fresh vegetables through salinity eustress and biofortification applications facilitated by soilless cultivation. Front Plant Sci 9:1254
19. Rouphael Y et al (2018b) Salinity as eustressor for enhancing quality of vegetables. Sci Hortic 234:361–369
20. Orsini F et al (2016) "Physiological quality" of organically grown vegetables. Sci Hortic 208:131–139
21. Khanam et al (2012) Phenolic acids, flavonoids and total antioxidant capacity of selected leafy vegetables. J Funct Foods 4(4):979–987
22. Colla et al (2006) Effect of salinity on yield, fruit quality, leaf gas exchange, and mineral composition of grafted watermelon plants. HortScience 41(3):622–627
23. Marin et al (2009) Antioxidant compounds in green and red peppers as affected by irrigation frequency, salinity and nutrient solution composition. J Sci Food Agric 89:1352–1359

24. Rouphael et al (2012) Improving melon and cucumber photosynthetic activity, mineral composition, and growth performance under salinity stress by grafting onto *Cucurbita* hybrid rootstocks. Photosynthetica 50:180–188
25. Sakamoto et al (2014) Effects of salinity and nutrients in seawater on hydroponic culture of red leaf lettuce. Environ Control Biol 52:189–195
26. Giuffrida et al (2017) Effects of salt stress imposed during two growth phases on cauliflower production and quality. J Sci Food Agric 97:1552–1560
27. El Shal RM et al (2022) Effect of nano-fertilizers on alfalfa plants grown under different salt stresses in hydroponic system. Agriculture (Basel) 12(8):1113
28. Lee SK et al (2010) Effect of silicon on growth and salinity stress of soybean plant grown under hydroponic system. Agrofor Syst 80:333–340
29. Krauss S et al (2006) The influence of different electrical conductivity values in a simplified recirculating soilless system on inner and outer fruit quality characteristics of tomato. J Agric Food Chem 54:441–448
30. Wu M et al (2004) Effects of nutrient solution EC, plant microclimate and cultivars on fruit quality and yield of hydroponic tomatoes (*Lycopersicum esculentum* L.). Acta Hort 659:541–547
31. Moya C et al (2017) Increased electrical conductivity in nutrient solution management enhances dietary and organoleptic qualities in soilless culture tomato. HortScience 52:868–872
32. Kim HJ et al (2008) Salt in irrigation water affects the nutritional and visual properties of romaine lettuce (*Lactuca sativa* L.). J Agric Food Chem 56:3772–3776
33. Colla G et al (2013) Effects of saline stress on mineral composition, phenolics acids and flavonoids in leaves of artichoke and cardoon genotypes grown in floating system. J Sci Food Agric 93:1119–1127
34. Klados, Tzortzakis (2014) Effects of substrate and salinity in hydroponically grown *Cichorium spinosum*. J Soil Sci Plant Nutr 14:211–222
35. Bonasia A (2017) Nutritional, biophysical and physiological characteristics of wild rocket genotypes as affected by soilless cultivation system, salinity level of nutrient solution and growing period. Front Plant Sci 8:300
36. Ntatsi G (2017) Salinity source alters mineral composition and metabolism of *Cichorium spinosum*. Environ Exp Bot 141:113–123
37. Petropoulos SA et al (2017) Salinity effect on nutritional value, chemical composition and bioactive compounds content in *Cichorium spinosum* L. Food Chem 214:129–136
38. Arzani, Ashraf (2016) Smart engineering of genetic resources for enhanced salinity tolerance in crop plants. Crit Rev Plant Sci 35(3):146–189
39. Ashraf, Munns (2022) Evolution of approaches to increase the salt tolerance of crops. Crit Rev Plant Sci 41(2):128–160
40. Kumar R et al (2024) Advances in genomic tools for plant breeding: harnessing DNA molecular markers, genomic selection, and genome editing. Biol Res 57(1):80
41. Chauhan PK et al (2023) Understanding the salinity stress on plant and developing sustainable management strategies mediated salt-tolerant plant growth-promoting rhizobacteria and CRISPR/Cas9. Biotechnol Genet Eng Rev 39:311–347
42. Raza A et al (2023) Smart reprograming of plants against salinity stress using modern biotechnological tools. Crit Rev Biotechnol 43(7):1035–1062
43. Magdy, Mansour (2023) Role of vacuolar membrane transport systems in plant salinity tolerance. J Plant Growth Regul 42:1364–1401
44. Khalid M et al (2022) Using exogenous melatonin, glutathione, proline, and glycine betaine treatments to combat abiotic stresses in crops. Int J Mol Sci 23:12913
45. Kumar V et al (2017) Plant small RNAs: the essential epigenetic regulators of gene expression for salt-stress responses and tolerance. Plant Cell Rep 37:61–75
46. Rashid MM et al (2021) Epigenetic regulation of salinity stress responses in cereals. Mol Biol Rep 1:12

Chapter 5
Physiological Interventions for Medicinal Plant Productions

Amit Kumar and Mahinder Partap

Abstract The global utilization of medicinal plants is steadily increasing due to their rich repository of bioactive compounds, known as secondary metabolites, which are pivotal in promoting health and treating various ailments. Beyond their traditional use in medicine, these plants play significant roles in culinary arts, food preservation, cosmetic industries, and herbal formulations. However, this growing demand has intensified challenges such as habitat loss, overharvesting, unauthorized cultivation, and the adverse effects of climate change, threatening the sustainability of medicinal plant resources. This chapter explores the innovative role of hydroponics as a sustainable and efficient alternative for cultivating medicinal plants. Hydroponic systems, by allowing precise control over environmental variables, including light intensity, pH, humidity, electrical conductivity, aeration, and nutrient composition-offer significant potential to optimize the yield and consistency of secondary metabolite production. These controlled conditions help overcome the limitations of traditional soil-based agriculture, providing benefits such as increased growth rates, improved biomass accumulation, enhanced resource efficiency, and year-round cultivation. Furthermore, this chapter delves into plant physiology-based approaches to strategically manipulate ambient conditions for stimulating specific metabolic pathways, thereby enhancing the biosynthesis of desirable secondary metabolites. Case studies and experimental insights are discussed to illustrate how hydroponics can be tailored to meet the specific growth and metabolite needs of various medicinal plant species. Overall, this chapter emphasizes hydroponics as a transformative cultivation platform, capable of supporting both the conservation of medicinal plants and the sustainable production of their valuable therapeutic compounds in an ecologically responsible manner.

A. Kumar
Fermentation and Phytofarming Technology Division, CSIR-Institute of Himalayan Bioresource Technology, Palampur, Himachal Pradesh, India

Academy of Scientific and Innovative Research (AcSIR), Ghaziabad, India

M. Partap (✉)
Agriculture Research Organisation (ARO), Volcani Institute, Rishon LeZion, Israel
e-mail: mahinder@volcani.agri.gov.il

P. Kumar, A. R. Warghat (eds.), *Transforming Medicinal Plant Agriculture*, Smart Agriculture 16, https://doi.org/10.1007/978-981-95-3963-5_5

Keywords Hydroponics · Medicinal plants · Secondary metabolites · Plant physiology

5.1 Introduction

Medicinal plants have played a central role in the development of healthcare systems for thousands of years, acting as reservoirs of therapeutic compounds critical to both traditional and modern medicine. These plants are recognized for their vast array of bioactive secondary metabolites, which possess a wide range of pharmacological activities. These compounds include, but are not limited to, alkaloids, flavonoids, terpenoids, saponins, steroids, phenolics, and various enzymes. Each of these compounds serves a specific function in plant defense, signaling, or adaptation and has been harnessed by humans for medicinal use. The biochemical diversity of medicinal plants is largely the result of evolutionary pressure, where plants have developed complex metabolic pathways to survive in varying environmental conditions. Over time, this evolutionary process has led to the synthesis of specialized molecules, many of which exhibit significant therapeutic properties. Today, this evolutionary legacy underpins numerous drug discovery and development efforts worldwide. Their continued reliance by vast populations underscores the global significance of medicinal plants. According to data from the World Health Organization (WHO), approximately 80% of the global population relies on traditional medicine for their primary healthcare needs, the majority of which are derived from plant sources [1]. This reliance is particularly evident in developing nations, where accessibility and affordability make plant-based remedies a practical choice. Even in more industrialized societies, there is a growing trend toward the use of natural products and herbal medicines, often perceived as safer and more holistic alternatives to synthetic drugs. The economic value of medicinal plants is also rapidly increasing. The global market, valued at USD 170 billion in 2022, is projected to surge to USD 600 billion by 2033 [2]. This rising demand is driven not only by the healthcare sector but also by the cosmetics, nutraceutical, and food industries, all of which utilize plant-derived bioactive compounds. However, the surge in demand presents several challenges. Traditional agricultural systems face mounting difficulties, including a reduction in arable land, degradation of soil quality, overuse of chemical fertilizers, and the prevalence of soil-borne pathogens. Additionally, climate change and environmental degradation are further exacerbating the instability and unreliability of conventional farming practices. These factors collectively highlight the urgent need for innovative, sustainable cultivation systems that can ensure a consistent and high-quality supply of medicinal plants.

In response to these challenges, hydroponics has emerged as a highly promising alternative approach for high value plant productivity. Hydroponics, a soilless method of plant cultivation, involves growing plants in a controlled environment using nutrient-rich water solutions or artificial substrates. This technique allows for the precise regulation of environmental conditions such as light intensity, temperature, humidity, pH, electrical conductivity, nutrient concentration, and airflow, all of which directly influence plant growth, development, and secondary metabolite production [3]. By eliminating soil, hydroponic systems avoid many of the issues associated with traditional farming, including soil-borne diseases and contaminants, while also significantly improving water and nutrient use efficiency. The concept of water-based plant cultivation is not new. Historical records reveal that ancient civilizations such as those in Babylon, Egypt, and the Aztec empire employed water gardening techniques reminiscent of modern hydroponics. The floating gardens—or "chinampas"—used by the Aztecs, for instance, are considered precursors to floating raft hydroponic systems [4]. These historical precedents underscore the enduring human interest in sustainable, space-efficient, and high-yield agricultural methods. Modern hydroponic systems offer a wide range of advantages. These include faster plant growth, higher yields per unit area, reduced use of pesticides and herbicides, and the potential for year-round cultivation irrespective of external climatic conditions [5]. This level of control and efficiency is particularly beneficial for high-value crops such as medicinal herbs, where consistency in active ingredient content is critical. Furthermore, hydroponic systems facilitate better monitoring of plant nutrition. The nutrient solution can be regularly analyzed and adjusted to correct deficiencies or imbalances, ensuring optimal growth conditions and maximizing the biosynthesis of targeted secondary metabolites [3, 6].

From a physiological perspective, plant responses to environmental conditions are intricately linked to their ability to produce secondary metabolites. Factors such as light quality and intensity, salt stress, temperature fluctuations, and nutrient availability can significantly impact photosynthesis, respiration, hormone regulation, and ultimately the production of bioactive compounds. Several studies have demonstrated that manipulating these variables in controlled environments can enhance the quality and quantity of desired phytochemicals [7, 8]. For example, adjusting the red-to-blue light ratio using LEDs has been shown to stimulate the accumulation of flavonoids and anthocyanins in various medicinal plants. Similarly, mild abiotic stress, such as controlled salinity, can act as an elicitor to trigger secondary metabolite biosynthesis, mimicking natural defense responses. The integration of omics technologies—including transcriptomics, genomics, proteomics, and metabolomics—has revolutionized our understanding of the molecular and genetic basis of secondary metabolism. These advanced tools allow researchers to identify key genes, transcription factors, and enzymes involved in the biosynthetic pathways of pharmacologically active compounds [9]. When applied in conjunction with hydroponic systems, omics approaches enable the design of targeted strategies for enhancing metabolite production. For instance, the knowledge of specific gene expression patterns under certain nutrient regimes or light conditions can inform adjustments to the growth environment, thereby optimizing yield and quality. Another promising approach within hydroponic cultivation is the use of elicitors—chemical or biological agents that stimulate the production of secondary metabolites. Elicitors can be biotic, such as microbial extracts, or abiotic, such as UV radiation or jasmonic acid. When applied under controlled hydroponic conditions, elicitors can effectively induce stress-like responses in plants, leading to increased biosynthesis of valuable compounds without compromising plant health [3]. Combining elicitor treatments with physiological and omics-based monitoring further refines this approach, offering precise and reproducible outcomes. In recent years, commercial agro-industries have increasingly adopted hydroponic methods not only for vegetables and ornamental crops but also for medicinal plants. This trend reflects the growing recognition of hydroponics as a reliable and scalable technology capable of producing consistent, high-quality botanical materials. The ability to standardize cultivation conditions and closely monitor all growth parameters makes hydroponics particularly suitable for pharmaceutical applications, where regulatory standards require rigorous quality control and traceability.

In conclusion, the rising global demand for medicinal plants, coupled with the limitations of traditional agriculture, necessitates the adoption of innovative cultivation methods. Hydroponics stands out as a powerful tool for sustainable and efficient medicinal plant production. It offers unparalleled control over growth conditions, reduces environmental impacts, and supports the enhancement of secondary metabolite content through physiological and molecular interventions. This chapter aims to explore in depth the application of hydroponic systems in the cultivation of medicinal plants. It will focus on the manipulation of environmental variables, nutrient management, elicitation strategies, and the integration of omics technologies to optimize the production of bioactive compounds. By bridging traditional knowledge with modern technology, hydroponics holds the potential to transform the future of medicinal plant agriculture.

5.2 Advantages of Hydroponics for Medicinal Plant Cultivation

Hydroponics offers numerous benefits for cultivating medicinal plants, enhancing the production of secondary metabolites. Hydroponics improves the purity, uniformity, efficiency, productivity, and concentration of secondary metabolites in medicinal plants (Fig. 5.1). Since secondary metabolites are frequently concentrated in plant organs, it is essential to adjust growing conditions to optimize the yield of the targeted organs and metabolites. For example, Image-driven characterization facilitates the quantification of colored metabolites, including anthocyanins and carotenoids, in plant leaves [10]. Hydroponics cultivates plants indoors under hygienic conditions, following stringent biosecurity measures that prevent pests and diseases, thereby obviating the need for chemical pest management agents. Moreover, hydroponic farming facilitates meticulous regulation of nutrient administration, minimizing surplus fertilizer waste on the plants. Assessing the quality of growth substrates such as perlite, vermiculite, water, and fertilizers helps avert pollution from heavy metals. Hydroponics facilitates the evaluation of essential contamination routes during production to guarantee the purity and safety of medicinal plants. Below, we

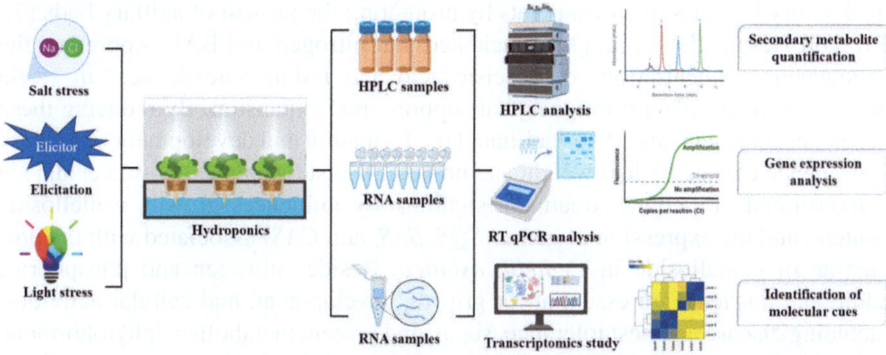

Fig. 5.1 Graphical illustration of hydroponics and associated technologies for enhanced plant production

highlight the factors and environmental conditions that affect the enhanced production of secondary metabolites in medicinal plants under hydroponics.

5.3 Nutrient Manipulation Strategies

Nutrients utilized in hydroponics are solubilized in water and predominantly exist in inorganic and ionic states. All needed ingredients for plant growth are provided through various chemical compositions, and creating a nutrient solution that offers an optimal ion ratio for plant growth is a crucial stage in hydroponic farming. A perfect nutrient recipe in hydroponics or any plant growth media includes macronutrients, micronutrients, vitamins, and other essential mineral ingredients. Nonetheless, nutrient imbalances can lead to deficits in hydroponically cultivated crops, hence severely limiting agricultural output. Moreover, the overuse of fertilizer can elevate production costs, induce plant toxicity, exacerbate environmental contamination, and diminish the nutritional value of crops [11]. In hydroponics, greater control of the nutrient solution depends on maintaining an ideal pH, electrical conductivity, and ion concentration. The effects of different nitrogen forms on plant morphology, physiology, and phytochemistry can be evaluated using hydroponic systems, owing to variations in soil nitrogen. Ammonium (NH_4^+) and nitrate (NO_3^-) are the two main forms of nitrogen that plants frequently employ to meet their nitrogen requirements. In *Melissa officinalis,* the NH_4^+ to NO_3^- ratio (0:100) enhances plant biomass, antioxidant capacity, and essential oil content [12]. The optimal ratio of $NH4^+$ to $NO3^-$ might stimulate plant development and increase steviol glycosides accumulation in stevia leaves [13]. The total steviol glycosides content was diminished in plants lacking the macronutrients nitrogen, phosphorus, sulfur, magnesium, or calcium due to impaired photosynthesis and lowered leaf yield [14]. Co-stimulation of nitrogen and 6-Benzylaminopurine (BAP) enhanced the leaf dry biomass of stevia plants by promoting the growth of axillary buds [7]. Tavakoli Hasanaklou et al. [13] elucidated that nitrogen and BAP suppressed the endogenous concentrations of abscisic acid and Indole-3-acetic acid in stevia plants, which are plant hormones that suppress bud expansion. By reducing these hormones, nitrogen, and BAP facilitated the formation and development of axillary buds, hence enhancing leaf weight and branching in stevia plants. Prasad et al. [15] demonstrated that copper treatment significantly influences growth, centelloside content, and the expression of genes *(SQS, βAS,* and *CAS)* associated with the production of centelloside in *Centella asiatica*. Besides nitrogen and phosphorus, additional nutrients are essential for growth, development, and cellular activities, including disease and pest tolerance, sugar, and protein metabolism, phytohormone production, the functioning of enzymes, photosynthesis, and the integrity of cell membrane structure [16]. Consequently, the alteration of macronutrient and micronutrient contents is crucial for plant growth and the production of phytochemicals.

5.4 pH and Electrical Conductivity Management

The pH of a nutrient solution is a critical factor influencing the accessibility of nutrients, absorption, and solubility. The ideal pH range for most of the crops is between 5.5 and 6.5, where nutrients are readily accessible [17]. For instance, elevated pH enhances the precipitation of calcium and magnesium and reduces the soluble capacity of iron and phosphate in the solution. Meanwhile, lower pH levels reduce the uptake of macronutrients such as nitrogen, calcium, and magnesium [18]. In *Araxacum officinale* and *Reichardia picroides*, low pH (pH 4) boosts the production of antioxidants, photosynthetic pigments, and overall plant growth [19]. Consequently, an optimal pH range must be sustained for adequate plant growth. Implementing an efficient nutrient solution approach is crucial for minimizing water and nutrient usage, reducing production costs, and enhancing cultivation.

The electrical activity (EC) value indicates the number of ions in the solution; thus, in most prior studies and standard practices, checking the nutrient solution is conducted by measuring EC many times every day. An ideal EC between 1.5 and 2.5 dS m^{-1} is necessary for plant growth [20]. Variations from this range may yield detrimental consequences: elevated EC induces salt stress, while diminished EC results in nutritional deficit in plants [21]. An EC of 3.5 mS cm^{-1} boosts the nutritional levels of nitrogen, phosphorus, potassium, and essential oil content in *Acmella oleracea* [22]. In Centella asiatica, EC 1.2 enhanced the plant growth and centellosides content [23]. The modification of the nutrient solution necessitates the application of the appropriate quantities of essential ions at various growth stages according to the crop's needs.

5.5 Light and Photoperiod Regulation

Light is essential for plant development, signaling, and metabolism. By adjusting critical light characteristics such as wavelength, intensity, and photoperiod, cultivators can influence plant growth and nutritional content. Light significantly influences plant physiology, photosynthesis, the circadian clock, and the photomorphogenesis process in plants. Light-emitting diodes (LEDs) are modern light sources characterized by energy savings, longevity, compactness, low weight, and great luminous efficacy. These benefits render LEDs an ideal illumination tool for the mechanical regulation of crop growth and development [24]. Red and blue light in the photosynthetically active spectrum (400–700 nm) substantially influence the secondary metabolic pathway and photosynthetic CO_2 assimilation. A study was conducted by [25] to examine the effect of low light intensity on *Glycyrrhiza uralensis*. Reduced light intensity enhances leaf area and concentrations of chlorophyll and glycyrrhizic acid, while reducing biomass and photosynthetic activity. *Artemisia annua* and *Hypericum perforatum* exposed to white, blue, and red light enhance artemisinin and coumaroylquinic acid accumulation in shoot and leaf tissue, respectively [26]. Elevated light intensity and prolonged photoperiod can elicit a stress response, prompting the synthesis of protective secondary metabolites. For example, elevated light intensities above 900 μmol m^{-2} s^{-1} augmented flavonoid content in *Orthosiphon aristatus* [27] and *Passiflora suberosa* [28]. Chou et al. [8] reported that exposure to red and green light with a lower photosynthetic photon flux density of 50 $\mu mol \cdot m^{-2} \cdot s^{-1}$ can enhance fresh and dry biomass in stevia and total steviol glycosides content and yield. Utilizing customized spectra, consisting of wavelength combos, can accurately regulate the metabolic processes of plants, thus increasing the yield of desired bioactive metabolites. Partap et al. [29] reported that a blue- and red-light combination enhances the morphology, antioxidant capacity, and physiological growth in lilium under the pin-trey soilless cultivation system. Furthermore, qRT-PCR data revealed up-regulation of flowering genes *FLOWERING LOCUS T* and *CONSTANS 9* genes while the flowering repressor *LOWERING LOCUS C* was down-regulated in blue and red-light combination. Nonetheless, our understanding of the synergistic and pleiotropic effects of altering various light factors on pharmacologically significant plant secondary metabolites is limited. Light manipulation presents challenges due to its differential impacts on secondary metabolite production, which vary depending on the plant species and their respective light sensitivities. These issues highlight the necessity for additional investigation to comprehend the influence of light on secondary metabolites' routes for dependable translation into hydroponics.

5.6 Elicitation Application

Elicitors are chemicals that trigger secondary pathways that lead to the enhanced synthesis of bioactive compounds in plants [30]. Elicitors are chemicals that function as signaling agents, detected by elicitor-specific receptors on the plant cell membrane, and stimulate the transcriptional activity of genes responsible for producing secondary metabolites [31]. Prior studies suggest that biotic and abiotic elicitors significantly enhance secondary metabolites in hydroponics [3, 32–34]. The concentration, duration, and timing of the elicitor applied also affect the production of phytochemicals. Application of Salicylic acid (50 μM) enhances yield, net photosynthesis rate, aucubin, and geniposidic acid content in *Bacopa monnieri* [34]. RNA sequencing data revealed that externally applied 0.10 $\mu mol \cdot L^{-1}$ IAA promotes plant growth and root formation in *Salvia miltiorrhiza* [35]. The application of 75 μM BAP and 150 mg/L ammonium nitrate in hydroponics increases the yield of stevioside (2.17 g/plant) and rebaudioside A (0.67 g/plant). Partap et al. [3] reported elevated levels of secondary metabolites and volatile oil content in *Valeriana jatamansi*. Application with 100 μM/L SA for 148 hours augmented the production of galanthamine and haemanthamine in *Hippeastrum papilio* [36]. Medicinal plants, capable of supplying therapeutically significant bioactive chemicals, present great opportunities for the advancement of natural therapeutics and diverse industrial applications. Elicitation approaches prove to be effective in augmenting the synthesis of medicinally beneficial metabolites, warranting further investigation into elicitor-mediated pathways across various plant systems under fluctuating climate conditions [31, 37].

5.7 Photosynthesis Performance and Efficiency

Chlorophyll concentration and fluorescence characteristics are recognized measures of the photosynthesis capacity and plant vitality, representing both the structural stability and functional efficacy of the photosynthetic system. Both red light and blue light have been documented to enhance carotenoid production [38]. Leaves of plants cultivated solely under LED lighting exhibited a markedly elevated concentration of chlorophyll a and β-carotene [39]. Partap et al. [3] reported that elicitor treatment enhances the physiological performance and yield of *Valeriana jatamansi* under aeroponic conditions. Mubeen et al. [33] also reported that elicitation treatment boosts photosynthetic pigment and lignin content in *Silybum marianum.* Salinity stress reduced the photosynthetic pigment accumulation in *Thymus vulgaris* [40]. The bioactive constituents present in medicinal plants typically encompass secondary metabolites, which possess therapeutic properties and are essential for plant adaptation to environmental conditions and stress resilience. Previous results indicate that prolonged exposure to elevated salt levels diminishes chlorophyll concentration by reducing the activity of *aminolevulinic acid synthase (ALA)*, an enzyme responsible for catalyzing the production of D-aminolevulinic acid, the initial precursor in chlorophyll production [41]. In *Salvia miltiorrhiza*, salt stress significantly enhanced osmolytes production and reduced Chl-b content while causing minimal alteration in Chl-a content, resulting in a marked overall decline in total chlorophyll concentration [42]. Ahmadirad et al. [7] demonstrated that the photosynthetic rate in stevia plants declined upon increased nitrogen treatment. A 40 mM NaCl salt stress diminishes photosynthetic qualities in *Cannabis sativa* due to reduced leaf gas exchange and nutrient solution assimilation in hydroponics. Salt stress also induces osmotic stress in plants, which leads to inhibition of photosynthetic enzymes, ultimately reducing photosystem II (Fv/Fm) [43]. Dutch Bucket Soil-Less cultivation system improves the morphology of *Chrysanthemum morifolium*, promotes antioxidant activity, and alters the gene expression patterns related to flavonoids and photosynthesis [44]. Table 5.1 represents the effect of elicitor treatment and salt stress on physiology and photosynthetic performance of hydroponically cultivated plants.

Table 5.1 Physiological intervention in hydroponically cultivated medicinal plants

Plant	Treatment	Response	Reference
Melissa officinalis	Different NH_4^+ (ammonium) to NO_3^- (nitrate) ratio	Enhanced chlorophyll, carotenoids, and macronutrient content	[12]
Valeriana jatamansi	Yeast extract and methyl jasmonate treatment	Enhance photosynthetic rate and stomatal conductance	[3]
Cannabis sativa	40 mM NaCl stress	Reduced photosynthetic efficiency	[7]
Picrorhiza kurroa	–	Enhance photosynthetic rate, stomatal conductance, and transpiration rate	[5]
Houttuynia cordata	–	Higher expression of photosynthetic genes (*HcatpB*, *HcpetC*, and *HcpetH*)	[9]
Salvia miltiorrhiza	NaCl salt stress treatment	A decrease in chlorophyll b content and a negligible change in chlorophyll a content	[42]
Silybum marianum	*Aspergillus Niger*, methyl jasmonate, and silver nanoparticles treatment	Boost chlorophyll content	[33]
Prunella vulgaris	0.20 mM phosphorus treatment	Influence total chlorophyll and carotenoid contents	[56]
Taraxacum officinale *Reichardia picroides*	Lower pH (4.0) treatment	Enhance chlorophylls (a, b, and total) and carotenoid content	[19]
Bacopa monnieri	Salicylic acid (50 μM) treatment	Enhance the net photosynthesis rate	[34]

5.8 OMICS Advancement

The omics (Genomics, transcriptomics, proteomics, and metabolomics) resources provide a thorough analysis of molecular biology inside a particular biological sample, devoid of any intrinsic bias [45]. OMICS resources have been utilized in the study of medicinal plants to identify critical molecular components, including key genes, transcription factors, and enzymes, associated with the metabolic pathways involved in secondary metabolites production [46]. Researchers identified the principal genes that are responsible for producing enzymes in the secondary metabolism biosynthesis route using transcriptomics and metabolomics analysis, resulting in novel opportunities for genetic manipulation utilizing a metabolic engineering framework. Yin et al. [9] employed multi-omics techniques (transcriptomics and metabolomics) alongside physiological data to elucidate the molecular mechanisms through which hydroponics influences the biomass of *H. cordata*. The study hypothesized that under hydroponic conditions, *H. cordata* may immediately assimilate nitrogen components via its roots. When the availability of nutritional nitrogen is sufficient, *H. cordata* can assimilate greater quantities of nitrogen, resulting in elevated nitrogen levels within the plant and influencing nitric oxide synthesis and metabolism. Further qPCR data also showed enhanced expression levels of photosynthesis-associated genes (*HcatpB, HcpetC*, and *HcpetH*). Nitrogen metabolism enhances photosynthesis in *H. cordata*, leading to improved root and leaf growth. External application of methyl jasmonate (MeJA) amplifies the pungent taste of soilless-grown Chinese chives by activating *AtuFMO1* genes associated with the biosynthesis of S-alk(en)ylcysteine sulfoxides (CSOs), thus elevating the production of flavor precursors (Methiin and Alliin). Integrative metabolomic and transcriptome analysis demonstrated that MeJA inhibits endogenous jasmonic acid production while enhancing sulfur absorption and CSO synthesis pathways. *AtuFMO1*, governed by transcription factors AtuPHL7 and AP2/ERF-ERF, is crucial in facilitating the sulfoxidation of S-alkyl-L-cysteine, a rate-limiting step in CSO synthesis [47].

The cooperative behavior of several biological components is essential for the proper functioning of a biological system; thus, the combination of omics information via rigorous statistical analysis provides comprehensive systems insight [48] that can be utilized to construct metabolic prototypes for predicting how the system operates. The reasons and mechanisms behind the extensive chemical diversity of plants remain unclear. Comprehending the method of refining faulty secondary metabolism via evolution may represent a significant breakthrough attainable through the fast-advancing omics study of many plant species [49].

5.9 Smart Hydroponics System

The latest advances in hydroponics have incorporated computational techniques that combine sensor technologies. The computational approach employs the Internet of Things, machine learning techniques, and deep learning techniques. These innovations enable the remote control of hydroponic systems by monitoring pH, electrical conductivity, and nutrient level, with data storage capabilities in the cloud [50]. Computational application offers multiple advantages in hydroponics, such as crop yield prediction and estimation [51], accurate nutrient delivery [52], and disease detection and prevention [53]. Establishing standard processes is crucial for optimizing technical advancements in hydroponics, particularly concerning uniform device operations. There are innovative research possibilities to investigate and analyze smart hydroponics systems for medicinal plant production using cutting-edge technologies. The integration of computation technology and hydroponics can address issues in the farming sector while meeting the needs of a growing global population. Fully harnessing the potential of cutting-edge technology in agriculture and actualizing its benefits for farmers, consumers, and the environment requires further research, development, and collaboration among participants.

5.10 Conclusion and Prospects

This book chapter highlighted the necessity of prioritizing hydroponics to improve the quality of biomass production in medicinal plants, given the issues of conventional agriculture, exacerbated by severe weather and suboptimal soil health. Hydroponics safeguards a stable phytochemical composition in medicinal plants by removing the influences of changeable weather, inconsistent soil conditions, and pathogens. Hydroponics has numerous applications due to its versatility and efficacy. Hydroponics is advantageous in urban environments with constrained space and can be implemented on balconies, rooftops, and gardens [54]. By altering growing factors, medium composition, elicitation, and salt stress treatment, growers can enhance the production rate and efficacy of secondary metabolites [12, 36]. The integration of functional genomics and hydroponics enables the precise manipulation of environmental conditions to achieve targeted biochemical profiles in medicinal plants [55]. Given the healthcare sector's huge need for secondary metabolites derived from medicinal plants for the formulation of secure drugs, alongside the constraints imposed on the cultivation of these plants due to adverse environmental factors such as temperature swings, salinity, and water shortages, it will soon be imperative to cultivate medicinal plants using hydroponic systems, owing to their many benefits.

References

1. Kumari R, Kotecha M (2016) A review on the standardization of herbal medicines. Int J Pharm Sci Res 7(2):97–106
2. Silveira D, Boylan F (2023) Medicinal plants: advances in phytochemistry and ethnobotany. Plants 12(8):1682

3. Partap M, Kumar P, Kumar A, Joshi R, Kumar D, Warghat AR (2020) Effect of elicitors on morpho-physiological performance and metabolites enrichment in *Valeriana jatamansi* cultivated under aeroponic conditions. Front Plant Sci 11:01263
4. Pachauri RK, Allen MR, Barros VR, Broome J, Cramer W, Christ R, Church JA, Clarke L, Dahe Q, Dasgupta P, Dubash NK (2014) Climate change 2014: synthesis report. In: Contribution of working groups I, II and III to the fifth assessment report of the intergovernmental panel on climate change. IPCC, p 151
5. Thakur K, Partap M, Kumar D, Warghat AR (2019) Enhancement of picrosides content in *Picrorhiza kurroa* Royle ex Benth. mediated through nutrient feeding approach under aeroponic and hydroponic system. Ind Crop Prod 133:160–167
6. Rattan S, Partap M, Kumar S, Warghat AR (2022) Nutrient feeding approach enhances the vegetative growth biomass, volatile oil composition, and myristicin content in hydroponically cultivated *Petroselinum crispum* (Mill.) Nyman. J Appl Res Med Aromat Plants 26:100359
7. Ahmadirad S, Tavakoli A, Mokhtassi-Bidgoli A, Mohasses Mostashari M (2024) Optimizing biomass and Steviol glycoside yield in hydroponically grown stevia (*Stevia rebaudiana* Bertoni) with ammonium nitrate and 6-Benzylaminopurine concentrations. Sugar Tech 26(2):595–607
8. Chou CT, Christabel V, Le MA TML, Wang ST (2025) Effects of light conditions on the leaf growth and Steviol glycoside yields of hydroponically cultivated stevia across growth stages. Horticulturae 11(3):316
9. Yin D, Song J, Xiang Z, Tian S, Li G, Qi S (2024) Hydroponic cultivation enhances biomass and medicinal safety of *Houttuynia cordata* Thunb: an integrated physiological and multi-omics analysis. Ind Crop Prod 220:119284
10. Gitelson AA, Keydan GP, Merzlyak MN (2006) Three-band model for noninvasive estimation of chlorophyll, carotenoids, and anthocyanin contents in higher plant leaves. Geophys Res Lett 33(11)
11. Rahman KA, Zhang D (2018) Effects of fertilizer broadcasting on the excessive use of inorganic fertilizers and environmental sustainability. Sustainability 10(3):759
12. Safaei F, Alirezalu A, Noruzi P, Alirezalu K (2024) Phytochemical and morpho-physiological response of *Melissa officinalis* L. to different NH4+ to NO3 ratios under hydroponic cultivation. BMC Plant Biol 1:968
13. Tavakoli Hasanaklou H, Ebadi Khazineh Ghadim A, Moradi F, Jahanbakhsh Ghodehkahriz S, Gholipouri A (2021) The effects of NH 4+ and NO 3− and plant growth regulators on the accumulation of nutrients, carbohydrates and secondary metabolites of *Stevia rebaudiana* Bertoni. Sugar Tech 23:65–76
14. Kafle GG, Midmore DJ, Gautam R (2017) Effect of nutrient omission and pH on the biomass and concentration and content of steviol glycosides in stevia (*Stevia rebaudiana* (Bertoni) Bertoni) under hydroponic conditions. J Appl Res Med Aromat Plants 7:136–142
15. Prasad A, Kumari M, Srivastava NK, Mathur AK, Mathur A (2018) Copper-induced modulation of biomass growth, physiological parameters, bioactive centellosides, and expression of biosynthetic pathway genes in an important medicinal herb, *Centella asiatica*. J Plant Growth Regul 37:471–480
16. Suryawanshi YC (2021) Hydroponic cultivation approaches to enhance the contents of the secondary metabolites in plants. In: Biotechnological approaches to enhance plant secondary metabolites. CRC Press, pp 71–88
17. Gillespie DP, Papio G, Kubota C (2021) High nutrient concentrations of hydroponic solution can improve growth and nutrient uptake of spinach (*Spinacia oleracea* L.) grown in acidic nutrient solution. HortScience 56(6):687–694
18. Velazquez-Gonzalez RS, Garcia-Garcia AL, Ventura-Zapata E, Barceinas-Sanchez JDO, Sosa-Savedra JC (2022) A review on hydroponics and the technologies associated for medium-and small-scale operations. Agriculture 12(5):646

19. Alexopoulos AA, Marandos E, Assimakopoulou A, Vidalis N, Petropoulos SA, Karapanos IC (2021) Effect of nutrient solution pH on the growth, yield and quality of *Taraxacum officinale* and *Reichardia picroides* in a floating hydroponic system. Agronomy 11(6):1118
20. Kannan M, Elavarasan G, Balamurugan A, Dhanusiya B, Freedon D (2022) Hydroponic farming–a state of art for the future agriculture. Mater Today Proc 68:2163–2166
21. Flores-Sánchez ID, Sandoval-Villa M, Soto-Hernández RM (2023) Effects of electrical conductivity and pruning on secondary metabolite contents in fruits of *Jaltomata procumbens* (cav.) JL Gentry. Nat Prod Commun 18(3):p.1934578X221150547
22. do Carmo APM, Freitas MSM, Machado LC, dos Santos Silva L, Petri DJC, Vimercati JC, Matos CRR, Mathias L, Vieira IJC, de Carvalho AJC (2024) Electrical conductivity of nutrient solutions affects the growth, nutrient levels, and content and composition of essential oils of *Acmella oleracea* (L.) RK Jansen from southeastern Brazil. J Agric Food Res 15:100968
23. Shawon MRA, Azad MOK, Ryu BR, Na JK, Choi KY (2023) The electrical conductivity of nutrient solution influenced the growth, centellosides content and gene expression of *Centella asiatica* in a hydroponic system. Agriculture 13(12):2236
24. Ferreira LT, de Araújo Silva MM, Ulisses C, Camara TR, Willadino L (2017) Using LED lighting in somatic embryogenesis and micropropagation of an elite sugarcane variety and its effect on redox metabolism during acclimatization. Plant Cell Tissue Organ Cult 128:211–221
25. Hou JL, Li WD, Zheng QY, Wang WQ, Xiao B, Xing D (2010) Effect of low light intensity on growth and accumulation of secondary metabolites in roots of *Glycyrrhiza uralensis* Fisch. Biochem Syst Ecol 38(2):160–168
26. Paponov M, Ziegler J, Paponov IA (2023) Light exposure of roots in aeroponics enhances the accumulation of phytochemicals in aboveground parts of the medicinal plants Artemisia annua and Hypericum perforatum. Front Plant Sci 14:p.1079656
27. Ibrahim MH, Jaafar HZ (2012) Primary, secondary metabolites, H_2O_2, malondialdehyde and photosynthetic responses of *Orthosiphon stimaneus* Benth. to different irradiance levels. Molecules 17(2):1159–1176
28. Ni YW, Lin KH, Chen KH, Wu CW, Chang YS (2020) Flavonoid compounds and photosynthesis in Passiflora plant leaves under varying light intensities. Plants 9(5):633
29. Partap M, Deekshith HN, Gupta H, Birsanta G, Kapoor P, Soni V, Bhargava B (2025) Morphological and biochemical analysis along with gene expression dynamics in Lilium 'brunello'under supplemented effect of blue and red light treatment in pin tray soil-less cultivation system. J Photoch Photobio B:113168
30. Partap M, Kumar A, Kumar P, Kumar D, Warghat AR (2025) Elicitors mediated enhancement of Picrosides and their pathway precursors in dedifferentiated cell suspension culture of *Picrorhiza kurroa* Royle ex Benth. J Plant Growth Regul 44(1):247–265
31. Kumar A, Partap M, Warghat AR (2025) Jasmonic acid: a versatile phytohormone regulating growth, physiology, and biochemical responses. J Plant Growth Regul 44(1):131–154
32. Mubeen B, Hasnain A, Mehboob R, Rasool R, Riaz A, Elaskary SA, Shah MM, Faridi TA, Ullah I (2022) Hydroponics and elicitation, a combined approach to enhance the production of designer secondary medicinal metabolites in *Silybum marianum*. Front Plant Sci 13:897795
33. Mubeen B, Hasnain A, Jie W, Zheng H, Peijnenburg WJ, Rozali SE, Rasool R, Naqvi SAH, Rao MJ, Sohail MA, Moustafa M (2023) Enhanced production of active photosynthetic and biochemical molecules in *Silybum marianum* L. using biotic and abiotic elicitors in hydroponic culture. Molecules 28(4):1716
34. Vinay CM, Shetty DN, Sanjay KU, Venkatasai NN, Rai PS (2025) Impact of salicylic acid elicitation on the growth, photosynthesis, and metabolites of *Bacopa monnieri* (L.) Pennell growing in a hydroponic control system. Sci Hortic 342:114056
35. Wang W, Zheng Y, Qiu L, Yang D, Zhao Z, Gao Y, Meng R, Zhao H, Zhang S (2024) Genome-wide identification of the SAUR gene family and screening for SmSAURs involved in root development in *Salvia miltiorrhiza*. Plant Cell Rep 43(7):165

36. Haist G, Sidjimova B, Denev R, Berkov S (2025) January) The effect of salicylic acid as an elicitor for Galanthamine production in *Hippeastrum papilio* (Ravenna) Van Scheepen hydroponic cultures. CR Acad Bulg Sci 78(1):64–72
37. Rattan S, Warghat AR (2023) Comparative analysis of salidroside and rosavin accumulation and expression analysis of biosynthetic genes in salicylic acid and methyl jasmonate elicited cell suspension culture of *Rhodiola imbricata* (Edgew.). Ind Crop Prod 198:116667
38. Ferrón-Carrillo F, Guil-Guerrero JL, González-Fernández MJ, Lyashenko S, Battafarano F, da Cunha-Chiamolera TPL, Urrestarazu M (2021) LED enhances plant performance and both carotenoids and nitrates profiles in lettuce. Plant Foods Hum Nutr 76:210–218
39. Mirgos M, Geszprych A, Przybył JL, Niedzińska M, Sujkowska-Rybkowska M, Gajc-Wolska J, Kowalczyk K (2025) Effect of lighting type on the nitrates concentration, selective bioactive compounds and yield of sweet basil (*Ocimum basilicum* L.) in hydroponic production. Agronomy 15(4):966
40. Salehi-Arjmand H, Akhoundi M, Akramian M, Khadivi A, Tunç Y (2025) Time course of physiological and biochemical responses of *Thymus vulgaris* to short-term salinity stress in hydroponics. Sci Rep 15(1):1–14
41. Coêlho MR, Rivas R, Ferreira-Neto JR, Bezerra-Neto JP, Pandolfi V, Benko-Iseppon AM, Santos MG (2021) Salt tolerance of *Calotropis procera* begins with immediate regulation of aquaporin activity in the root system. Physiol Mol Biol Plants 27:457–468
42. Gengmao Z, Quanmei S, Yu H, Shihui L, Changhai W (2014) The physiological and biochemical responses of a medicinal plant (*Salvia miltiorrhiza* L.) to stress caused by various concentrations of NaCl. PLoS One 9(2):e89624
43. Yep B, Gale NV, Zheng Y (2020) Aquaponic and hydroponic solutions modulate NaCl-induced stress in drug-type *Cannabis sativa* L. Ind Crop Prod 11:1169
44. Partap M, Chandel A, Sharma D, Verma K, Parmar P, Soni V, Srivatsan V, Bhargava B (2025) Morphological traits and biochemical and volatile compound composition along with gene expression profiling in *Chrysanthemum morifolium* Ramat. Grown under soil and soil-less conditions. ACS Agr. Sci Technol 5(5):701–713
45. Sharma S, Kumar P, Sharma R, Warghat AR (2023) In vitro propagation and omics breakthroughs for understanding specialized metabolite production in high-value Himalayan Fritillaria species. Ind Crops Prod 205:117541
46. Rattan S, Kumar P, Kaur E, Sood A, Acharya V, Warghat AR (2022) Comparative transcriptome and tissue-specific expression analysis of genes reveal tissue-cultured plants as an alternative source for phenylethanoids and phenylpropanoids in *Rhodiola imbricata* (Edgew.). Gene 836:146672
47. Wang C, Zhang J, Li J, Chai Q, Xie J (2025) Integrated metabolomics and transcriptomics analysis reveals the potential mechanism by which Methyl jasmonate enhances the pungent flavor of soilless-cultivated Chinese chives *(Allium tuberosum)*. BMC Plant Biol 25(1):375
48. Riano-Pachon DM, Nagel A, Neigenfind J, Wagner R, Basekow R, Weber E, Mueller-Roeber B, Diehl S, Kersten B (2009) GabiPD: the GABI primary database—a plant integrative 'omics' database. Nucleic Acids Res 37(suppl_1):D954–D959
49. Rai A, Saito K, Yamazaki M (2017) Integrated omics analysis of specialized metabolism in medicinal plants. Plant J 90(4):764–787
50. Srivastava U, Mathur A (2023) Recent advancements in prevalent practices for plant cultivation by hydroponics. Def Life Sci J 8:255–268
51. Rafael GJ, Pierre M, Frank E, Heidi W, Sibylle D, Daniel C, Matthew R, Gemma M, Daniel M, Guillermo G, Gustavo S (2022) Evidence for increasing global wheat yield potential. Environ Res Lett 17(12)
52. Gul Z, Bora S (2023) Exploiting pre-trained convolutional neural networks for the detection of nutrient deficiencies in hydroponic basil. Sensors 23(12):5407
53. Guo Y, Gao J, Tunio MH, Wang L (2022) Study on the identification of mildew disease of cuttings at the base of mulberry cuttings by aeroponics rapid propagation based on a BP neural network. Agronomy 13(1):106

54. Venkatasai NN, Shetty DN, Vinay CM, Sekar M, Muthusamy A, Rai PS (2025) A comprehensive review of factors affecting growth and secondary metabolites in hydroponically grown medicinal plants. Planta 261(3):1–18
55. Dsouza A, Dixon M, Shukla M, Graham T (2025) Harnessing controlled-environment systems for enhanced production of medicinal plants. J Exp Bot 76(1):76–93
56. Yu M, Chen Y, Zhu Z, Liu L, Zhang L, Guo Q (2016) Effect of phosphorus supply on plant productivity, photosynthetic efficiency and bioactive-component production in *Prunella vulgaris* L. under hydroponic condition. J Plant Nutr 39(12):1672–1680

Chapter 6
Smart Urban Agriculture for Enhancing Resilience and Sustainability

Reza Azadi Gonbad and Nasrin Farhadi

Abstract Urban smart agriculture is a solution for food security and sustainable development in cities. Urban smart farming (USF) integrates digital technologies (e.g., IoT, AI, and robotics) into urban agriculture optimizes resource use, enhances productivity, and reduces environmental impact. This paper explores the role of USF in addressing food security challenges in rapidly urbanizing environments. Through a systematic literature review, we analyze key technologies, benefits, and implementation challenges. Results highlight USF's potential to transform urban food systems while supporting the United Nations' Sustainable Development Goals. Findings indicate that USF can considerably reduce water usage and increase crop yields tenfold in limited urban spaces. However, high initial costs and policy gaps are the critical challenges in development of urban smart farming.

Keywords Artificial intelligence (AI) · Food security · Internet of Things (IoT) · Vertical farming

R. Azadi Gonbad (✉)
Tea Research Center, Horticultural Science Research Institute (HSRI), Agricultural Research, Education and Extension Organization (AREEO), Lahijan, Iran

N. Farhadi
Greenhouse and Controlled Environments Research Center, Horticultural Science Research Institute (HSRI), Agricultural Research, Education and Extension Organization (AREEO), Karaj, Iran

P. Kumar, A. R. Warghat (eds.), *Transforming Medicinal Plant Agriculture*, Smart Agriculture 16, https://doi.org/10.1007/978-981-95-3963-5_6

6.1 Introduction

6.1.1 The Imperative of Urban Smart Farming: A Sustainable Solution for Future Cities

Urban smart farming (USF) has emerged as a transformative solution to address the complex challenges of food security. With the United Nations projecting that 68% of the world's population will live in cities by 2050, traditional food systems will no longer be able to meet the nutritional needs of this rapidly growing urban population. Meanwhile, climate change and global warming pose serious threats to conventional agricultural production, making the need for sustainable alternatives more urgent than ever [1, 2].

The modern technologies such as the Internet of Things (IoT), artificial intelligence (AI) and automated control systems, enables the urban smart farming to agricultural production in the heart of cities. These systems reduce carbon emissions by 70% and water consumption by 90% compared to traditional methods [3, 4]. Moreover, utilizing underused urban spaces such as rooftops and basements can increase food production per unit area by up to tenfold. The USF establishes sustainable food security by reducing dependence on imports, enabling year-round production, and providing citizens with direct access to fresh nutritious produce. Additionally, this technology improves air quality through increased urban green spaces, creates tech-driven jobs and promotes a circular economy. In the face of challenges like supply chain disruptions (clearly demonstrated during the COVID-19 pandemic) urban smart farming serves as a resilience-building solution for cities [5]. Its intelligent monitoring systems allow for the prediction and management of potential crises [6].

Given these advantages, it is clear that the development of urban smart farming is no longer optional but an unavoidable necessity for the cities of the future [7]. However, fully realizing this potential requires investment in research and development, establishment of appropriate regulatory frameworks and increased public awareness. By merging agricultural knowledge with digital technologies, this innovation presents a new model of coordination relationship between cities and nature that can solve the many of the critical challenges of food supply for increased population. Urban smart farming is not just an agricultural revolution; it is a fundamental strategy of sustainable urban living in the twenty-first century [8, 9]. Compared to traditional agriculture, this system offers numerous advantages including weather independence, 70–95% reduction in water usage, year-round production capability, and significantly lower carbon footprint by eliminating transportation requirements. These characteristics make urban smart farming a crucial solution for addressing food security challenges in urban areas [10]. Table 6.1 highlighted the key differences between urban smart agriculture and traditional farming.

The present study ultimately seeks to provide useful insights for urban planners, technology developers, and community influencers to accelerate the adoption of smart farming solutions as part of sustainable urban development strategies. By

Table 6.1 The differences between urban smart agriculture and traditional agriculture

Aspect	Urban smart farming	Traditional agriculture
Location	Integrated into cities (rooftops, vertical farms, indoors)	Rural areas, large open fields
Space efficiency	High (vertical stacking, compact systems)	Low (requires large land areas)
Water usage	90–95% less (closed-loop hydroponics/aeroponics)	High (flood irrigation, evaporation loss)
Energy dependency	Renewable energy-powered (solar, smart grids)	Relies on fossil fuels for machinery
Crop yield	10–40x higher per square (year-round production)	Seasonal, weather-dependent yields
Labor	Automated (AI, robotics, IoT monitoring)	Manual-intensive
Climate resilience	Fully controlled environments (no weather risks)	Vulnerable to droughts/floods/pests
Supply chain	Hyper-local (reduces transport emissions/costs)	Long-distance distribution networks
Chemical use	Minimal (precision nutrients, no pesticides)	Pesticides/fertilizers common
Carbon footprint	Negative (absorbs CO_2, renewable energy)	High (machinery, transport emissions)
Cost structure	High initial investment, lower long-term costs	Low startup, high operational costs
Biodiversity impact	Increases urban biodiversity	Can reduce natural biodiversity
Food safety	Sterile, traceable (blockchain-enabled)	Risk of contamination (soil/water)
Scalability	Easily expandable in urban settings	Limited by land availability

increasing awareness in this field, the research will contribute to building more resilient, food-secure, and environmentally sustainable cities for future generations.

6.1.2 Definition of Urban Smart Farming (USF)

Urban smart farming represents an emerging model in food production which offering an integrated technology-driven approach of agriculture in urban environments. This advanced system fundamentally transforms traditional food production methods by combining digital technologies with sustainable agricultural principles [11, 12].

The SUF utilizes Internet of Things (IoT) technology through a network of complex sensors that continuously monitor critical plant growth parameters including light, temperature, humidity, and nutrient levels. The collected data is processed by artificial intelligence and machine learning systems, which employ predictive algorithms to optimize growing conditions [13].

A distinctive feature of this system is its implementation of innovative cultivation methods such as hydroponics, aquaponics, and aeroponics, which eliminates the need for fertile soil. In these controlled environments, plants use precisely calibrated nutrient solutions resulting significantly improving yields while reducing resource consumption [14]. Vertical farming serves as the backbone of this system, maximizing urban spaces through efficient vertical utilization to enable mass production in limited areas. Combination of vertical farming with renewable energy sources like solar power systems; creates a fully sustainable food production model [15].

On a broader scale, this system represents not just a food production method, but a vital component of future smart city ecosystems. It simultaneously addresses environmental, economic and social challenges in urban areas. This integrated approach presents a new model for combining digital technologies with sustainable food production that could revolutionize urban agriculture. As cities continue to grow, urban smart farming stands poised to become an essential element of sustainable urban development strategies worldwide [16, 17].

6.2 Key Technologies in Urban Smart Farming

6.2.1 IoT and Data-Driven Agriculture

The integration of Internet of Things (IoT) technologies has emerged as a basis of urban smart farming, revolutionizing how cities approach food production. By deploying networks of interconnected sensors, these systems enable real-time monitoring of critical growth parameters, including soil moisture (accurate to ±2%), ambient light intensity (measured in Photo-synthetically Active Radiation), and microclimate conditions [18]. This data-driven pattern facilitates precision agriculture at unique scales, with automated delivery systems achieving 90% efficiency in water and nutrient distribution in contrast to traditional farming's 30%–40% efficiency rates. In vertical farming applications, IoT-enabled dynamic LED lighting systems have demonstrated 40% yield increases while reducing energy consumption by 35% [19, 20].

The environmental implications are profound: smart irrigation algorithms have reduced urban agricultural water demand by 70%–95%, with aeroponic systems reaching near-zero water waste through closed-loop recycling. Advanced predictive models, combining IoT data with machine learning, now provide 14–21 day early warnings for pest infestations and plant diseases, potentially preventing 80% of crop losses. When integrated with renewable energy micro-grids, these systems have shown capacity to reduce carbon footprint of urban food production by half [20, 21].

6.2.2 Artificial Intelligence

Artificial intelligence combines computer vision and data processing to enable intelligent management of urban farms (Fig. 6.1). AI-based systems can monitor plant health with over 90% accuracy through satellite simulation and IoT sensor data analysis, detecting diseases weeks before visible symptoms appear. AI-powered robots like Agrobot harvest crops at three times human speed, while intelligent spray drones reduce pesticide use by 60%. Smart urban farms could meet up to 30% of cities' food needs by 2030, achieving 20%–40% greater productivity with 70% lower carbon emissions. Additionally, AI-powered contaminant detection systems are taking food safety to unprecedented levels [2, 4].

However, high initial costs, workforce training requirements and cybersecurity concerns are the main challenges in using artificial intelligence in urban farming. Addressing these issues requires developing edge AI for on-site data processing, using generative AI for climate simulations, and integrating blockchain for supply chain transparency. These achievements will transform urban agriculture from a marginal activity into a vital component of future city infrastructure. To fully realize this potential, collaboration between governments, research institutions, and the private sector is essential. Investment in applied research and cost-effective solutions can make sustainable urban food security achievable [12, 22].

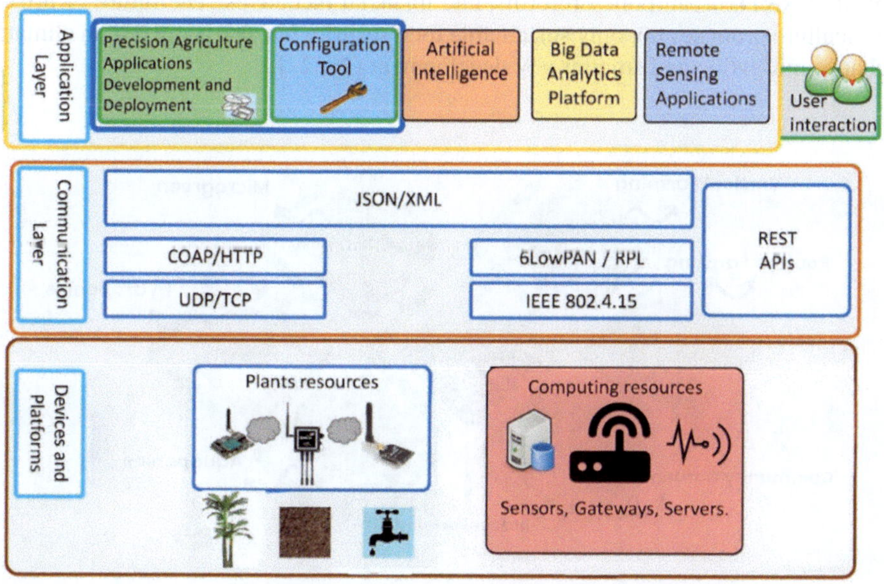

Fig. 6.1 Conceptual framework of an IoT-based urban farm [6]

6.2.3 Advanced Cultivation Systems

In Fig. 6.2, the different types of urban farming are shown. Smart urban farming depends on advanced cultivation systems especially vertical farming, which utilizes multi-tiered structures with intelligent LED lighting and hydroponic/aeroponic systems to achieve yields 100 times greater than traditional methods. Hydroponic and aeroponic systems have fundamentally changed urban agriculture by eliminating soil dependence. These methods accelerate plant growth by up to 40% compared to conventional farming. Aquaponic systems have further enhanced urban food production by creating symbiotic fish-plant ecosystems that simultaneously yield protein and vegetables. The development of container farms and rooftop greenhouses has enhanced the access to advanced agricultural technology. These systems can be operational within 24 h, providing ideal solutions for areas lacking arable land [23]. Rooftop greenhouses offer additional benefits, reducing building energy consumption by 20% while absorbing 50% of rainwater [24]. From a sustainability perspective, these systems create circular economies by converting food waste into compost and energy. By eliminating long supply chains, they reduce carbon footprints while supporting urban biodiversity through natural habitats [25].

Despite challenges like high initial costs and technical complexity, emerging solutions such as solar-powered farms, low-cost units, and beginner-friendly apps are lowering barriers to entry. Projections indicate smart urban farms could supply 40% of city vegetable needs by 2035. Accelerating this revolution requires investment in R&D, urban policy reforms, and financial incentives. The future of urban agriculture promises not only sustainable food security but also a key role in climate change mitigation and smarter city development [26, 27].

Fig. 6.2 Different types of urban farming

6.2.4 Robotics and Automation

Modern urban agriculture is undergoing a technological revolution through the integration of advanced robotics and automation systems. These innovations are fundamentally determining how cities approach for food production, enabling high levels of efficiency, and sustainability in constrained urban environments. At the forefront of this transformation are intelligent robotic systems that handle the complete agricultural cycle. Automatic planting machines equipped with computer vision achieve millimeter-perfect seed placement and optimizing use of every square centimeter of valuable urban growing space [6]. This precision continues through the plant growth cycle with AI-powered monitoring robots using hyperspectral imaging to detect nutrient deficiencies or disease symptoms days before human observers might notice. In addition, robotic arms that can identify and pick ripe produce without damage, operating at triple human speed with near-perfect accuracy, have enhanced the harvesting process [28, 29].

The operational benefits of these automated systems are profound. Labor requirements have been reduced by 60–70%, while production continues uninterrupted under optimal conditions. Resource efficiency has considerably increased with automated nutrient delivery systems that reducing fertilizer use by 35% on average. Food safety has improved as minimized human contact contamination risks by 80%, and closed-loop water systems recycle up to 95% of irrigation supplies. These facilities represent a new model where urban farms operate more like advanced manufacturing plants than traditional agricultural operations. Current systems work best for leafy greens and certain fruits but for complex crops still requiring human intervention [30].

Despite to amazing effects of integrated automation in smart urban farming however, it must be considered that this system needs the high investments for sophisticated technology, technical maintenance requirements, energy demands, and workforce retraining. This robotic transformation makes urban farming as a commercially viable solution to food security challenges, capable of producing consistent and high-quality yields year-round regardless of external climate conditions. As the technology matures and costs decrease, automated urban agriculture will increasingly become an integral component of smart city infrastructure, potentially forming closed-loop systems that share energy and resources with surrounding urban ecosystems [31].

6.2.5 Renewable Energies

Market projections indicate that 70% of urban farms will primarily use renewable energy by 2035, according to International Renewable Energy Agency (IRENA). Smart urban farming is undergoing a fundamental transformation through the adoption of renewable energy systems. By controlling solar, wind, bioenergy, and kinetic energy technologies, urban farms are self-sufficient and eco-friendly food production centers that significantly reduce reliance on fossil fuels [2].

The most prominent applications of renewable energy include rooftop systems that profit from power LED grow lights, climate control and automation systems. Solar thermal technology further enhances efficiency by heating water for hydroponic systems and reducing energy costs by 30%. Small-scale wind turbines complement energy needs, particularly in high-rise vertical farms [32–34]. The conversion of organic waste into biogas through anaerobic digestion represents another breakthrough for greenhouse heating. Algae bioenergy systems, which generate power while absorbing CO_2, represent another exciting point. Emerging technologies such as piezoelectric floors in high-traffic urban farms are even harvesting energy from foot traffic to power sensors and lighting systems [12, 33]. Generally, these integrated renewable solutions offer multiple advantages:

- Energy independence, with some facilities achieving complete self-sufficiency.
- 40–60% reduction in long-term energy costs
- Carbon-neutral operations in fully renewable-powered farms.
- Enhanced resilience during power outages or fuel shortages.

Despite the current progress, challenges remain, particularly regarding high initial costs and space limitations. Innovative solutions are addressing these barriers especially hybrid energy systems combining solar, wind, and battery storage. Singapore's Sustenir Agriculture exemplifies this progress, operating solar-powered vertical farms that use 95% less energy than conventional greenhouses while producing premium crops like kale and strawberries [35].

6.2.6 Blockchain and Product Tracking

Blockchain technology enhances supply chain transparency and provides digital certification, building consumer trust [36].

6.2.7 Augmented Reality (AR)

Emerging technologies like augmented reality contribute to urban farmer education through growth simulations and operational guidance via wearable devices. This integrated technological framework represents a significant leap forward in our ability to develop self-sufficient, environmental-friendly food production systems within urban landscapes while addressing pressing challenges of population growth and resource scarcity. The full realization of this potential requires interdisciplinary collaboration among agricultural scientists, IT specialists, robotics engineers, and clean energy experts [37, 38].

6.3 Benefits of Urban Smart Farming

Urban smart farming represents a new approach in food production, offering a comprehensive solution to pressing urban challenges through technological innovation. By integrating advanced agricultural systems into city landscapes, this approach delivers multifaceted benefits that extend far beyond traditional farming methods. Urban smart farming offers transformative benefits that address key challenges in modern cities.

6.3.1 Enhanced Food Security and Local Production

Urban smart farming provides food systems by enabling hyper-local production within city limits. Establishing vertical farms, hydroponic systems in basements and rooftop greenhouses shorten supply chains from farm-to-table. During the COVID-19 pandemic, cities with established urban farming infrastructure maintained 30–40% better fresh produce availability during supply chain disruptions according to FAO reports. The controlled environments allow year-round production regardless of external weather conditions, with some facilities achieving 15–20 harvest cycles annually for leafy greens compared to 2–3 cycles in conventional agriculture [39–42].

6.3.2 Resource Conservation

Smart farming systems through precision technology conserve natural resources. IoT-enabled hydroponic systems demonstrate 90–95% water savings by recirculating nutrient solutions, also aeroponic systems reduce water usage by up to 98% compared to soil farming. Energy consumption is optimized through AI that adjusts LED lighting spectra and intensity based on real-time plant growth data, resulting 30–40% energy reductions. Urban vertical farms use 70% less fertilizer through closed-loop nutrient delivery systems that prevent nutrient solution runoff. Regarding space utilization, a 1000 m^2 vertical farm can produce yields equivalent to 5000 m^2 of traditional farmland [41, 43].

6.3.3 Environmental Regeneration

These systems actively improve urban ecosystems through multiple pathways. Rooftop farms reduce building cooling demands by 15–20% through evapotranspiration, while absorbing 50% more storm water than conventional roofs according to

Environmental Protection Agency data [44, 45]. By eliminating agricultural runoff, they prevent the annual discharge of 2–3 million tons of nitrogen pollutants into waterways. The cultured systems of urban farms create urban biodiversity hotspots, with some facilities reporting 30–50% increases in local pollinator populations [24, 46, 47].

6.3.4 Economic Transformation

Urban smart farming generates novel economic value through multiple channels. It creates more jobs with high-skilled tech than traditional farming. Local economies in the USF benefit directly from import substitution, a trend that is further amplified as the global urban farming market grows and attracts significant venture capital investment [1, 8].

6.3.5 Public Health Advancements

Urban smart farming delivers measurable health benefits through multiple pathways. Locally harvested products retaining 30–40% more nutrients compared to transported ones also communities with urban farm showed higher consumption of fresh produce. Most important of all, urban-grown produce 98–100% lower pesticide residues that considerably improves the quality of urban residents' food basket. Green farming spaces reduce urban heat by 2–4 °C while improving air quality. Therapeutic gardening programs in these facilities have demonstrated 30–50% reductions in stress hormone levels among participants [48].

6.3.6 Climate Resilience

Smart urban agriculture provides critical climate adaptation benefits. Controlled environment agriculture such as USF maintains production during extreme weather events that damage traditional farms. Decentralized production networks in urban smart farming reduce single-point failure risks. These systems utilize flood-resistant designs and can operate as an alternative energy during grid outages. Urban farms also mitigate heatwaves impacts by reducing the urban heat by 1–2 °C [31, 43, 46].

6.3.7 Technological Innovation

Urban farms drive cross-sector technological advancement through their integration of AI, robotics, and bioscience. Machine vision systems achieve 99% accuracy in identifying plant stress indicators weeks before human detection. Automated harvesting robots now perform delicate operations like strawberry picking with 95% success rates [11]. The sector's energy innovations include hybrid renewable systems that combine solar, biogas and kinetic energy operate at 90% energy self-sufficiency. The involved technological in USF are useful for other industries such as urban farming-developed LED spectra are now used in medical therapies and wastewater treatment sensors derived from hydroponic monitoring have applications in industrial process control [43].

6.4 Challenges and Limitations

6.4.1 Technical Challenges

Urban smart farming faces significant technical barriers, primarily the high initial setup costs. Advanced equipment requirements including full-spectrum LED lighting systems, precision environmental sensors and sophisticated hydroponic systems demand substantial capital investment. These systems also consume considerable energy due to their continuous need for precise climate control (temperature, humidity, and lighting) and if relying on non-renewable energy sources their economic and environmental durability gets weak. Furthermore, the technical complexity necessitates specialist forces for maintenance and troubleshooting which resulting workforce shortages in many urban areas [18].

6.4.2 Economic and Financial Barriers

The extended return on investment period (often up to 7 years) presents a major deterrent, particularly for private investors. Higher operational costs typically make smart farming products more expensive than conventional alternatives, creating market penetration challenges. The sector also suffers from immature business models, with many projects remaining dependent on government subsidies [20].

6.4.3 Regulatory and Policy Challenges

Municipal laws and land-use regulations frequently restrict urban farming development. The absence of clear legal frameworks for permitting, quality standards, and food safety compliance creates additional limitation. Insurance services for these novel agricultural operations remain underdeveloped due to their emerging nature. At the macro level, the lack of appropriate incentive policies and subsidies has decreased the growth of USF as well as conflicting regulations between different government agencies is created implementation obstacles [44, 49].

6.4.4 Social-Cultural Barriers

Many conventional farmers and distributors consider USF systems as threats to their livelihoods. Due to limited awareness about the benefits and quality concerns of USF products, consumers are hesitant to consume these products. The transition to products that may differ in appearance from conventional alternatives requires extensive consumer education and gradual market development, as taste preferences and purchasing habits slow to change [47].

6.4.5 Environmental Limitations

While offering numerous ecological benefits, smart urban agriculture confronts its own environmental challenges. If the USF powered by non-renewable energy, these systems' carbon footprints may increase unexpectedly. The management of nutrient-rich wastewater in hydroponic systems demands careful handling to prevent groundwater contamination. Additionally, the production of growing media and specialized equipment carries its own environmental costs that must be accounted for in sustainability assessments. Electronica waste from obsolete equipment requires specialized disposal solutions [50, 51].

6.4.6 Technological and Operational Challenges

Strong dependency on stable internet connectivity and digital infrastructure creates vulnerabilities to network outages and cyber threats. Crop diversity limitations (particularly for strategical crops like wheat and rice) constrain system capabilities. The technical integration of various subsystems (environmental controls, irrigation, and lighting) presents ongoing engineering challenges. Moreover, the rapid rate of technological obsolescence in this field requires continuous system upgrades, adding to

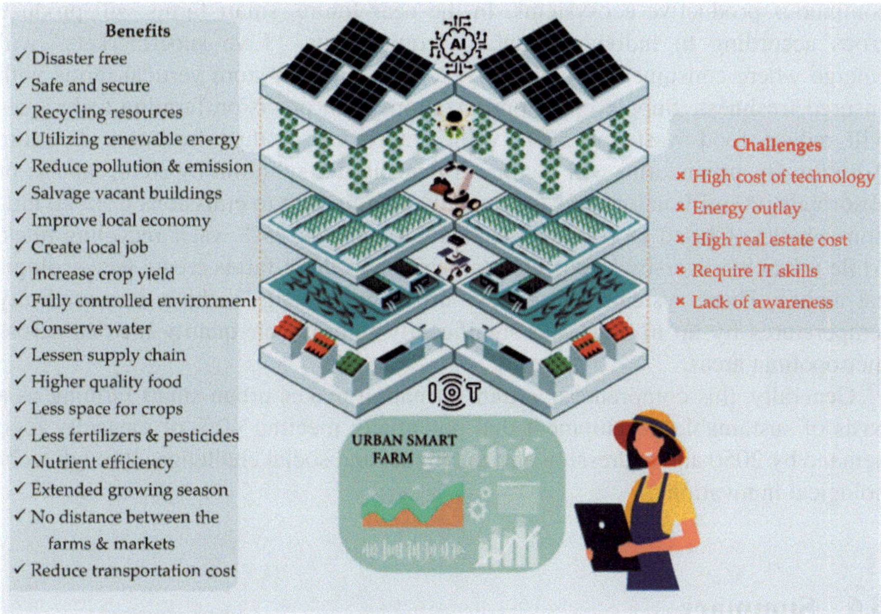

Fig. 6.3 Urban smart farm's benefits and challenges [8]

operational complexity and costs [52, 53]. In Fig. 6.3 the advantages and challenges of urban smart farming are reported next to each other.

6.5 Future Trends

By 2045, urban agriculture will undergo a revolutionary transformation through three key technological advancements. Next-generation AI systems utilizing self-learning algorithms will autonomously manage all plant growth parameters with 98% accuracy. Robotic automation will replace 70% of human labor in planting, maintenance, and harvesting operations, with specialized robots achieving 95% success rates in delicate fruit harvesting. Nanotechnology advancements will introduce molecular-scale greenhouses allowing precise manipulation of plant growth conditions, potentially tripling crop yields compared to conventional methods. Urban farming will become an organic component of city structure by 2045. Renewable micro-grids will supply 100% of vertical farming operations and autonomous drone and vehicle networks will enable hyper-local food distribution with under two-hour delivery. Architectural innovations will consider 30% of new building facades designed as multifunctional vertical farms by 2035, serving dual purposes as both food production sites and building insulation systems that reduce energy consumption. These integrated systems will transform urban landscapes into continuous productive ecosystems. In the near future, smart farms will produce crops according to individual dietary requirements. "Live supermarkets" will emerge where consumers directly harvest crops from in-store vertical farms with ensured freshness. Supply chain optimizations with precision farming techniques will reduce food waste by 50%, while improving product freshness duration. Additionally, urban farms will become critical climate solutions by 2050, capable of absorbing 100 million tons of CO_2 annually—equivalent to emissions from 20 million vehicles. Closed-loop water systems will achieve 95% water recycling rates, while urban biodiversity is expected to increase as these farms create new ecological niches. These systems will act as "urban green lungs" which decreased city temperatures by up to 3 °C and significantly improve air quality metrics across metropolitan areas.

Generally, this comprehensive transformation makes urban smart farming as a basis of sustainable development that potentially meeting 40% of megacity food demand by 2050 and addressing environmental and social challenges through technological innovation.

6.6 Summary

Ultimately, urban smart farming offers not just a solution for food security, but also a comprehensive model for sustainable urban development capable of simultaneously addressing environmental, economic, and social challenges. Cities that lay the groundwork for this transformation will emerge as pioneers of sustainable development in the second half of the twenty-first century, setting new standards for urban quality of life. This transformation will undoubtedly reshape the future of urban living and human consumption patterns.

References

1. Yuan X et al (2025) Advanced technologies in plant factories: exploring current and future economic and environmental benefits in urban horticulture. Hortic Res 12:1–18
2. Sikka R et al (2024) Advancing agriculture in smart cities: renewable energy and artificial intelligence-powered IoT. E3S Web Conf 500:1–10
3. Evan F, Jayadi R (2023) A prototype system design for microgreens IoT-based smart vertical farm. J Syst Manag Sci 12:1–12
4. Rawal S (2017) IoT based smart irrigation system. Int J Sci Res Eng Manag 1:1–10
5. Schoen V et al (2021) We have been part of the response: the effects of COVID-19 on community and allotment gardens in the global north. Front Sustain Food Syst 5:1–15
6. Atalla S et al (2023) An intelligent recommendation system for automating academic advising based on curriculum analysis and performance modeling. Mathematics 11(5):1098
7. Çakmakçı R, Salık MA, Çakmakçı S (2023) Assessment and principles of environmentally sustainable food and agriculture systems. Agriculture 13:1–20

8. Moghayedi A et al (2022) Effects of urban smart farming on local economy and food production in urban areas in African cities. Sustainability 14:10836
9. De Sousa R et al (2024) Challenges and solutions for sustainable food systems: the potential of home hydroponics. Sustainability 16:1–18
10. Khan N et al (2024) Cultivating change: empowering communities among elderly through social innovation and entrepreneurship in smart urban farming. J Hum Earth Future 5:1–12
11. Pai HA et al (2023) Enhancing urban farming through smart systems and water conservation: an IoT-based approach. In: 2023 3rd international conference on technological advancements in computational sciences (ICTACS). IEEE, pp 1–10
12. Rajendiran G, Rethnaraj J (2023) Future of smart farming techniques: significance of urban vertical farming systems integrated with IoT and machine learning. Open Access J Agric Res 8:1–12
13. Shuhaimi FN, Jamil N, Hamzah R (2020) Evaluations of internet of things-based personal smart farming system for residential apartments. Bull Electr Eng Inform 9:1–10
14. Insigne A, Balba N (2024) Optimizing urban farming with smart aquaponics systems. Asia Pac J Manag Sustain Dev 12:1–15
15. Luo Y, Ball P (2025) Adaptive production strategy in vertical farm digital twins with Q-learning algorithms. Sci Rep 15:1–14
16. Hic C et al (2016) Food surplus and its climate burdens. Environ Sci Technol 50:1–12
17. Benke K, Tomkins B (2017) Future food-production systems: vertical farming and controlled-environment agriculture. Sustainability 9:1–15
18. Thilakarathne N et al (2025) Internet of things enabled smart agriculture: current status, latest advancements, challenges and countermeasures. Heliyon 11:1–15
19. Sai M et al (2022) Internet of things for smart agriculture—state of the art and challenges. Ecol Eng Environ Technol 23:1–12
20. Dutta M et al (2025) Internet of things-based smart precision farming in soilless agriculture: opportunities and challenges for global food security. IEEE Access 13:1–20
21. Kim J et al (2024) Sustainable and inflatable aeroponics smart farm system for water efficiency and high-value crop production. Appl Sci 14:1–15
22. Ghandar A et al (2021) A decision support system for urban agriculture using digital twin: a case study with aquaponics. IEEE Access 9:1–15
23. Bakirov K et al (2024) Advances in the design and optimization of smart irrigation systems for sustainable urban vertical farming. Sci J Astana IT Univ 6:1–15
24. Özsoy T (2023) Enabling circularity for food safety: the rooftop farming model. Int J Agric Environ Food Sci 7:1–10
25. Hamidon MH et al (2019) Design and development of smart vertical garden system for urban agriculture initiative in Malaysia. J Teknol 81:1–10
26. Aisyah AP (2024) Assistance in urban farming for the cultivation of chili peppers in a greenhouse based on internet of things systems. ETHOS J Penelit Pengabdi Masy 12:1–12
27. Ramli M et al (2023) Design of a smart portable farming kit for indoor cultivation using the raspberry pi platform. Pertanika J Sci Technol 31:1–18
28. Shamshiri R et al (2018) Advances in greenhouse automation and controlled environment agriculture: a transition to plant factories and urban agriculture. Comput Electron Agric 153:1–12
29. Sugeru H, Musa P, Miska M (2020) Design and implementation of ecoholistic vertiminaponic with automated control and internet of things based monitoring system for sustainable urban farming. IOP Conf Ser Earth Environ Sci 500:1–12
30. Konar M (2024) Integrating IoT and AI for automated crop with smart urban farming. In: 2024 international conference on emerging innovations and advanced computing (INNOCOMP). IEEE, pp 1–10
31. Balogun A et al (2022) Digitalization for transformative urbanization, climate change adaptation, and sustainable farming in Africa: trends, opportunities, and challenges. J Integr Environ Sci 19:1–20

32. Abdelhamid MA et al (2025) Design and evaluation of a solar-powered smart irrigation system for sustainable urban agriculture. Sci Rep 15:1–14
33. Chatterjee A, Debnath S, Pal H (2020) Implication of urban agriculture and vertical farming for future sustainability. In: Urban horticulture - necessity of the future. Springer, pp 1–20
34. Oh S, Lu C (2022) Vertical farming—smart urban agriculture for enhancing resilience and sustainability in food security. J Hortic Sci Biotechnol 97:1–15
35. Pueyo-Ros J, Comas J, Corominas L (2023) Ediblecity: an R package to model and estimate the benefits of urban agriculture. Open Res Eur 3:1–15
36. Gulyas BZ, Edmondson J (2021) Increasing city resilience through urban agriculture: challenges and solutions in the global north. Sustainability 13:1–20
37. Kirwan RF et al (2023) Scalable agritech growbox architecture. Front Internet Things 2:1–15
38. EkaPutri R et al (2024) Smart tower farming system based on the internet of things in greenhouse system. Tikrit J Agric Sci 24:1–15
39. Pulighe G, Lupia F (2020) Food first: COVID-19 outbreak and cities lockdown a booster for a wider vision on urban agriculture. Sustainability 12:1–20
40. Barthel S et al (2019) Global urbanization and food production in direct competition for land: leverage places to mitigate impacts on SDG2 and on the earth system. Anthropol Rev 6:1–20
41. Jindal D (2023) Harmonizing urbanization and food security: a novel AI-driven approach for sustainable urban agriculture. Int J Sci Res 12:1–10
42. Lal R (2020) Home gardening and urban agriculture for advancing food and nutritional security in response to the COVID-19 pandemic. Food Secur 12:1–20
43. Hachem-Vermette C (2024) Enhancing urban climate resistance through the application of selected strategies and technologies. Discov Cities 5:1–12
44. Deksissa T et al (2021) Integrating urban agriculture and stormwater management in a circular economy to enhance ecosystem services: connecting the dots. Sustainability 13:1–20
45. Prudencio L, Null S (2018) Stormwater management and ecosystem services: a review. Environ Sci Pol 84:1–15
46. Altieri M et al (2025) Unleashing the potential of urban agroecology to reach biodiversity conservation, food security and climate resilience. Agriculture 15:1–20
47. Villavicencio-Valdez GV et al (2023) Urban agroecology enhances agrobiodiversity and resilient, biocultural food systems. The case of the semi-dryland and medium-sized Querétaro City, Mexico. Front Sustain Food Syst 7:1–15
48. Ebenso B et al (2022) Nature-based one health approaches to urban agriculture can deliver food and nutrition security. Front Nutr 9:1–15
49. Christmann A, Graf-Drasch V, Schäfer R (2024) Smart urban agriculture. Bus Inf Syst Eng 66:1–15
50. Jasmee K, Jumaat S (2025) Solar-powered Peltier thermoelectric water cooling and pH monitoring. Int J Electr Eng Appl Sci 7:1–12
51. Richter I et al (2023) Willing to be the change: perceived drivers and barriers to participation in urban smart farming projects. J Urban Aff 45:1–20
52. Chojnacka KE et al (2024) Green urban transition: interdisciplinary insights on green façades design in hot climates as one of crucial strategies for low-carbon development. Clean Technol Environ Policy 26:1–15
53. Agilesh R et al (2021) IoT based smart agriculture. J Manuf Eng 16:1–12

Chapter 7
Elicitation in Hydroponics for Medicinal Plants Production

Nasrin Farhadi and Mohammad Moghaddam

Abstract Cultivation in the hydroponic system is a technological and economical method in terms of water consumption, so that the development of this system as an alternative to traditional agriculture having a suitable irrigation pattern for the cultivation and production of different plant species. The use of this system as an efficient way for plant production has also shown favorable results to produce medicinal plants. Hydroponic system due to its possibility to control environmental conditions is more efficient than conventional agriculture to production of medicinal plants. Also, the reported results indicate the success of the hydroponic system for the production of medicinal plants at a commercial level. This system, with its greater control of different environmental factors, gave an appropriate opportunity to change the quality and quantity of medicinal plants by application of different biotic and abiotic elicitors. This chapter explores and highlights the use of different elicitors to enhancing the quantity and quality of medicinal plants under hydroponic system.

Keywords Elicitors · Secondary metabolites · Signaling · Soilless culture

N. Farhadi
Greenhouse and Controlled Environments Research Center (GCER), Horticultural Science Research Institute (HSRI), Agricultural Research, Education and Extension Organization (AREEO), Karaj, Iran

M. Moghaddam (✉)
Department of Horticultural Science and Landscape Architecture, Faculty of Agriculture, Ferdowsi University of Mashhad, Mashhad, Iran
e-mail: m.moghadam@um.ac.ir

P. Kumar, A. R. Warghat (eds.), *Transforming Medicinal Plant Agriculture*, Smart Agriculture 16, https://doi.org/10.1007/978-981-95-3963-5_7

7.1 Introduction

Hydroponic cultivation of medicinal plants refers to the soilless farming techniques that plants grow in the nutrient-rich water solutions and provide the precise control of environmental condition. This innovative approach has gained prominence as a sustainable alternative to traditional agriculture, particularly in response to the growing global demand for natural remedies and herbal medicines [1]. By mitigating common challenges associated with soil-based farming, such as disease prevalence and resource scarcity, hydroponics enhances the efficiency and quality of medicinal plant production, making it a notable focus within agricultural science and commercial farming sectors [2]. The significance of hydroponics in medicinal plant production lies in its ability to deliver higher yields with consistent active compound concentrations, which are crucial for the therapeutic efficacy of these plants. Research has shown that species like echinacea and basil thrive in hydroponic systems, benefiting from tailored nutrient solutions that stimulate secondary metabolite production [3]. The technique has considerably evolved since its early experimental stages, with advances in technology and methodology leading to its adoption by commercial growers worldwide, particularly in regions with limited arable land. Additionally, concerns over the environmental impact of energy consumption in hydroponic systems and the limited variety of crops that can be effectively cultivated are important considerations for sustainability advocates [4]. The future of hydroponic medicinal plant production looks promising, with ongoing innovations aimed at enhancing efficiency and accessibility, while addressing the evolving needs of the market and the environment [5]. As the industry matures, it presents opportunities for integrating smart technologies and sustainable practices that further enhance the appeal of hydroponic systems. By embracing these advancements, hydroponics may play a pivotal role in the future of agriculture, contributing to the global shift toward sustainable food systems and the increasing popularity of herbal remedies among consumers [6].

7.2 Hydroponic Culture

Hydroponic culture is a method of growing plants without soil, instead utilizing nutrient-rich water solutions to supply essential nutrients directly to plant roots. This innovative agricultural approach has gained significant attention for its ability to maximize space efficiency, resource use, and crop yields, making it particularly relevant in the context of urban farming and sustainable agriculture [7]. These methods can be categorized into two major types: solution culture types and medium culture types, each employing distinct techniques suitable for various plant species and growing conditions [7–9] (Fig. 7.1).

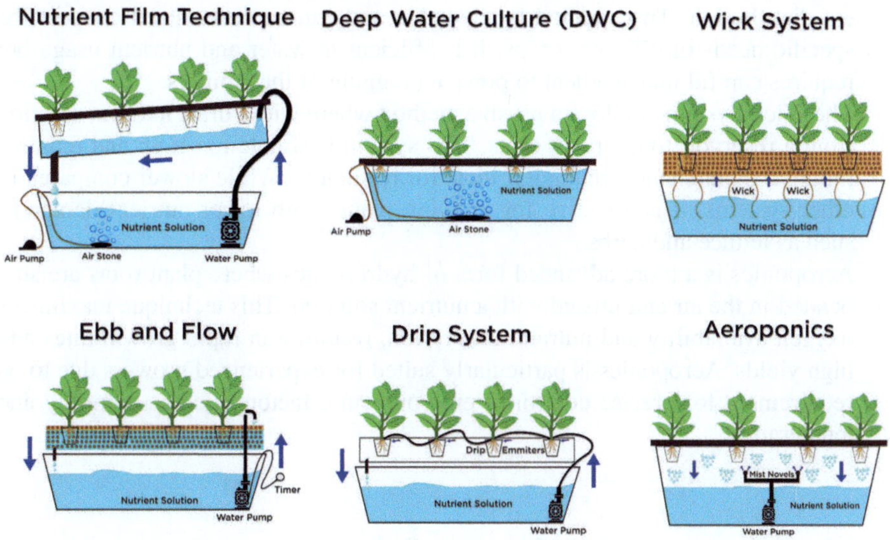

Fig. 7.1 Different types of hydroponic cultures

7.2.1 Hydroponic Types

Solution Culture Types

- The Nutrient Film Technique (NFT) is a widely utilized hydroponic system where a thin layer of nutrient solution flows continuously over the roots of the plants, which are supported in a sloped growing tray. This method provides both nutrition and aeration, promoting healthy root development and faster plant growth. The NFT system is particularly efficient in resource utilization, consuming less water and nutrients while yielding maximum results, making it ideal for crops like lettuce and herbs.
- In Deep Water Culture (DWC) plant roots are suspended in a nutrient solution that is continuously oxygenated, typically using air stones or diffusers. This method allows for rapid plant growth due to the constant supply of nutrients and oxygen, making it suitable for a wide range of crops, including larger fruiting plants like tomatoes.
- Ebb and Flow System also known as flood and drain systems, periodically flood the growing area with nutrient solution before draining it back into the reservoir. This cycling allows for excellent oxygenation and nutrient absorption by the roots, supporting vigorous plant growth. These systems are versatile and can accommodate various types of plants, from leafy greens to larger fruit-bearing crops.

Medium Culture Types

- The Drip System delivers nutrient solution directly to the base of each plant through a network of tubes and emitters, allowing for precise control over nutri-

ent distribution. This method is adaptable and can be customized to meet the specific needs of different crops. It is efficient in water and nutrient usage but requires careful management to prevent clogging of the emitters.
- The Wick System employs a passive method where wicks draw nutrient solution from a reservoir to the plant roots. This system is simple to set up and requires minimal maintenance, making it ideal for beginners. While slower compared to other systems, it is effective for growing plants with lower nutrient demands, such as lettuce and herbs.
- Aeroponics is a more advanced form of hydroponics where plant roots are suspended in the air and misted with a nutrient solution. This technique maximizes oxygen availability and nutrient absorption, resulting in rapid growth rates and high yields. Aeroponics is particularly suited for experienced growers due to its requirement for precise control of environmental factors such as humidity and temperature.

7.2.2 Advantages of Hydroponic Culture

Unlike traditional agriculture, which requires expansive land areas, hydroponic systems can be implemented in smaller spaces, making them particularly advantageous in urban settings where land is limited. Vertical hydroponic systems further enhance space efficiency by allowing crops to be stacked vertically, thus maximizing production per square foot and addressing the challenges posed by urbanization and land scarcity [1, 6, 8]. Hydroponic culture is recognized for its efficient use of resources, particularly water and nutrients which utilizes a closed-loop water circulation method and consequently reduces water usage compared to traditional soil farming. This system minimizes water loss due to evaporation or runoff, making hydroponics a highly water-efficient cultivation method [7]. Additionally, the nutrient solution in hydroponics can be carefully formulated to provide plants with an optimal balance of essential elements, enhancing nutrient uptake and minimizing deficiencies. So that the hydroponic system has higher yields and faster growth rates. Plants grown hydroponically can yield 30% to 50% more than those cultivated in soil under comparable conditions [10]. The controlled environment of hydroponics allows for optimal conditions, such as precise nutrient delivery and climate control, which contribute to quicker maturation and robust plant development. As a result, crops like leafy greens often thrive in hydroponic systems, leading to increased productivity and efficiency [1]. The environmental impact of hydroponic culture is generally more favorable than that of traditional farming. Hydroponics eliminates soil-borne diseases and pests, reducing the need for pesticides and herbicides. This approach not only leads to cleaner food production practices but also contributes to environmental protection by minimizing fertilizer runoff and soil erosion. Moreover, the ability to grow crops year-round, regardless of external weather conditions, helps stabilize food supply and prices, further enhancing food security [6, 11]. Hydroponic culture plays a crucial role in sustainable agriculture by addressing key challenges such as soil degradation and water scarcity. As arable land becomes increasingly limited, hydroponics presents a viable alternative that can produce food in a resource-efficient manner. Furthermore, hydroponics can be integrated into urban environments, providing local food sources and reducing food miles, which lowers the carbon footprint associated with transporting agricultural products. This localized production contributes to a more sustainable food system that aligns with the growing demand for fresh, locally grown produce [12–14].

7.2.3 Challenges and Limitations of Hydroponic Culture

Hydroponics, while offering numerous advantages, presents various challenges and limitations that must be acknowledged for successful implementation. Establishing a hydroponic system requires substantial initial investment due to the specialized equipment necessary, such as pumps, grow lights, and nutrient delivery systems, which can accumulate quickly. The installation costs vary significantly based on the scale of the system, and material and labor expenses constitute a major portion of the financial outlay [1, 6–8]. Ongoing operational costs, including regular maintenance, nutrient replenishment, water supply, and further strain resources can deter potential growers. Effective operation of hydroponic systems necessitates technical expertise and skilled labor. Monitoring nutrient levels, adjusting pH, and ensuring appropriate lighting and temperature conditions require a considerable understanding of the system's mechanics, which can pose a barrier for new growers [13]. Additionally, training staff to address the intricate demands of hydroponics is often necessary, adding to the overall costs and complexity of managing these systems. Achieving the correct balance of nutrients in a hydroponic setup is critical yet challenging. Improper nutrient ratios or fluctuations in pH can lead to deficiencies or toxicities that adversely affect plant health and productivity. Regular testing and adjustments are essential, as hydroponic systems lack the buffering capacity of soil, making them more susceptible to rapid changes in nutrient levels and pH [15]. Hydroponics heavily relies on technology, including pumps, timers, and artificial lighting, which necessitates a continuous electricity supply. Power outages or equipment failures can severely impact plant growth if not addressed promptly. The energy consumption associated with advanced climate control systems, like heating, cooling, and lighting, can also contribute to elevated operational costs [16]. While hydroponics offers potential sustainability benefits, it raises concerns about its environmental impact. The reliance on nutrient solutions and energy-intensive systems can challenge sustainability claims, necessitating a careful evaluation of the ecological footprint associated with hydroponic practices [17].

7.3 Medicinal Plants Production

Medicinal plant production involves the cultivation and harvesting of plants recognized for their therapeutic properties, serving as a vital component of both traditional and modern healthcare systems worldwide. The significance of medicinal plants is rooted in their long history of use, dating back thousands of years, where they have provided natural remedies for various ailments before the advent of contemporary pharmaceuticals [6]. Today, the global market for medicinal plants is valued in the billions, reflecting an increasing consumer preference for natural and herbal alternatives to synthetic medications. The cultivation of medicinal plants is characterized by diverse practices tailored to meet the specific needs of both herbal and pharmaceutical industries [7, 8]. Despite the growing interest in medicinal plant production, significant challenges remain, including over-exploitation of wild populations, environmental sustainability concerns, and the need for robust quality control measures [18]. The ecological impacts of unsustainable harvesting practices threaten both biodiversity and the livelihoods of communities reliant on these resources. Furthermore, the erosion of traditional knowledge related to medicinal plants poses risks to the sustainable management and cultural heritage associated with their use [19]. Looking forward, the future of medicinal plant production is expected to leverage advancements in agricultural techniques and biotechnological innovations, potentially enhancing the yield and quality of these essential crops. Collaborative efforts among farmers, researchers, and policymakers will be crucial in addressing the complex challenges facing this industry, ensuring the preservation of biodiversity and the equitable sharing of benefits derived from medicinal plants [18–20].

7.3.1 Elicitation and Secondary Metabolite Production

Elicitation has emerged as a practical strategy for enhancing the production of secondary metabolites. Elicitors can trigger complex biochemical pathways that enhance the production of these compounds. Although the precise mechanisms of elicitation are still under investigation, the utilization of elicitors represents a promising approach for large-scale production of valuable phytochemicals in medicinal plants. Elicitors can be classified into biotic and abiotic categories, each affecting secondary metabolite production through different pathways (Fig. 7.2) [21]. The primary metabolic pathways involved are the salicylic acid and jasmonic acid pathways, both of which are integral to the plant's immune response. A balanced activation of these pathways is essential; excessive activation of SA can lead to increased susceptibility to pests, while overactivation of jasmonic acid may hinder growth by diverting resources toward defense production instead of promoting growth [22]. The efficacy of elicitors is influenced by several factors including dosage, method of application, compatibility with other treatments, and environmental conditions

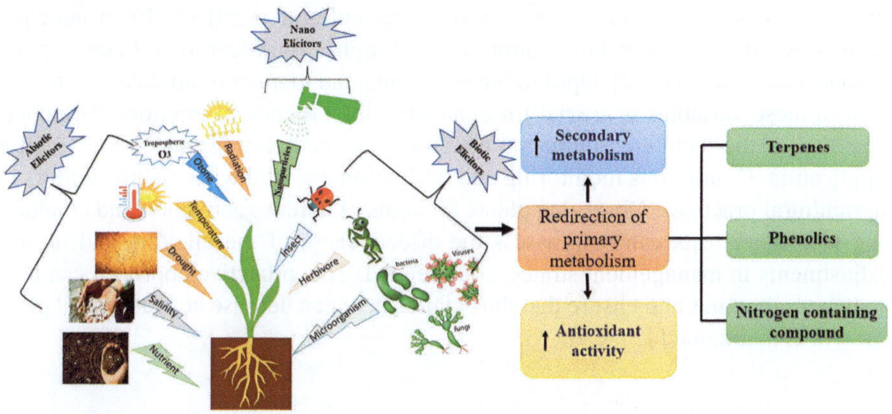

Fig. 7.2 Influence of exogenous elicitors on the production of secondary metabolite in medicinal plants [21]

such as temperature and humidity. Incorrect application can either fail to activate plant defenses or induce excessive responses that may harm plant health. Continuous monitoring and evaluation of plant responses to elicitors are recommended to optimize their use and enhance secondary metabolite production [21].

The application of elicitors has been shown to improve both primary and secondary metabolic processes in plants, contributing to optimal growth and development. A well-balanced interaction between these two metabolic pathways is vital for plants to cope with changing environmental conditions [21]. The success of elicitor application is influenced by several factors, including dosage, compatibility with other products, environmental conditions, and continuous monitoring of plant responses. The concentration and type of used elicitor can significantly impact the induction of secondary metabolites in different plant species. Additionally, environmental factors such as temperature, humidity, and light must be considered to optimize the timing and manner of elicitor application [23].

7.3.2 Challenges of Elicitor Application

The application of elicitors in the production of medicinal plants presents various challenges that must be addressed to optimize their effectiveness. One of the most critical aspects of using elicitors is determining the appropriate dosage and method of application. Incorrect dosages may fail to adequately activate the plant's defense mechanisms or, conversely, could induce an excessive response that adversely affects plant growth and development [24]. Elicitors must also be compatible with other applied treatments and fertilizers. Negative interactions between elicitors and these products can reduce their efficacy or compromise the effectiveness of other agricultural inputs [20]. It is essential to conduct compatibility assessments to ensure successful integration into existing agricultural practices. Environmental factors, such as temperature, humidity, and light, significantly influence plant responses to elicitors [22]. Optimizing the timing and manner of application in relation to these variables is crucial for enhancing the efficacy of elicitors. An understanding of the local environmental conditions is necessary for successful elicitor application. Continuous monitoring and evaluation are vital when using elicitors in agricultural practices. Observing plants for signs of defense activation and conducting laboratory tests can help assess the effectiveness of the elicitors and inform adjustments in management strategies as needed. This proactive approach can lead to better outcomes and ensure that the balance between defense activation and plant growth is maintained [21–24].

7.3.3 Hydroponic Production of Medicinal Plants

Hydroponic production of medicinal plants is an innovative agricultural practice that utilizes soilless systems to cultivate herbs and plants with therapeutic properties. This method offers several advantages over traditional soil-based cultivation, including increased control over growing conditions, higher yields, and the potential for enhanced medicinal quality through precise nutrient management [16, 25]. The hydroponic method provides a controlled environment that allows growers to manipulate variables such as light, temperature, humidity, and nutrient levels to optimize plant growth. This control can lead to increased production efficiency and improved quality of medicinal compounds [24]. Additionally, hydroponic uses significantly less water compared to conventional farming, making it an environmentally sustainable option. The hydroponic cultivation of medicinal plants appears promising, especially in light of increasing consumer demand for natural remedies and sustainable agricultural practices [26]. As research continues to explore the potential of hydroponics in enhancing the bioactive properties of medicinal plants, it may soon become an integral part of commercial agriculture. This evolution in farming practices not only supports the therapeutic benefits of plants but also addresses environmental concerns associated with traditional farming methods [6, 27]. Hydroponics has gained popularity as a viable method for cultivating a variety of medicinal plants, offering controlled environments and optimal nutrient absorption [28]. The production of echinacea, thyme, basil, lavender, mint, rosemary, and chamomile has been successfully achieved under hydroponic conditions with the highest quantity and quality. The thyme plants require minimal fertilization and full sunlight that easily provide in hydroponic system. Growing lavender hydroponically allows for better control over nutrient delivery and can enhance the production of essential oils. Rosemary requires adequate lighting and specific growing conditions to ensure its medicinal potency when grown hydroponically [21–24]. Hydroponic cultivation can enhance the quality and consistency of chamomile flowers, providing a reliable source for its medicinal properties. By harnessing the benefits of hydroponic cultivation, these medicinal plants can be produced with greater efficiency and purity, allowing for a steady supply of natural remedies for various health needs [29, 30]. The future of hydroponic medicinal production is promising, driven by increasing demand for natural remedies and the need for sustainable agricultural practices. Hydroponics offers a viable solution for the cultivation of medicinal plants, allowing for year-round production in controlled environments that optimize growth conditions [30]. This method significantly reduces the variability associated with traditional open-field cultivation, leading to more consistent biomass production and active ingredient content in medicinal herbs. The successful production of differ medicinal plant species regards to quantity and quality were reported in *Ligularia fischeri* [30], *Echinacea purpurea* [31], *Crocus sativus* [32], *Helichrysum odoratissimum* [33], *Moringa oleifera* [34]. However, the future of hydroponic medicinal production will also be influenced by ongoing technological advancements in cultivation techniques and nutrient management [1, 2, 22].

7.3.4 Elicitation in Hydroponic Cultures of Medicinal Plant

Elicitation in hydroponics refers to the application of elicitors to enhance the production of secondary metabolites in plants grown in soilless systems. Elicitation in hydroponics has gained significant attention due to its potential to enhance the production of secondary metabolites, which are vital for plant defense and have various applications in agriculture and medicine [30]. The hydroponic condition due to precise management of nutrient delivery and plant health, facilities optimized responses to elicitors. The integration of elicitation techniques within hydroponic systems presents a promising avenue for enhancing plant productivity and the concentration of secondary metabolites [35–38]. Regular monitoring and adjustment of elicitor types and concentrations can lead to significant improvements in crop quality, thereby aligning with the goals of both commercial and organic growers [22, 36, 37]. Moreover, the controlled environment of hydroponics allows for the efficient application of elicitors to maximizing their potential benefits in terms of increasing the quantity and quality of active substances [38]. Recent studies, confirmed the enchantment of medicinal plant quality regards to induced secondary metabolites production under hydroponic conditions in response to applied elicitors. Under hydroponic culture of *M. pulegium* applied salinity and salicylic acid synergistically enhanced antioxidant capacity [39] and essential oil accumulation in oil glands [40] (Fig. 7.3). Methyl jasmonate as an effective elicitor in the hydroponic cultures, through increasing secondary metabolite accumulation enhance antioxidant capacity of hydroponically grown Chinese chives [41]. The change of light intensity and quality in the hydroponically grown of *Hypericum perforatum*, besides to enhancing flower yield, significantly increased the content of hypericin, pseudohypericin, and hyperforin of harvested flowers [27]. Applied elicitation in the hydroponic system efficiently enhanced rhinacanthin production in *Rhinacanthus nasutus* roots [42]. Salicylic acid at the proper concentration increased flavonolignans accumulation in the fruits of hydroponically cultured *Silybum marianum* [43]. Additionally,

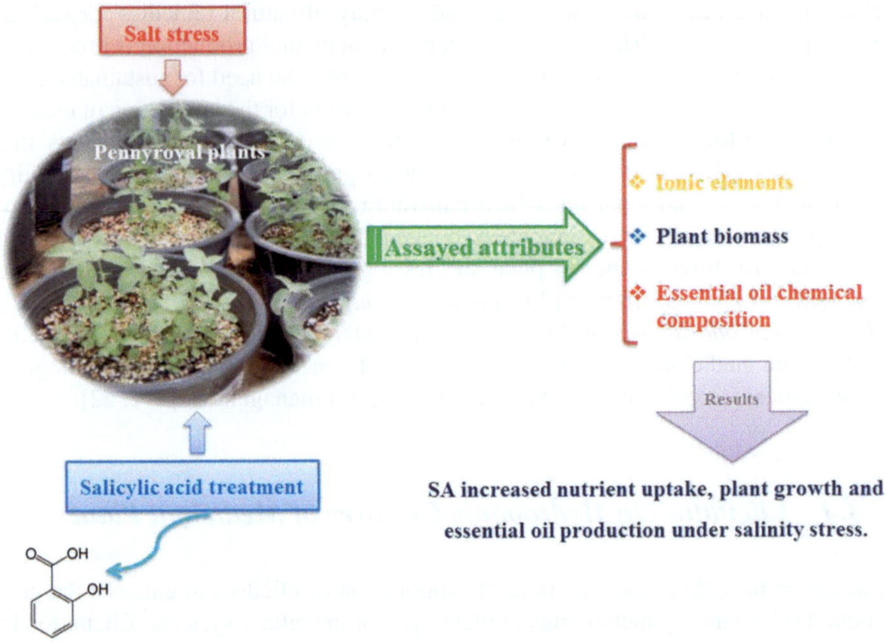

Fig. 7.3 Under hydroponic culture of *M. pulegium* applied salinity and salicylic acid synergistically enhanced essential oil accumulation in oil glands [36]

as climate change continues to affect agricultural productivity, the role of elicitors in hydroponics will be critical for developing sustainable farming practices. By improving the resilience of plants through elicitation, farmers can produce crops more efficiently, even in adverse conditions. This approach not only contributes to food security but also aligns with environmentally friendly agricultural practices by reducing the need for chemical fertilizers and pesticides [44, 45]. As reported by Chrysargyris et al. [46] and Farhadi et al. [47] applied appropriate elicitors under stress condition in the hydroponic culture the accumulation of active substances increased withoutt reduction of growth and crop performance.

7.3.5 Challenges and Future of Elicitation in Hydroponics

Elicitation under hydroponic conditions presents several challenges and limitations that can impact the effectiveness and efficiency of this agricultural practice. The correct dosage and method of application of elicitors are crucial for maximizing their efficacy in hydroponic systems. Incorrect applications may fail to activate plant defenses adequately, while excessive application can negatively affect plant growth by inducing an overstimulation of defense responses [22, 37]. Striking a balance between activating defense mechanisms and promoting growth is essential to avoid compromising plant productivity. Another significant challenge is the compatibility of elicitors with other applied treatments and fertilizers [28]. Therefore, careful consideration is required when integrating elicitors into existing hydroponic management practices. Environmental conditions such as temperature, humidity, and light levels play a critical role in how plants respond to elicitors. These factors can significantly influence the timing and effectiveness of elicitor application [38]. In hydroponic setups, where these conditions can be controlled, adjustments may still be necessary to optimize the benefits of elicitors. Continuous monitoring and evaluation of plant responses to elicitors are vital for assessing their effectiveness. This includes observing for signs of activated defenses and performing laboratory tests when necessary. The requirement for ongoing assessment can be resource-intensive and may complicate the management of hydroponic systems [48]. Elicitors must be designed to activate multiple pathways in a balanced manner, ensuring that plant defenses are enhanced without compromising growth [22]. Applied salinity and 5-aminolevulinic acid at an optimum level considerably enhanced rutin content in seeds of *Fagopyrum esculentum* however high concentration of these elicitors although increased rutin accumulation but negatively affected plant growth and seed yield [47] (Fig. 7.4). Achieving this balance requires advanced technical knowledge and may involve extensive research and development. In addition, the economic feasibility of using elicitors in hydroponic systems can be a limiting factor. However, elicitors have the potential to enhance crop quality and resilience, the costs associated with their application and the need for specialized knowledge may deter widespread adoption among growers [21]. As the agricultural industry seeks sustainable management strategies, these challenges must be addressed to facilitate the effective integration of elicitors into hydroponic practices [8, 49].

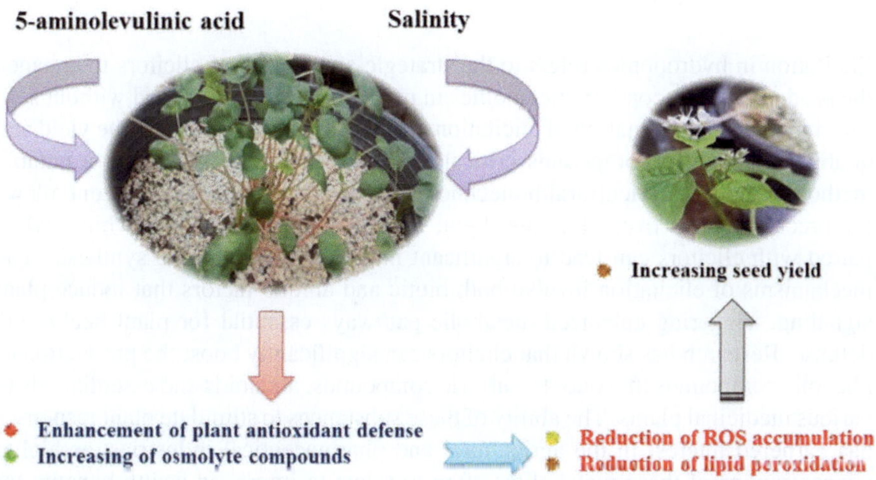

Fig. 7.4 Applied 5-aminolevulinic acid significantly affected the physiological and phytochemical traits of hydroponic culture of *Fagopyrum esculentum* under salinity stress [47]

The study of elicitation under hydroponic conditions is poised to become a significant area of research, particularly in the context of sustainable agriculture and food security amidst climate change challenges. Researchers are exploring innovative formulations of biomaterials and their applications as elicitors, which could enhance plant growth and resilience in extreme conditions [45]. The exploration of different microorganisms as plant growth-promoting bacteria (PGPB) has expanded, allowing for a broader understanding of their effects on various plant species. This ongoing research is expected to lead to more effective strategies for utilizing PGPB in hydroponic systems. The future of elicitor research in hydroponics includes identifying new elicitors derived from both biotic and abiotic sources. As the understanding of plant defense mechanisms deepens, researchers can harness elicitors such as methyl jasmonate and silver nanoparticles [7, 21], which have demonstrated significant impacts on photosynthetic pigment production and overall plant growth. The integration of various elicitors could optimize the balance between plant defense responses and growth, leading to enhanced crop productivity without compromising plant health [16].

The incorporation of advanced biotechnological techniques, such as genetic engineering and nanotechnology, could further enhance the efficacy of elicitors in hydroponic environments [9, 37]. The use of nanoparticles, for instance, could provide targeted delivery systems for elicitors, improving their absorption and overall effectiveness in stimulating plant defenses [21, 22]. Additionally, employing a combination of elicitors in hydroponic systems can be tailored to the specific needs of different plant species, maximizing the benefits of elicitation while minimizing potential drawbacks associated with overactivation of defense pathways [20, 48].

7.4 Summary

Elicitation in hydroponics refers to the strategic application of elicitors to enhance the production of secondary metabolites in medicinal plants cultivated without soil. This innovative combination of elicitation and hydroponics maximizes the yield and quality of bioactive compounds critical for therapeutic use, making it a notable method in modern agricultural biotechnology. The hydroponic environment allows for precise control over nutrients, light, and other growth factors, which, when paired with elicitors can lead to significant increases in metabolite synthesis. The mechanisms of elicitation involve both biotic and abiotic factors that induce plant signaling, triggering enhanced metabolic pathways essential for plant health and defense. Research has shown that elicitors can significantly boost the production of phenolic compounds, flavonoids, sulfuric compounds, alkaloids and essential oils in various medicinal plants. The ability of these substances to stimulate plant responses has garnered interest in the agricultural and pharmaceutical industries, as higher concentrations of these metabolites often translate to improved health benefits for consumers. While the integration of elicitation and hydroponics presents substantial benefits, it also raises notable challenges, such as the need for ongoing research to optimize the conditions and concentrations of used elicitors. The potential for variations in metabolite production due to environmental factors necessitates careful management and control within hydroponic systems. Additionally, the ecological implications of commercial elicitation practices, including the sustainability of sourcing these compounds, contribute to ongoing debates within the field of agricultural biotechnology. In summary, elicitation in hydroponics represents a promising frontier in the cultivation of medicinal plants, aiming to meet the increasing demand for high-quality bioactive compounds. By addressing both the opportunities and challenges associated with this approach, researchers and cultivators can pave the way for more sustainable and effective production practices in the medicinal herbs.

References

1. Fussy A, Papenbrock J (2022) An overview of soil and soilless cultivation techniques-chances, challenges and the neglected question of sustainability. Plants 11:1153
2. Strzemski M et al (2020) The impact of different cultivation systems on the content of selected secondary metabolites and antioxidant activity of carlina acaulis plant material. Molecules 25:146
3. Vahl M et al (2025) Influence of hydroponic cultivation subsystems on vegetative growth, vigour, and metabolite profiles in vanilla. J Hortic Sci Biotechnol 1–17
4. Lam VP, Kim SJ, Park JS (2020) Optimizing the electrical conductivity of a nutrient solution for plant growth and bioactive compounds of *Agastache rugosa* in a plant factory. Agron 10:76
5. Dass SM, Chai T-T, Wong F-C (2022) Investigation on the effects of hydroponic-planting versus soil-planting using *Plectranthus amboinicus* as a plant model. J Trop Life Sci 12:53–61
6. Zeljkovic SC et al (2022) Optimizing growing conditions for hydroponic farming of selected medicinal and aromatic plants. Food Chem 375:131845
7. Erere A et al (2021) Hydroponics research trends: a review and bibliometric analysis (2008–2018). Nat Volatiles Essent Oils 8:3240–3251
8. Macwan J et al (2020) Review on soilless method of cultivation: hydroponics. Int J Recent Sci Res 11:37122–37127
9. Waiba KM et al (2020) Soil-less vegetable cultivation: a review. J Pharmacogn Phytochem 9:631–636
10. Hosseini H et al (2021) Nutrient use in vertical farming: optimal electrical conductivity of nutrient solution for growth of lettuce and basil in hydroponic cultivation. Horticulturae 7:283
11. Menegaes JF et al (2020) Germination and production of edible flowers of nasturtium in hydroponic cultivation. Braz J Dev 6:13381–13394
12. Dewir YH et al (2022) Effects of growing substrate, mode of nutrient supply, and saffron corm size on flowering, growth, photosynthetic competence, and cormlet formation in hydroponics. HortTechnology 32:234–240
13. Ai PH et al (2021) Comparison of chrysanthemum flowers grown under hydroponic and soil-based systems: yield and transcriptome analysis. BMC Plant Biol 21:1–17
14. Sakamoto M, Suzuki T (2020) Effect of nutrient solution concentration on the growth of hydroponic sweetpotato. Agron 10:1708
15. Kumar P, Saini S (2020) Nutrients for hydroponic systems in fruit crops. In: Solankey SS et al (eds) Urban horticulture—necessity of the future. IntechOpen, London, UK, pp 1–11
16. Velazquez-Gonzalez RS et al (2022) A review on hydroponics and the technologies associated for medium- and small-scale operations. Agriculture 12:646. https://doi.org/10.3390/agriculture12050646

17. Savvas D, Gruda N (2018) Application of soilless culture technologies in the modern greenhouse industry—a review. Eur J Hortic Sci 83:280–293
18. Giurgiu RM et al (2017) A study of the cultivation of medicinal plants in hydroponic and aeroponic technologies in a protected environment. Acta Hortic 1170:671–678
19. Dsouza A et al (2025) Harnessing controlled-environment systems for enhanced production of medicinal plants. J Exp Bot 76:76–93
20. Amarie E-O, Onochie P (2024) A review of hydroponic systems for medicinal herb cultivation: practices and prospects. Proceedings 105:133. https://doi.org/10.3390/proceedings2024105133
21. Humbal A, Pathak B (2023) Influence of exogenous elicitors on the production of secondary metabolite in plants: a review ("VSI: secondary metabolites"). Plant Stress 8:100166. https://doi.org/10.1016/j.stress.2023.100166
22. Abdul Malik NA, Kumar IS, Nadarajah K (2020) Elicitor and receptor molecules: orchestrators of plant defense and immunity. Int J Mol Sci 21:963. https://doi.org/10.3390/ijms21030963
23. Partap M et al (2020) Effect of elicitors on morpho-physiological performance and metabolites enrichment in *Valeriana jatamansi* cultivated under aeroponic conditions. Front Plant Sci 11:01263. https://doi.org/10.3389/fpls.2020.01263
24. Mubeen B et al (2022) Hydroponics and elicitation, a combined approach to enhance the production of designer secondary medicinal metabolites in *Silybum marianum*. Front Plant Sci 13:897795. https://doi.org/10.3389/fpls.2022.897795
25. Li QS et al (2018) Growth responses and root characteristics of lettuce grown in aeroponics, hydroponics, and substrate culture. Horticulturae 4:35
26. Schroeder F-G, Lozoya DR, Ruser P (2020) Hydroponic forcing of saffron (*Crocus sativus* L.). Acta Hortic 1273:281–288
27. Karimi M, Ahmadi N, Ebrahimi M (2022) Red LED light promotes biomass, flowering and secondary metabolites accumulation in hydroponically grown *Hypericum perforatum* L. (cv. Topas). Ind Crops Prod 175:114239
28. Sarmah R, Bora S, Sarmah R (2020) Quality blooming of marigold in hydroponics. Int J Curr Microbiol Appl Sci 9:1792–1799
29. Gautam S, Bhagyawant SS, Srivastava N (2018) Antioxidant responses and isoenzyme activity of hydroponically grown safflower seedlings under copper stress. Indian J Plant Physiol 23:342–351
30. Wei H (2018) Effect of different cultivation systems on the accumulation of nutrients and phytochemicals in *Ligularia fischeri*. Hortic Plant J 4:24–29
31. Ahmadi F et al (2021) Optimizing hydroponic culture media and NO_3^-/NH_4^+ ratio for improving essential oil compositions of purple coneflower (*Echinacea purpurea* L.). Sci Rep 11:8009
32. García-Rodríguez MV, López-Córcoles H, Alonso GL (2017) Effect of the hydroponic growing of forced *Crocus sativus* L. on the saffron quality. Acta Hortic 1184:287–292
33. Zantanta N et al (2022) Comparing crop yield, secondary metabolite contents, and antifungal activity of extracts of *Helichrysum odoratissimum* cultivated in aquaponic, hydroponic, and field systems. Plants 11:2696
34. Managa LR, du Toit ES, Prinsloo G (2021) Variations in the leaf metabolite profile between hydroponic and field grown *Moringa oleifera* Lam. genotypes. Biochem Syst Ecol 97:104302
35. Akiyama H et al (2017) Evaluation of the safety and efficacy of *Glycyrrhiza uralensis* root extracts produced using artificial hydroponic and artificial hydroponic-field hybrid cultivation systems. J Nat Med 71:265–271
36. Verdoliva SG et al (2021) Controlled comparisons between soil and hydroponic systems reveal increased water use efficiency and higher lycopene and β-carotene contents in hydroponically grown tomatoes. Sci Hortic 279:109896
37. Halder M, Sarkar S, Jha S (2019) Elicitation: a biotechnological tool for enhanced production of secondary metabolites in hairy root cultures. Eng Life Sci 19(12):880–895. https://doi.org/10.1002/elsc.201900058

38. Venkatasai NN (2025) A comprehensive review of factors affecting growth and secondary metabolites in hydroponically grown medicinal plants. Planta 261:48. https://doi.org/10.1007/s00425-025-04619-y
39. Farhadi N, Ghassemi-Golezani K (2020) Physiological changes of *Mentha pulegium* in response to exogenous salicylic acid under salinity. Sci Hortic 267:109325. https://doi.org/10.1016/j.scienta.2020.109325
40. Ghassemi-Golezani K, Farhadi N (2022) The efficacy of salicylic acid levels on photosynthetic activity, growth, and essential oil content and composition of pennyroyal plants under salt stress. J Plant Growth Regul 41(5):1953–1965
41. Wang C et al (2022) Effect of methyl jasmonate treatment on primary and secondary metabolites and antioxidant capacity of the substrate and hydroponically grown Chinese chives. Front Nutr 9:859035
42. Suksawat T, Panichayupakaranant P (2024) Enhanced rhinacanthin production in *Rhinacanthus nasutus* roots using a hydroponics and elicitation system. J Young Pharm 16(2):216–222. https://doi.org/10.5530/jyp.2024.16.28
43. Ahmed HS et al (2020) Salicylic acid increases flavonolignans accumulation in the fruits of hydroponically cultured *Silybum marianum*. Saudi Pharm J 28:593–598
44. Thakur K et al (2019) Enhancement of picrosides content in *Picrorhiza kurroa* Royle ex Benth. mediated through nutrient feeding approach under aeroponic and hydroponic system. Ind Crops Prod 133:160–167
45. Mehdizadeh L, Moghaddam M (2023) Hydroponic system for cultivation of medicinal plants. In: Kumar NS, Singh R (eds) Biosynthesis of bioactive compounds in medicinal and aromatic plants. Food bioactive ingredients. Springer, Cham. https://doi.org/10.1007/978-3-031-35221-8_10
46. Chrysargyris A et al (2021) Copper tolerance and accumulation on *Pelargonium graveolens* L'Her. grown in hydroponic culture. Plants 10:1663
47. Farhadi N et al (2022) Effect of 5-aminolevulinic acid on phytochemical and biochemical traits of *Fagopyrum esculentum* under salinity stress. J Soil Sci Plant Nutr 22:3254–3267
48. Chen QY et al (2021) An intelligent hydroponic device for *Astragalus membranaceus* Bge. var. mongolicus (Bge.) Hsiao. J Sens 2021:4967954
49. Kheyri Z et al (2022) Inoculation efficiency of different mycorrhizal species on growth, nutrient uptake, and antioxidant capacity of *Calendula officinalis* L.: a comparative study. J Soil Sci Plant Nutr 22:1160–1172

Chapter 8
Integration of Robotics in Soilless Cultivation Systems for Medicinal Plant Production

Meenakshi Thakur and Avika Subba

Abstract Robotics is rapidly emerging as a transformative force in the soilless cultivation of medicinal plants (MPs), offering precise, efficient, and sustainable alternatives to conventional farming. In this context, growers are increasingly relying on hydroponic, aeroponic, aquaponic, and vertical farming technologies to mitigate the impacts of climate change, soil degradation, and labour shortages. This chapter explores how robotics—integrated with artificial intelligence, sensor networks, and automation—enables real-time monitoring and management of critical variables such as nutrient supply, temperature, humidity, and plant growth. Robotic platforms perform key operations including transplanting, pruning, harvesting, and disease detection, with higher accuracy and reduced labor inputs. Drawing on global case studies, this chapter illustrates successful applications such as soft robotic grippers in Japan, AI-based nutrient assessment tools in Iran, and collaborative robots in container-based vertical farms. These examples showcase how robotics enhances consistency, phytochemical quality, and year-round production of plants in controlled environments. This chapter also addresses current limitations, including high capital investment, energy consumption, and technological complexity, particularly for small-scale growers. To overcome these barriers, this chapter emphasizes the need for affordable, scalable, and crop-specific robotic solutions. By aligning robotics with circular economy practices and smart farming technologies, this chapter

M. Thakur (✉)
Regional Cum Facilitation Centre Northern Region-I, NMPB, Ministry of Ayush, GOI, New Delhi, India

Research Institute in Indian Systems of medicine (RIISM), Mandi, Himachal Pradesh, India

National Medicinal Plants Board, New Delhi, India

A. Subba
Regional Cum Facilitation Centre Northern Region-I, NMPB, Ministry of Ayush, GOI, New Delhi, India

National Medicinal Plants Board, New Delhi, India

National Medicinal Plants Board, New Delhi, India

P. Kumar, A. R. Warghat (eds.), *Transforming Medicinal Plant Agriculture*, Smart Agriculture 16, https://doi.org/10.1007/978-981-95-3963-5_8

presents a forward-looking framework for the future of medicinal plant cultivation in soilless systems.

Keywords Artificial intelligence · Hydroponics · Medicinal plants · Phenotyping · Robotics · Soilless farming · Sustainable cultivation

8.1 Introduction

Medicinal plants (MPs) have always been the integral part of traditional medicine systems, viz., Ayurveda, Unani, and Traditional Chinese Medicine. In recent years, there has been a growing global interest in medicinal plants such as Tulsi (*Ocimum sanctum*), Ashwagandha (*Withania somnifera*), Brahmi (*Bacopa monnieri*), Turmeric (*Curcuma longa*), Stevia (*Stevia rebaudiana*), Mint (*Mentha arvensis*), Gotu Kola (*Centella asiatica*), and Amla (*Phyllanthus emblica*). These plants are increasingly recognized for their therapeutic properties and are extensively utilized across the pharmaceutical, nutraceutical, and herbal product sectors [1–4]. However, the sustainable cultivation of medicinal plants is facing increasing challenges due to the multifaceted impacts of climate change, soil degradation, and weather unpredictability. Rising global temperatures and changing patterns of rainfall are disrupting plant growth cycles, reducing yields, and affecting the phytochemical composition of medicinal plants, which is critical for their therapeutic value [5, 6]. In addition, soil health—which plays a vital role in supporting plant nutrition, microbial interactions, and water retention—is rapidly declining due to over-cultivation, erosion, excessive use of agrochemicals, and contamination from industrial pollutants [7, 8]. These factors contribute to the loss of organic matter, reduction in soil biodiversity, and depletion of essential nutrients, all of which directly impact the growth and quality of medicinal plants. Such challenges are particularly detrimental for small and marginal farmers, who typically lack the infrastructure, irrigation facilities, and insurance mechanisms to cope with climate-related risks [9].

To overcome these problems, soilless farming methods like hydroponics, aeroponics, aquaponics, and vertical farming are becoming popular. These systems allow precise control over growing conditions—such as water, nutrients, temperature, and light—which helps in producing high-quality medicinal plants with consistent levels of active compounds [10, 11]. Although soilless systems offer many advantages—such as saving water, reducing diseases, and enabling year-round cultivation—they also require constant monitoring and skilled labor. This is where robotics and smart automation play a key role. Robots and smart devices can take over repetitive tasks like sowing/planting, nutrient supply, pruning/thinning, temperature control, diseases detection, and harvesting (Fig. 8.1). This makes farming more efficient, reduces labor costs, and improves accuracy [12]. Artificial intelligence (AI)-based systems can also predict plant needs, control environment, and maintain consistent quality which is important for producing quality herbal products [13, 14]. This chapter will explore how robotics is changing the way we grow

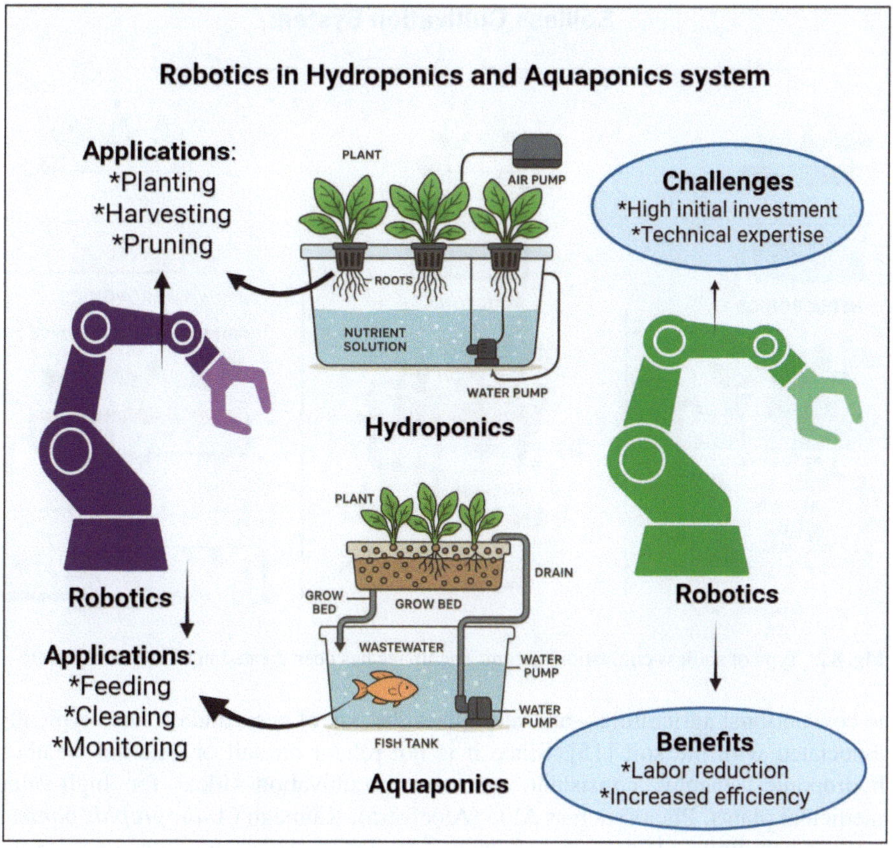

Fig. 8.1 Representation of robotics in soilless system. The image has been created in biorendor software

plants in soilless systems. It will cover the latest technologies, real-world examples, benefits, and challenges—helping farmers, researchers, and industry professionals understand how to adopt these innovations for a more sustainable future in agriculture.

8.2 Soilless Agriculture: The Need for Transition

8.2.1 Hydroponics: Efficient and Controlled Cultivation

Hydroponics is a method of growing plants without soil, in which essential nutrients are supplied directly to the plant roots through a water-based solution (Fig. 8.2). This technique significantly cuts down water usage—by as much as 90% compared

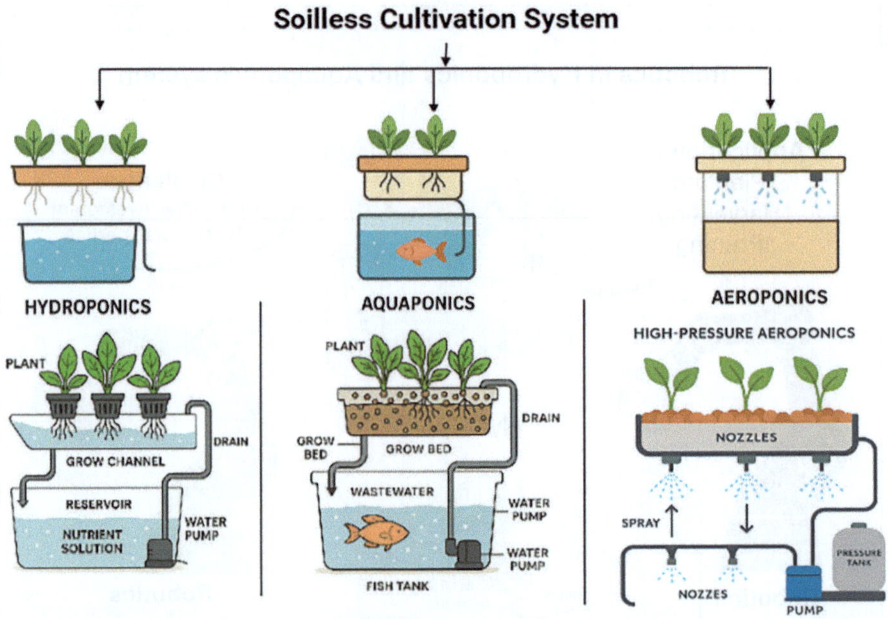

Fig. 8.2 Type of soilless cultivation system. The image has been created in biorendor software

to conventional agriculture—and minimizes the risk of pests and diseases typically associated with the soil [15]. Since it is not reliant on soil or external weather, hydroponics enables consistent, year-round cultivation—ideal for high-value medicinal plants. Plants such as Aloe (Aloe vera), Kalmegh (*Andrographis paniculata*), Chamomile (*Matricaria chamomilla*), Tagar (*Valeriana jatamansi*), Sage (*Salvia officinalis*), Ashwagandha (*Withania somnifera*), Tulsi (*Ocimum sanctum*), Brahmi (*Bacopa monnieri*), Stevia (*Stevia rebaudiana*), Kutki (*Picrorhiza kurroa),* and Parsley (*Petroselinum crispum*) have shown improved growth and better medicinal properties when grown in such systems [1, 16–19]. Globally, hydroponic systems cover approximately 35,000 ha, though challenges remain regarding energy consumption and production costs [20]. Hydroponic cultivation typically uses inert substrates like sand, gravel, rock wool, perlite, vermiculite, or coco peat to support plant roots. This allows for optimal nutrient uptake while eliminating the need for soil [21].

8.2.2 Aeroponics: Advanced Soilless Innovation

Aeroponics builds upon the principles of hydroponics by eliminating the need for any growing medium. In this technique, plant roots are exposed to the air inside a closed, dark chamber and are regularly sprayed with a fine mist containing essential nutrients, ensuring optimal growth conditions (Fig. 8.2). This method maximizes oxygen availability to roots, improves nutrient absorption efficiency, and reduces water use than hydroponics [11]. Aeroponic systems use 95% less water than conventional farming, require less space, and are ideal for vertical integration. Medicinal plants like Kutki (*Picrorhiza kurroa*), Valeriana (*Valeriana jatamansi*), and Brahmi (*Bacopa monnieri*) have been successfully cultivated under aeroponics system [1, 18, 22, 23]. They are especially beneficial for growing medicinal plants under precisely controlled environments [14].

8.2.3 Aquaponics: Sustainable Symbiosis

Aquaponics is a sustainable farming method that combines fish farming (aquaculture) with soilless plant cultivation (hydroponics) in a single, interconnected system. In this setup, the waste generated by fish or other aquatic animals provides natural nutrients for the plants, while the plants, in turn, filter and clean the water, which is then recirculated back to the fish tanks. This creates an efficient, eco-friendly cycle that supports both plant and animal growth (Fig. 8.2). Fish waste acts as a natural fertilizer, supplying essential nutrients to the plants. In turn, the plants filter and clean the water, which is then cycled back to the aquaculture tanks for reuse [24]. Unlike hydroponics, aquaponics reduces the need for synthetic fertilizers and creates a more sustainable, low-input production model [25]. Plants like Gold carpet (*Helichrysum cymosum*), Basil (*Ocimum basilicum*), Artemisia (*Artemisia annua*), and Kunth (*Cuphea hyssopifolia*) have successfully been cultivated in this system [26–29]. These studies highlight the suitability of aquaponics for producing medicinal plants sustainably. This system not only supports environmental sustainability but also meets the growing demand for organic and chemical-free medicinal plants by maintaining natural nutrient cycles [30].

8.2.4 Vertical Farming and Controlled Environment Agriculture (CEA)

Soilless techniques like hydroponics, aeroponics, and aquaponics are often integrated into vertical farming systems, where crops are grown in stacked layers to maximize space. These systems are often housed in enclosed structures that use artificial lighting, CO_2 control, and automated nutrient supply to optimize plant growth [31]. This spatial arrangement optimizes land use, enabling high-density crop production even in urban or resource-limited settings [32]. Vertical farms typically operate within Controlled Environment Agriculture (CEA) facilities or plant factories, which leverage advanced technologies such as artificial LED lighting, climate control, CO_2 enrichment, and automated nutrient supply systems to create ideal growing conditions [33]. CEA systems allow growers to meticulously manage environmental parameters—light spectra, temperature, humidity, photoperiods, and nutrient concentrations—thus minimizing variability and maximizing consistency in yield and quality. This precision is particularly crucial in the cultivation of medicinal plants, where consistent phytochemical profiles are vital for their efficacy in pharmaceutical and nutraceutical applications [34]. Basil (*Ocimum basilicum*), Ashwagandha (*Withania somnifera*), and Gotu kola (*Centella asiatica*) have shown increased concentration of secondary metabolite content when cultivated under these systems, enhancing their therapeutic potential and market value [35, 36]. Moreover, integrating vertical farming with CEA greatly lowers the reliance on chemical pesticides and fertilizers, conserves up to 95% more water than conventional farming, and overcomes challenges related to seasonal changes, pest infestations, and soil depletion [37, 38]. These advantages make it a highly sustainable solution for year-round plants production, especially in areas with poor soil quality or limited arable land. Moreover, automation and data-driven decision-making—often powered by IoT, AI, and machine learning—enhance productivity and reduce human error in CEA systems. These smart controls not only stabilize the cultivation environment but also optimize resource use and minimize operational costs over time [39]. All of the methods examined may offer potential enhancements in terms of environmental sustainability [40].

8.3 Robotics in Agriculture: From Evolution to Soilless Cultivation

8.3.1 Evolution of Agricultural Robotics: A Brief Timeline

The evolution of agricultural robotics began in the 1960s with the introduction of remote-controlled tractors and basic automation tools, marking the first step toward mechanized farming [41] (Fig. 8.3). By the 1980s, precision agriculture emerged through Global Positioning System (GPS)-based field mapping, significantly enhancing spatial data analysis and farm planning [42]. In the 1990s, progress in technology paved the way for autonomous tractors and sprayers, transforming agricultural operations like planting and spraying, and significantly decreasing the need for human labor [43]. The 2000s witnessed a shift toward task-specific robots, particularly in horticulture, with machines developed to perform harvesting of horticultural crops in structured settings [44]. The 2010s brought about the integration of drones for aerial monitoring and the application of swarm robotics for tasks such as pollination and weeding. These systems began leveraging AI and machine learning (ML) for improved data analysis and operational decision-making [45, 46]. By the 2020s, robotics had evolved into sophisticated systems capable of performing real-time weed detection, disease monitoring, and were increasingly incorporated into vertical farming and CEA [47, 48] (Fig. 8.3). Looking ahead, future developments are likely to lead to fully autonomous farms that combine robotics, AI, and genetic engineering to enable highly productive, resource-efficient farming systems [49].

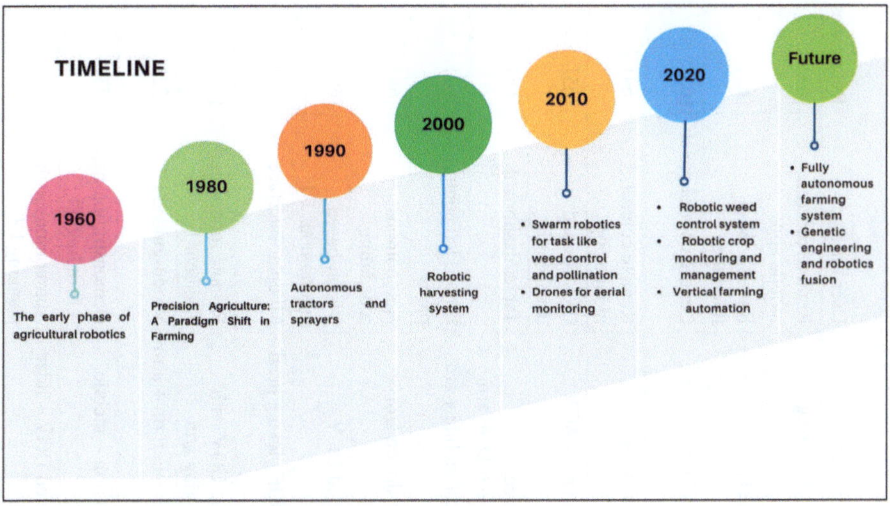

Fig. 8.3 Development and Milestones in Agricultural Robot History

8.3.2 Types of Agricultural Robots

Agricultural robots are broadly classified into individual robotic systems and multi-robot systems, based on their design, task execution capacity, and level of coordination. These categories are especially important in soilless cultivation systems, where accurate control and optimal use of space are essential.

8.3.2.1 Individual Robotic Systems

In soilless farming systems, individual robotic units are designed to operate independently, carrying out specific agricultural tasks without the need for human involvement (Table 8.1). These robots are equipped with precision sensors, actuators, control systems, and wireless communication technologies, allowing them to execute critical tasks such as nutrient solution supply, seedling, transplantation, canopy monitoring, pruning, and harvesting within controlled environments [50, 51]. Unlike in traditional soil-based systems, hydroponic robots operate in highly structured environments such as vertical farms, Nutrient Film Technique (NFT) setups, and aeroponic chambers. Their compact designs and high maneuverability make them ideal for navigating narrow rows and multi-tier growing systems. These robots use AI-driven vision systems and robotic arms to detect plant health indicators, monitor root development, and apply precise interventions like misting or nutrient adjustments [52, 53].

An autonomous robot called CropScout, is equipped with multispectral imaging sensors to check crop health in hydroponic greenhouses. By using Wi-Fi to send real-time data to nutrition management systems, it makes it easier to make prompt

Table 8.1 Type and role of robotics in agriculture

Sr. No.	Robot Name (Company)	Type	Application	Locomotion	User Interface	References
1.	Agrobot SW6010 (Agrobot)	Individual	Automated strawberry harvesting in hydroponic beds	2-wheel steering	Remote control, mobile app	[61]
2.	HV-100 (harvest automation)	Multi-UGV fleet	Tray spacing & logistics in nursery hydroponic systems	4-wheel steer	Radio + Wi-fi interface	[62]
3.	SwarmBot (SwarmFarm robotics)	Multi-UGV system	Mechanical weeding & nutrient dosing in polyhouse systems	4-wheel steer	Central control panel	[63]
4.	Agras T20 + ground bot (DJI)	Multi-drone/ UGV	Pest scouting + precision spraying in greenhouse hydroponics	UAV + AGV	Remote control, mobile app	[64]
5.	Iron ox robotic arm (iron ox)	Individual	Seeding, transplanting, and harvesting leafy & medicinal crops	Stationary + AGV	Cloud-based dashboard	[65]
6.	Mobile robotics platform (MRP)	Individual/UGV hybrid	Monitoring & selective harvest in vertical hydroponic farms	Wheeled UGV + 6-DOF arm	Tablet/dashboard	[66]
7.	Hydroponic vision arm (ATech)	Individual	Fruit picking in compact hydroponic setups	Vertical mobile + arm	Embedded control panel	[67]
8.	Cable-driven monitoring bot	Individual	Biomass & structural monitoring in vertical hydroponics	Cable-driven arm	Autonomous scheduling	[68]
9.	SARAL-bot	Individual	Leaf detection/removal & plant care in hydroponic strawberries	Wheeled UGV	Vision-based autonomous UI	[69]
10.	Robofruit arm	Individual	Modular strawberry harvester for greenhouse environments	2.5-DOF picking head	Modular interface	[70]
11.	AHPPEBot	Individual	Autonomous tomato harvesting in greenhouses	Mobile UGV with harvesting arm	Remote monitoring dashboard	[71]
12.	Harvey	Individual	Sweet pepper picking in protected cropping environments	Mobile platform + arm	Vision-guided UI	[72]
13.	Containerized Cobot system	Multi-arm collab	Automated transplanting & harvesting in shipping-container farms	Cobot arm + mobile base	Demonstration-based scheduling UI	[73]
14.	Fujinaga's Lidar robot	Individual	Path planning & harvesting among raised beds with lidar navigation	Wheeled UGV + lidar	Autonomous path planning UI	[74]

decisions about lighting, ventilation, and nutrient dosage [54]. Similarly, WeedBot has been adapted for substrate-based soilless systems, using machine vision to identify and remove weed growth mechanically, which is particularly useful in systems where manual weeding is inefficient or infeasible [55]. Another prominent example is GrapeBot, a modular robotic system specifically developed for managing fruit-bearing crops grown in hydroponic trellis setups. It autonomously determines grape ripeness based on visual and environmental parameters and adjusts harvesting protocols accordingly, supporting high-value crop production in controlled environments [56].

These robots often integrate wireless communication protocols such as Wi-Fi, LoRaWAN, and GSM to interact with centralized control systems and other devices within a smart greenhouse network [57]. Integration with middleware platforms and cloud-based data processing enables real-time analytics and remote control. Moreover, embedded GPS and Simultaneous Localization and Mapping (SLAM) technologies allow robots to navigate and localize within indoor hydroponic setups with minimal collision risks [58, 59]. Design variations are tailored to specific tasks—lightweight robots are employed for foliar monitoring or pest detection, while larger units equipped with multi-jointed arms handle harvesting or nutrient line maintenance [53, 60]. By performing precise, site-specific interventions, these robots reduce human labor, conserve resources, and enhance yield uniformity. Their application in soilless systems plays a pivotal role in the advancement of climate-resilient and sustainable food production technologies.

8.3.2.2 Multi-Robot Systems

The integration of multi-robot systems (MRS) within soilless cultivation has advanced rapidly in response to the increasing demand for sustainable, resource-efficient food production (Table 8.1). Hydroponic systems, characterized by controlled environments and nutrient-enriched substrates, benefit significantly from coordinated robotic platforms that enhance automation in planting, monitoring, harvesting, and input management [50, 75]. In hydroponic greenhouses and vertical farming units, modular and mobile ground robots perform repetitive yet precision-dependent tasks such as seeding, transplanting, and nutrient supply. These robots operate in tandem with unmanned aerial vehicles (UAVs) for environmental sensing and plant health assessment. The UAVs collect multispectral and thermal data that inform the decision-making algorithms of ground robots, which then execute tasks like localized irrigation and robotic harvesting with minimal human intervention [51, 60].

For instance, in closed-loop hydroponic facilities, drones equipped with high-resolution cameras assess vegetative indices such as Normalized Difference Vegetation Index (NDVI), chlorophyll fluorescence, and canopy temperature, which are indicative of plant stress or nutrient deficiencies. These data are relayed to ground-based units that adjust nutrient dosing or initiate targeted interventions, such as mechanical trimming or root-zone aeration [52, 59]. Such integration has been demonstrated to improve yield consistency and reduce resource wastage in recirculating hydroponic systems. Collaborative systems such as GrapeBot and Agrobot have been adapted for hydroponic fruiting crops, including vine tomatoes and strawberries. GrapeBot employs environmental sensors and machine vision to detect fruit ripeness and autonomously navigates through hydroponic racks to harvest or prune vines with precision [56]. Similarly, Agrobot's modular harvesting arms have been used to pick strawberries in soilless systems by detecting fruit color and softness, optimizing harvest timing and minimizing crop damage [53, 76].

Additionally, UAV-ground robot coordination enhances input-use efficiency. Moisture sensors on drones assess evapotranspiration patterns within the hydroponic setup, enabling real-time feedback loops to adjust nutrient film thickness or misting frequency in aeroponic systems [75]. Robots equipped with pH, (electrical conductivity (EC), and dissolved oxygen (DO) sensors regulate reservoir parameters and ensure optimal nutrient availability across growth stages. In vertical hydroponic farming systems, mobile robots are designed to navigate tight pathways and manage tasks across various growing levels. These camera-assisted robots can observe root health and detect algae formation in nutrient flow channels, either sending notifications or autonomously activating cleaning and disinfection measures [51]. UAVs enhance the functionality of ground-based robotic systems by offering an aerial perspective, enabling swift detection of anomalies in crop health or system performance. Achieving this level of precision relies on efficient and coordinated communication between robotic units. Technologies such as Wi-Fi, LoRaWAN, and 5G are increasingly utilized to ensure seamless real-time data sharing and control synchronization. Wi-Fi is suitable for intra-greenhouse operations, while LoRa enables long-range communication in multi-facility hydroponic farms [57]. High-precision navigation, often supported by RTK-GPS and Simultaneous Localization and Mapping (SLAM), ensures accurate robot pathing and collision avoidance in dense hydroponic layouts [58]. The integration of these robots in hydroponics not only reduces labor dependency and operating costs but also contributes to closed-loop sustainability by enabling data-driven decisions in real time. This level of automation enhances the resilience and scalability of urban and peri-urban food systems, positioning hydroponics as a cornerstone of future smart agriculture.

8.3.3 Robot Structures for Agricultural Purposes

The structural design of robots plays a crucial role in determining their field performance and task suitability (Fig. 8.4). Robotic platforms are adapted to specific farming environments, ranging from open fields to greenhouse setups. Tracked robotic platforms are commonly chosen for muddy or rough terrains because they offer better traction and can support heavier loads. In contrast, wheeled robots are more suitable for even, flat surfaces where speed are prioritized [77, 78]. Legged robots, though still experimental, offer versatility in irregular or fragile crop zones, enabling minimal crop disturbance. UAVs or drones, serve as aerial platforms for crop health monitoring, pest detection, and aerial imaging [79]. Autonomous ground vehicles (AGVs) are another key structure used for scouting, transporting goods, and performing repetitive tasks without human supervision [80].

Effective robot operation depends on an array of sensor technologies (Fig. 8.4). Visual sensors including thermal cameras, multispectral imaging units, and RGB cameras are essential for detecting plant health, identifying stress indicators, and guiding autonomous actions. Depth sensors support accurate spatial mapping, which is essential for navigation and obstacle avoidance. AI-enabled visual processing systems detect diseases, pests, or nutrient deficiencies in real-time and send feedback to control units. Geospatial tools such as GPS and inertial measurement units (IMUs) assist in precise localization, and real-time kinematic (RTK) positioning enhances accuracy to the centimeter level [81]. Soil monitoring technologies, including moisture sensors, nutrient analyzers, and gas detectors, help manage irrigation, fertilization, and soil health. pH sensors using glass electrodes, along with soil compaction sensors, play a vital role in monitoring root-zone conditions and ensuring optimal plant growth and productivity [82, 83].

Robotic manipulators or arms are critical in tasks such as selective harvesting and pruning (Fig. 8.4). These robotic arms, controlled by AI, are fitted with soft grippers or suction devices to handle fragile crops gently and prevent damage. Integrated force sensors adjust grip strength based on fruit softness, while computer vision helps detect ripeness and size. Advanced path planning ensures that robots take the most efficient route through a field, maximizing productivity and minimizing errors. These advanced manipulators have made it possible for robots to handle biological materials with high precision and care, even in dynamic field environments.

8.3.4 Applications of Robotics in Modern Agriculture

The scope of robotics in modern agriculture has expanded dramatically to include seeding, planting, pruning, weeding, and harvesting (Fig. 8.5). Robots equipped with RGB-D cameras and deep learning algorithms are used to detect ripe fruits and perform nondestructive harvesting, particularly beneficial during peak labor shortages [84, 85]. Pruning robots use AI and visual feedback to identify unproductive branches, cutting them precisely without harming the plant. Autonomous weeding robots distinguish between crops and weeds using AI and eliminate weeds through mechanical tools or lasers, significantly reducing the need for herbicides [86].

Seeding robots employ variable-rate planting techniques that adjust seeding depth and spacing based on real-time soil data (Fig. 8.5). Post-harvest, robots sort horticultural crops using hyperspectral imaging and deep learning to ensure uniformity and quality. These robotics systems run on battery packs, hybrid power, or

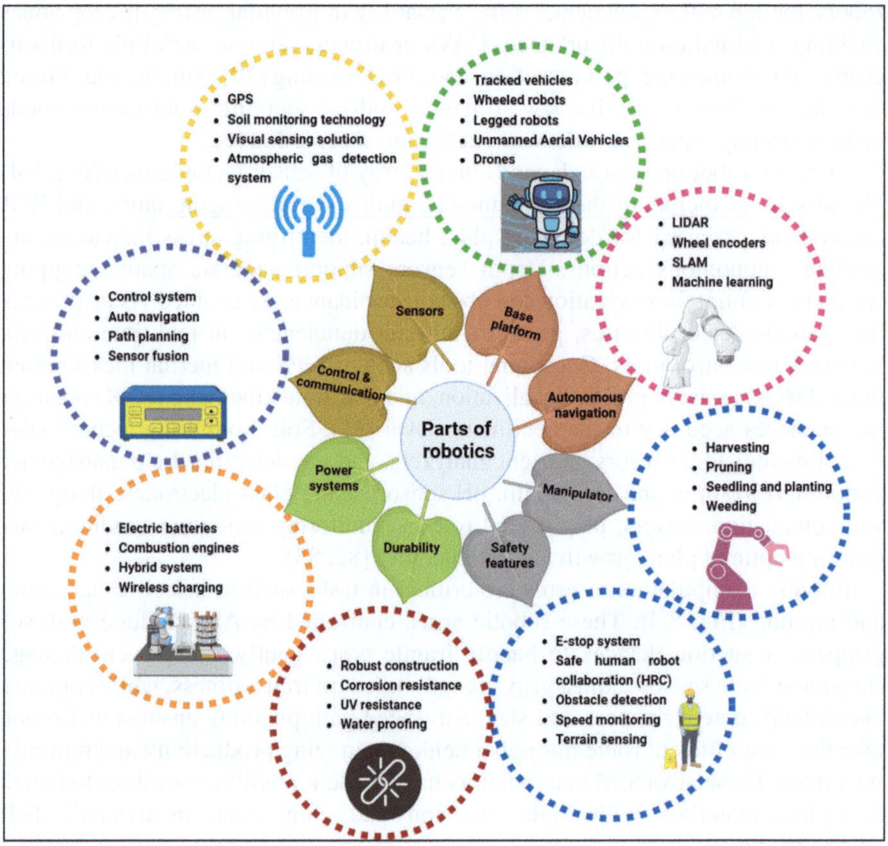

Fig. 8.4 Structural Variants of Agricultural Robots

wireless charging units to ensure uninterrupted operation [87]. The integration of sensor systems, AI, and automated control allows these robots to make real-time decisions, optimize resource use, and support sustainable farming practices. Robotic manipulators in soilless setups are equipped with nutrient and pH sensors to manage environmental conditions such as humidity, light, and temperature (Fig. 8.5). These parameters are critical in hydroponics and aeroponics, where minor variations can significantly affect plant growth. Robots ensure consistent seed spacing, early detection of nutrient deficiencies, and optimal root-zone management, thereby improving crop health and yield. The adoption of such systems in CEA has enabled scalability, reduced water consumption, and eliminated many of the risks associated with soil-borne pathogens [88, 89].

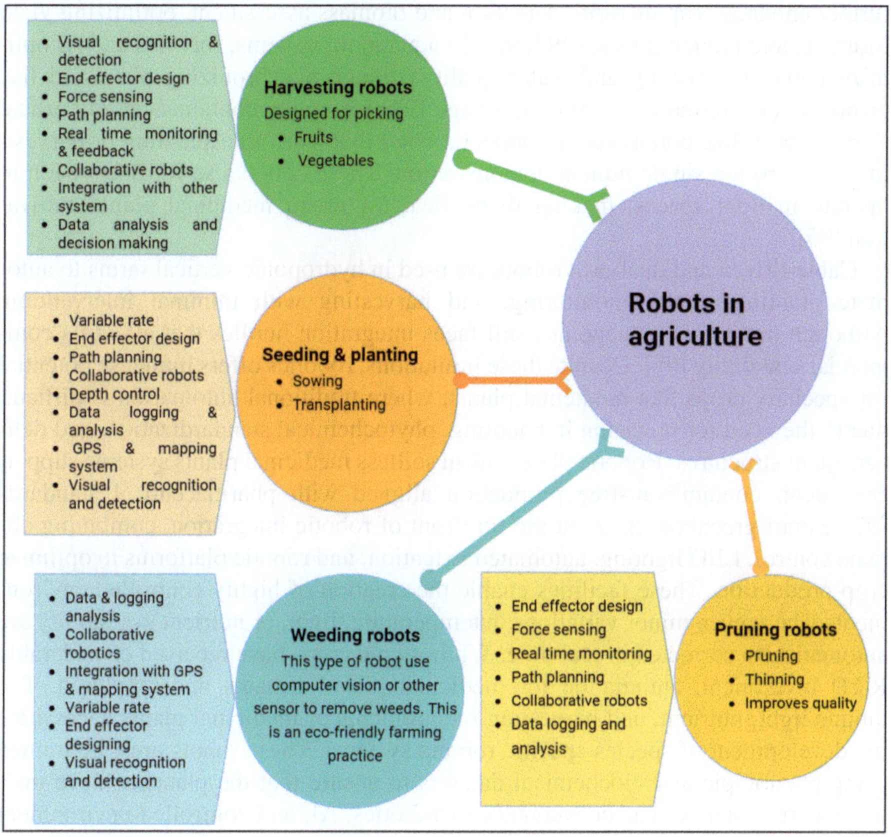

Fig. 8.5 Integration of Robotic Arms in Agricultural Tasks

8.4 Integration of Robotics in Soilless Medicinal Plant Cultivation

The agricultural robotics market is expected to reach USD 26.35 billion by 2032, driven by factors such as rising food demand, labor shortages, and the adoption of precision technologies [90]. Smart farming systems increasingly rely on AI-integrated robots capable of managing pH, light, temperature, and nutrient supply with minimal human oversight. These systems, however, involve high capital investments and energy costs [91]. Mechanical innovations like multi-jointed grippers and cable-driven manipulators have improved harvesting precision, achieving over 95% success rates in delicate plant handling. Deep learning vision systems further enhance crop maturity detection and biomass assessment, optimizing yield forecasts and reducing losses [92, 93]. In aquaponic systems, robotics and AI optimize nutrient recycling and water quality, though synchronizing plant and fish growth cycles remains a major challenge [94] Compact, container-based vertical farms now utilize collaborative robots (cobots) that learn transplanting or harvesting tasks from a single human demonstration. These cobots use RGB-D vision to operate in tight spaces, making them ideal for urban medicinal plants cultivation [95].

Cable-driven and dual-arm robots are used in hydroponic vertical farms to automate planting, health monitoring, and harvesting with minimal intervention. Although promising, aquaponics still faces integration hurdles that affect its commercial scalability [96]. Despite these limitations, robotics offers immense potential for specialty crops like medicinal plants, where traditional automation is difficult due to the need for precision in handling, phytochemical standardization, and delicate plant structures. Robotic platforms in soilless medicinal plants systems support consistent, contaminant-free production aligned with pharmaceutical standards [97]. Smart greenhouses are at the forefront of robotic integration, combining climate control, LED lighting, automated irrigation, and robotic platforms to optimize crop production. These facilities enable the creation of highly controlled environments where even minor variations in temperature, light, or nutrient availability are automatically corrected in real time. While staple crops have received considerable R&D investment, automation for medicinal plants remains underexplored. The unique light, nutrient, and harvesting requirements of medicinal plants necessitate the development of species-specific robotic systems. These robots must be trained using phenotypic and biochemical datasets to ensure that the plants achieve their therapeutic potency. The convergence of robotics, AI, and controlled-environment agriculture is essential for reducing labor, improving consistency, and scaling the standardized cultivation of medicinal plants [93, 98].

8.5 Techno-Economic Feasibility of Robotics in Soilless Cultivation of Medicinal Plants

8.5.1 Advantages

8.5.1.1 Environmental and Economic Advantages of Hydroponics

Hydroponic cultivation represents a breakthrough in sustainable agriculture, enabling significant water and nutrient savings. The study of Delaide et al. [99] highlighted the use of NUTRISENSE software for optimizing nutrient supply, especially under salinity stress, promoting recirculation and resource reuse. Han et al. [100] found that deploying Frequency Domain Reflectometry (FDR) sensors in hydroponic strawberry farms reduced water consumption by 1.7 times and fertilizer inputs by 41%. Similarly, Yep and Zheng [101] demonstrated that vertical hydroponics could reduce water use by up to 88% compared to traditional systems, while improving yield per unit area. The application of circular economy models, such as nutrient-rich drainage reuse, cuts freshwater use by 30% and reduces nitrate emissions by 40% [102]. Nutrient recovery systems like electrodialysis-forward osmosis enable the recycling of essential ions, maintaining balance and reducing waste [103, 104]. These innovations, when coupled with robotic monitoring and dosing systems, significantly enhance environmental performance while ensuring optimal growth conditions for medicinal plants.

8.5.1.2 Profitability and Productivity Through Controlled Soilless Systems

From an economic point of view, hydroponic and robotic integration offers high returns, particularly in areas with constrained agricultural land or water. The study of Caplan et al. [105] reported a Net Present Value (NPV) of over $177,000 and an Internal Rate of Return (IRR) of more than 30% in a Brazilian hydroponic setup, emphasizing economic viability. Controlled environments, as noted by Resh [106] and Blackmore et al. [41], allow for pest-free conditions, less input waste, and consistent yields. Kumar et al. [107] observed that early transplanting of *Bacopa monnieri* using robotic arms improved uniformity and allowed earlier harvesting, increasing overall system profitability by 7%. Moreover, robotic monitoring tools, or "scouts," helped reduce operational costs by 20% and improved early stress detection [108]. This makes these systems especially suitable for high-value medicinal species, which require precise environmental control and timely harvests.

8.5.1.3 Automation and Robotics in Enhancing Efficiency

Robotic automation has become a cornerstone of efficiency in soilless systems, enabling streamlined nutrient dosing, environmental monitoring, and disease control. Bac et al. [109] showcased robotic platforms equipped with AI and vision sensors that manage multiple tasks simultaneously, significantly reducing human labor dependency. In Nutrient Film Technique (NFT) systems, automation replaces manual watering and weeding [106], while vertical setups benefit from robots adaptable to canopy structure and height [110]. When integrated with IoT and drones, robots can enable hyper-local crop management by detecting stress markers and responding with precision interventions [111].

This efficiency extends to aquaponic systems, where robotics improves resilience and productivity in medicinal plant cultivation. Love et al. [112] reported a 75% increase in fish biomass with concurrent gains in crop yield. Automated calcium and magnesium supplementation, as noted by Somerville et al. [113], enhanced plant health, while decoupled systems enabled better control over pH and nutrient supply, raising yields by 36% [114]. Fernandez et al. [115] observed that AI-driven fish locomotion monitoring informs automated environmental alterations, improving plant and aquatic synergy. Although some crops underperform in aquaponics, tomatoes maintain comparable yields to hydroponics, underlining the need for species-specific automation protocols [116].

8.5.1.4 Robotics and Circular Economy Integration

Robotic technologies enhance the implementation of circular economy strategies in hydroponics, focusing on resource recovery and reuse. According to Goddek and Korner [116], reusing nutrient-enriched drainage water can reduce freshwater usage by 30% and lower nitrate emissions by 40%, contributing to environmental sustainability. Robotic systems equipped with sensors and dosing actuators can maintain pH, EC, and nutrient balance in real time [117]. Nutrient recovery methods, including electrodialysis and forward osmosis, facilitate continuous recycling within the system, thereby lowering reliance on additional nutrient inputs [118]. The study of Singh et al. [110] emphasized that integrating robotics with circular farming practices enhances organic compliance and supports the United Nation (UN) Sustainable Development Goals (SDGs) through reduced resource footprints.

8.5.1.5 Labor Reduction and Economic Efficiency

Incorporating robotics into hydroponic farming greatly reduces the need for manual labor and enhances consistency in operations. The study of Han et al. [100] showed that automating pH and nutrient monitoring in lettuce cultivation significantly minimized the requirement for human intervention. Carey et al. [119] conceptualized 'hydroponic capital' to describe how automation compensates for environmental and labor constraints. In the UAE, Al-Kodmany [120] reported that robotic systems reduced import dependence and associated logistics costs for medicinal herbs. Kumar et al. [121] demonstrated the use of adjustable robotic arms for transplanting in vertical farming systems, enabling higher throughput. Collaborative robots further reduce strain by handling tasks like nutrient distribution, tray lifting, and movement of materials, thus contributing to reduced labor costs and enhanced production consistency [122].

8.5.2 Limitations

Despite the evident benefits, there are multiple challenges associated with adopting robotics in soilless cultivation, especially for medicinal plants. One of the primary limitations is the lack of standardization across plant species and growing systems. Medicinal plants exhibit diverse morphologies and specific phytochemical needs, requiring robots to be customized for different tasks such as harvesting delicate leaves or detecting root-bound stress in plants grown aeroponically [123]. This variability hinders the use of a single robotic solution across multiple crops, increasing cost and complexity. Another major concern is resource intensiveness. The deployment of robotic systems requires investment in high-precision sensors, computing infrastructure, and software integration. These requirements can make adoption difficult for small and medium-scale growers, especially in developing regions [41]. Additionally, technical limitations—such as low adaptability to fluctuating indoor microclimates, delayed decision-making during real-time operations, and challenges in sensor calibration for non-soil substrates—further restrict field-level scalability.

There is also a knowledge gap due to limited field trials and research focused on long-term performance data in real-world greenhouse or vertical farming scenarios. The absence of such evidence contributes to growers' hesitation in adopting these systems [111]. Moreover, current robotic platforms are often not fully compatible with existing soilless cultivation infrastructure, which requires redesigning layouts or retrofitting systems—adding to the upfront costs and technical burden. Ultimately, the substantial costs associated with purchasing and maintaining robotic systems continue to be a major obstacle to their widespread adoption. Without institutional or governmental support, many growers may find it financially unviable. This opens opportunities for international and cross-sector collaboration aimed at developing open-source robotic platforms or public-private partnerships for technology dissemination and skill development.

8.6 Case Study

The integration of robotics and smart technologies into soilless cultivation systems—particularly hydroponics and aquaponics—is rapidly transforming agriculture system. Across the globe, innovative applications are being developed to address challenges such as labor shortages, nutrient deficiencies, harvesting, and environmental constraints, enabling precise, efficient, and sustainable production. In South Korea, a robotic system using Faster R-CNN and centroid-based tracking was deployed in hydroponic tomato greenhouses for real-time fruit monitoring. This vision-based technology, which achieved 90.2% detection accuracy, can be effectively adapted to identify optimal harvest stages in other crops [124]. Similarly, researchers in Iran developed a robotic model utilizing artificial neural networks (ANN) and support vector regression (SVR) to diagnose and treat iron deficiencies in hydroponically grown spinach. With 96% accuracy, this system holds promise for precision nutrient management in iron-sensitive plant species [125].

In China, a mobile robot equipped with AI-driven perception and a soft gripper enabled nondestructive harvesting of strawberries, achieving a 96.4% mean average precision [126]. This gentle handling approach can be scaled to medicinal herbs which are prone to damage during manual harvesting. Japan developed a robotic arm utilizing shape memory alloy (SMA) actuators specifically designed for transplanting seedlings within tissue culture vessels [127]. In arid regions like Qatar, IoT-enabled vertical hydroponic systems were adopted to remotely manage temperature, humidity, and nutrient levels [128]. Such systems are ideal for cultivating climate-sensitive medicinal plants under extreme environmental conditions, showcasing how digital automation can make cultivation viable where conventional methods fail. In aquaponics, robotics and digital technologies are being leveraged to optimize resource use, plant health, and environmental control. Simulation-based models developed in countries like the Faroe Islands, the Netherlands, and Namibia were used to fine-tune system parameters such as plant density and transpiration [129]. The studies conducted in Jordan and Qatar, focused on energy-efficient cooling and heating mechanisms within aquaponic setups to stabilize microclimates [130].

A multi-country initiative deployed IoT-based sensors across five geographically diverse sites to monitor aquaponic parameters like pH, dissolved oxygen, and nutrient content in real time. Predictive analytics supported maintenance and yield decisions, making these systems more reliable for consistent medicinal plant output [131]. In Spain and Latin America, small-scale aquaponic systems were introduced for educational and self-reliance purposes, encouraging pesticide-free cultivation of medicinal herbs and promoting community engagement in traditional medicine [132]. Urban growers have also begun using digital twin technology to simulate aquaponic behavior under different scenarios. These virtual tools enable adaptive decision-making and real-time parameter control, which are vital for ensuring uniformity in sensitive crops [133]. Moreover, deep learning models such as YOLO v4 have been applied to monitor fish locomotion—an indirect indicator of ecosystem health. Such closed-loop systems, where fish behavior informs plant care, are particularly suited for precision cultivation of high-value medicinal species [134]. Overall, these case studies highlight the transformative impact of robotics, AI, and IoT in soilless medicinal plant cultivation. By enhancing precision, automation, and environmental adaptability, these technologies are paving the way for a new era of sustainable, high-quality, and scalable medicinal plant production across diverse agro-climatic zones.

8.7 Gaps and Future Directions

Robotic technologies are poised to revolutionize nutrient management in soilless systems, particularly for medicinal plants that demand precise nutrient inputs for optimal growth and quality. These systems can greatly benefit from advanced sensors and automation capable of detecting imbalances and adjusting nutrient supply in real-time. The root-zone environment is another crucial aspect, especially in hydroponics and aquaponics, where factors like oxygen levels and the buildup of harmful exudates affect plant health. Robotic systems have the potential to improve root-zone conditions by enhancing aeration and removing phytotoxic substances, yet this area remains largely untapped. In addition, soilless systems often rely on nutrient recycling from wastewater, a process that is both complex and resource-intensive. Robotics could streamline this by automating nutrient recovery and treatment, although energy-efficient and crop-specific robotic solutions still need development. Economic feasibility remains a major barrier; while robotics offer efficiency, their high cost and uncertain scalability in medicinal plant farming call for more research into affordable, adaptable platforms suited for different scales of operation.

Another promising area is microbial community management. Microbes play a vital role in nutrient cycling and plant health, especially in controlled environments. Robotic interventions could support monitoring and managing these microbial populations, but current understanding of how robotics interact with plant microbiomes is limited and warrants further study. Emerging technologies like artificial intelligence, nanotech, and microbial biotechnology could work synergistically with robotics to further enhance precision, sustainability, and quality in medicinal plant cultivation. The integration of these tools could lead to the development of advance robotic systems capable of adapting to dynamic environmental and crop-specific needs. Future research must focus on multifunctional robots that can perform a variety of tasks such as transplanting, nutrient dosing, pruning, and harvesting—particularly suited to the specific requirements of medicinal plants. Swarm robotics, involving multiple smaller units working collaboratively, may prove especially effective in managing tasks in space-limited or repetitive operational setups. Improvements in imaging technologies—like hyperspectral and 3D vision—combined with AI-based analytics will aid in accurate detection of plant health issues, growth stages, and diseases.

Furthermore, energy-efficient robotics powered by renewable sources, such as solar energy or smart battery systems, will support continuous operation in greenhouses and vertical farms. The integration of IoT and machine learning will enable real-time feedback, predictive maintenance, and data-driven decision-making. Designing user-friendly interfaces is equally important to ensure these systems can be adopted even by operators with limited technical expertise, making collaboration between developers, scientists, and growers essential. Robotic innovation should also address labor-intensive tasks such as transplanting and harvesting, while exploring new solutions for artificial pollination and precision micro-nutrient supply. Such advancements can further reduce labor needs and enhance the quality of pharmaceutical-grade crops. In conclusion, integrating robotics into the soilless cultivation of medicinal plants presents a promising path toward more sustainable, efficient, and high-quality production systems. With the right blend of technological advancement, user collaboration, and policy support, robotics could redefine the future of medicinal agriculture by aligning productivity with precision and environmental responsibility.

References

1. Singh A, Prasad R (2024) Advances in medicinal plant research: from Ethnopharmacology to industrial applications. Springer Nature
2. Liu C, Tseng CY, Liu C (2018) Overview of traditional herbal medicine and its modern applications. J Tradit Complement Med 8(3):331–336. https://doi.org/10.1016/j.jtcme.2017.09.001
3. Sharifi-Rad J, Sureda A, Tenore GC, Daglia M, Sharifi-Rad M, Valussi M et al (2021) Biological activities of essential oils: from plant chemoecology to traditional healing systems. Molecules 26(1):12. https://doi.org/10.3390/molecules26010012
4. World Health Organization (2020) WHO global report on traditional and complementary medicine 2019. World Health Organization. https://www.who.int/publications/i/item/9789241515436
5. Intergovernmental Panel on Climate Change (IPCC) (2022) Climate change 2022: impacts, adaptation and vulnerability. https://www.ipcc.ch/report/ar6/wg2/
6. Myers SS, Smith MR, Guth S, Golden CD, Vaitla B, Mueller ND et al (2017) Climate change and global food systems: potential impacts on food security and under nutrition. Annu Rev Public Health 38:259–277. https://doi.org/10.1146/annurev-publhealth-031816-044356
7. Lal R (2015) Restoring soil quality to mitigate soil degradation. Sustainability 7(5):5875–5895. https://doi.org/10.3390/su7055875
8. Food and Agriculture Organization (FAO) (2021) Status of the World's soil resources. https://www.fao.org/3/i5199e/I5199E.pdf
9. Altieri MA, Nicholls CI, Henao A, Lana MA (2015) Agroecology and the design of climate change-resilient farming systems. Agron Sustain Dev 35(3):869–890. https://doi.org/10.1007/s13593-015-0285-2
10. Resh HM (2022) Hydroponic food production: a definitive guidebook for the advanced home gardener and the commercial hydroponic grower, 8th edn. CRC Press
11. Mahalingam M, Adhikari T, Choudhary R (2023) Hydroponics and vertical farming for sustainable agriculture: an emerging trend. J Clean Prod 397:136388. https://doi.org/10.1016/j.jclepro.2023.136388
12. Bhargava V, Shukla A, Singh M, Gaur A (2022) Recent developments in robotic systems for precision agriculture: a review. Comput Electron Agric 192:106582. https://doi.org/10.1016/j.compag.2021.106582
13. Wolfert S, Ge L, Verdouw C, Bogaardt MJ (2017) Big data in smart farming—a review. Agric Syst 153:69–80. https://doi.org/10.1016/j.agsy.2017.01.023
14. Kamilaris A, Kartakoullis A, Prenafeta-Boldú FX (2017) A review on the practice of big data analysis in agriculture. Comput Electron Agric 143:23–37. https://doi.org/10.1016/j.compag.2017.09.037
15. Chen X, Wang L, Zhang Y, Li M (2022) Water and nutrient dynamics in hydroponic systems: a review. Agric Water Manag 263:107404. https://doi.org/10.1016/j.agwat.2021.107404
16. Sgherri C, Cecconami S, Pinzino C, Navari-Izzo F, Izzo R (2020) Hydroponically-grown medicinal plants: phytochemical content and antioxidant capacity. Ind Crop Prod 145:112105. https://doi.org/10.1016/j.indcrop.2020.112105
17. Sharma R, Rana JC, Lal B (2020) Comparative analysis of medicinal quality in soil and hydroponically grown plants. J Med Plants Res 14(7):329–336. https://doi.org/10.5897/JMPR2020.7039
18. Thakur M, Mehta A, Kumar V (2019) Cultivation of high-value medicinal plants using aeroponic and hydroponic techniques. Ind J Horticult Technol 4(1):44–50
19. Rattan R, Kapoor S, Katoch V (2022) Enhanced phytochemical yield in medicinal plants under controlled hydroponic environments. Plant Physiol Rep 27(2):273–280
20. Luna-Maldonado AI, Valdez-Cepeda RD, Rodríguez-González A, Cota-Larrea M (2019) Global outlook of hydroponic technologies: land use and energy considerations. Renew Agric Food Syst 34(4):293–300. https://doi.org/10.1017/S1742170518000210

21. Wong YC, Phang SM, Poon CY (2020) Substrate media in hydroponics: a review on physical and chemical properties. J Plant Nutr 43(5):656–671
22. Pratap R, Singh V, Rawat D (2020) Aeroponic farming for enhanced yield and quality in medicinal plants. Asian J Agric Exten Econ Sociol 38(1):1–9
23. Mehta A, Thakur M, Sharma R (2023) Impact of aeroponics on secondary metabolite accumulation in Bacopa monnieri. J Herbal Med 38:100643
24. Rakocy JE, Masser MP, Losordo TM (2006) Recirculating aquaculture tank production systems: aquaponics—integrating fish and plant culture, SRAC Publication No. 454. Southern Regional Aquaculture Center
25. Goddek S, Joyce A, Kotzen B, Burnell G (2015) Aquaponics: the basics. In: Aquaponics food production systems. Springer, pp 1–24. https://doi.org/10.1007/978-3-030-15943-6_1
26. Mapfumari G, Mwale M, Mutsamba EF (2024) Evaluation of medicinal plant growth in aquaponic systems. Afr J Plant Sci 18(2):33–42
27. Hasan MR, Chowdhury SH, Sarker BC (2022) Suitability of Ocimum basilicum for aquaponic production: a performance study. Aquac Rep 22:100967
28. Kumar D, Rajak SK, Mishra R (2021) Artemisia annua cultivation in integrated aquaponic systems: an eco-friendly approach. Ind J Plant Sci 10(4):16–22
29. Ramirez-Godina F, García-Gaytán V, Morales-Cabrera M (2021) Evaluation of Cuphea hyssopifolia under aquaponic conditions. Hortic Bras 39(3):319–324
30. Somerville C, Cohen M, Pantanella E, Stankus A, Lovatelli A (2014) Small-scale aquaponic food production: integrated fish and plant farming, FAO fisheries and aquaculture technical paper no. 589. Food and Agriculture Organization
31. Despommier D (2013) The vertical farm: feeding the world in the 21st century. Picador
32. Kalantari F, Tahir OM, Joni RA, Fatemi E (2017) Opportunities and challenges in sustainability of vertical farming: a review. J Landsc Ecol 11(1):35–60. https://doi.org/10.1515/jlecol-2017-0004
33. Kozai T, Niu G, Takagaki M (2016) Plant factory: an indoor vertical farming system for efficient quality food production. Academic Press
34. Beacham AM, Vickers LH, Monaghan JM (2019) Vertical farming: a summary of approaches to growing skywards. J Hortic Sci Biotechnol 94(3):277–283. https://doi.org/10.1080/14620316.2019.1574212
35. Dou H, Niu G, Gu M, Masabni J (2019) Effects of light quality on growth and phytochemicals of two basil cultivars. HortScience 54(11):1832–1839. https://doi.org/10.21273/HORTSCI14170-19
36. Srivastava S, Thakur M, Mehta A (2023) Comparative analysis of bioactive compounds in Ashwagandha and Gotu Kola grown under controlled environment agriculture. J Med Plant Stud 11(2):45–51
37. Al-Kodmany K (2018) The vertical farm: a review of developments and implications for the vertical city. Buildings 8(2):24. https://doi.org/10.3390/buildings8020024
38. Kozai T (2013) Resource use efficiency of closed plant production system with artificial light: concept, estimation and application to plant factory. Proc Japan Acad B 89(10):447–461. https://doi.org/10.2183/pjab.89.447
39. Zheng Y, Wang L, Zheng X (2020) Smart farming: technologies for automated CEA systems. Sensors 20(23):6671. https://doi.org/10.3390/s20236671
40. Jurkenbeck K, Heumann A, Spiller A (2019) Sustainability in vertical farming: challenges and potential. J Clean Prod 237:117783. https://doi.org/10.1016/j.jclepro.2019.117783
41. Blackmore BS, Stout BA, Wang M, Runov B (2005) Robotic agriculture—the future of agricultural mechanisation? In: 5th European conference on precision agriculture. Wageningen Academic Publishers, Uppsala, pp 621–628
42. Zhang N, Wang M, Wang N (2002) Precision agriculture—a worldwide overview. Comput Electron Agric 36(2–3):113–132. https://doi.org/10.1016/S0168-1699(02)00096-0
43. Bechar A, Vigneault C (2016) Agricultural robots for field operations: current status and future trends. Biosyst Eng 149:94–111. https://doi.org/10.1016/j.biosystemseng.2016.06.014

44. Arad B, Ben-Shahar O, Werman M (2020) Agricultural robotics for harvesting: fundamental research and field applications. Robot Auton Syst 134:103606. https://doi.org/10.1016/j.robot.2020.103606
45. Liakos KG, Busato P, Moshou D, Pearson S, Bochtis D (2018) Machine learning in agriculture: a review. Sensors 18(8):2674. https://doi.org/10.3390/s18082674
46. Tokekar P, Hook JV, Mulla D, Isler V (2016) Sensor planning for a symbiotic UAV and UGV system for precision agriculture. IEEE Trans Robot 32(6):1498–1511. https://doi.org/10.1109/TRO.2016.2610158
47. Mogili UR, Deepak BBVL (2018) Review on application of drone systems in precision agriculture. Proc Comput Sci 133:502–509. https://doi.org/10.1016/j.procs.2018.07.063
48. Mahmud MS, Hasan MK, Rahman A, Kaiser MS (2022) A review on intelligent monitoring systems in agriculture using unmanned aerial vehicles. Inf Process Agric 9(2):264–281. https://doi.org/10.1016/j.inpa.2021.02.004
49. Kamilaris A, Prenafeta-Boldú FX (2018) Deep learning in agriculture: a survey. Comput Electron Agric 147:70–90. https://doi.org/10.1016/j.compag.2018.02.016
50. Shamshiri RR, Kalantari F, Ting KC, Thorp KR, Hameed IA, Weltzien C, Ahmad D, Ehsani R (2018) Advances in greenhouse automation and controlled environment agriculture: a transition to plant factories and urban agriculture. Int J Agric Biol Eng 11(1):1–22. https://doi.org/10.25165/j.ijabe.20181101.3210
51. Satti SR, Shamshiri RR, Abdul-Hamid MA, Abdul-Rahman A (2021) Design considerations of robotics in plant production systems: a review. Appl Eng Agric 37(3):395–404. https://doi.org/10.13031/aea.14493
52. Moghimi A, Gharari S, Hosseinpour S (2020) Development of an intelligent greenhouse monitoring robot using sensor fusion and AI. J Agric Robot 2(1):45–55
53. Zhang H, Li Y, Wang J, Liu Z (2023b) Multi-task robotic systems for smart agriculture: from sensing to harvesting. IEEE Access 11:11568–11579. https://doi.org/10.1109/ACCESS.2023.3242569
54. Design R (2023) CropScout: a multispectral imaging robot for autonomous greenhouse monitoring. AgriTech Robotics Monthly 12(3):18–22
55. Tran QH, Pham CT, Nguyen VH (2023) WeedBot: a robot for mechanical weed removal in controlled environments. J Precis Agric Eng 5(2):77–86
56. Peng Y, Wang Y, He L, Zhang D (2022) GrapeBot: development of a robotic system for autonomous grape harvesting in hydroponic setups. Biosyst Eng 215:12–24. https://doi.org/10.1016/j.biosystemseng.2022.01.003
57. Pincheira F, Soto FA, Parra D, Barros J (2023) Wireless communication in smart greenhouses: a comparative study. Sensors 23(5):2671. https://doi.org/10.3390/s23052671
58. Roldán JJ, Joossen G, Sanz D, del Cerro J, Barrientos A (2019) Mini-UAV based sensory system for measuring environmental variables in greenhouses. Sensors 19(15):3337. https://doi.org/10.3390/s19153337
59. Shamshiri RR, Ismail WIW, Zakaria AJ, Hameed IA (2024) Smart navigation technologies for autonomous robots in controlled-environment agriculture: a review. Comput Electron Agric 214:108001. https://doi.org/10.1016/j.compag.2024.108001
60. Gao Y, Zhang C, Wang J, Li M (2021) Development of a robotic system for automatic maintenance tasks in hydroponic greenhouses. Biosyst Eng 202:72–85. https://doi.org/10.1016/j.biosystemseng.2021.10.004
61. Henríquez D, Torres R, Valenzuela M (2019) Development of an autonomous robot for precision agriculture in vineyards. J Field Robotics 36(7):1245–1260. https://doi.org/10.1002/rob.21817
62. Peng J, Wang Z, Huang Y (2022) Automation technologies for robotic fruit harvesting: a comprehensive review. J Agric Sci Technol 24(5):455–472
63. Farm (2024) Farm automation technologies: an overview. Retrieved from https://www.farm.com/

64. Oliveira L, Dias J, Silva M (2021) Autonomous navigation and manipulation systems for agricultural robotics. Comput Electron Agric 187:106280. https://doi.org/10.1016/j.compag.2021.106280
65. Joseph R, Singh A, Kapoor P (2022) Multi-functional AI-driven robotic systems for controlled-environment agriculture. J Agric Eng Res 85(3):201–215
66. Ren L, Zhou J, Li X (2023) AI-driven plant disease diagnosis and robotic application in smart agriculture. Front Plant Sci 14:1178402. https://doi.org/10.3389/fpls.2023.1178402
67. Design and development of an autonomous robot for precision hydroponics farming (2024) Smart Agric Technol 5:100215. https://doi.org/10.1016/j.atech.2024.100215
68. Chen Q, Liu S, Wang H (2022) Vision-based robotic systems for precision agriculture: a review. Robot Auton Syst 154:104126. https://doi.org/10.1016/j.robot.2022.104126
69. Ahmed A, Singh M, Zhao Y (2025) Advances in robotic automation for high-efficiency greenhouse farming. Automation in Agriculture 18(2):101–120
70. Parsa J, Mahdavian SM, Farahani R (2023) Sensor-based automation for greenhouse crop management: a smart agriculture perspective. Sensors 23(11):4556. https://doi.org/10.3390/s23114556
71. Li Y, Zhang H, Wang T (2024) Deep learning-powered robotics in agricultural applications: opportunities and challenges. arXiv preprint. arXiv:2403.12345. https://arxiv.org/abs/2403.12345
72. Lehnert C, English A, McCool C, Tow AW, Perez T, Upcroft B (2017) Autonomous sweet pepper harvesting for protected cropping systems. IEEE Robot Automat Lett 2(2):872–879. https://doi.org/10.1109/LRA.2017.2655410
73. Mahalingam S, Ramesh N, Kulkarni A (2023) Design of an AI-assisted robotic arm for plant phenotyping in hydroponics. arXiv preprint arXiv:2306.11234. https://arxiv.org/abs/2306.11234
74. Fujinaga T, Yamamoto S, Ito K (2025) Integration of machine vision with robotics in smart agriculture. Comput Electron Agric 212:107013. https://doi.org/10.1016/j.compag.2024.107013
75. Moshayedi H, Yang G (2024) Multi-robot collaboration and autonomous decision-making for smart agriculture: a review. Comput Electron Agric 213:107093. https://doi.org/10.1016/j.compag.2023.107093
76. Qu C, Li Z, Wang X (2022) Development of an autonomous strawberry harvesting robot for soilless cultivation systems. Smart Agricultural Technology 3:100087. https://doi.org/10.1016/j.atech.2022.100087
77. Moshayedi H, Zhang Y, Yang G (2021) Development and challenges of ground-based robots in agricultural applications: a review. Comput Electron Agric 186:106171. https://doi.org/10.1016/j.compag.2021.106171
78. Zhang C, Wang Z, Wu Z (2020) A survey on agricultural robots: capabilities, design, and applications. Smart Agricultural Technology 1:100005. https://doi.org/10.1016/j.atech.2020.100005
79. Huang Y, Hoffmann WC, Lan Y (2022) Unmanned aerial vehicles for pest management and crop monitoring. Agric Eng Today 46(3):15–22
80. Patel K, Jain R, Mehta M (2023) Design and deployment of autonomous ground vehicles in precision farming. J Smart Agric Syst 4(2):65–72
81. Deshmukh A, Sharma R, Joshi A (2022) Precision agriculture using RTK-GPS and sensor fusion for location-specific management. Comput Electron Agric 33(4):211–224
82. Sharma V, Rana V, Rawat P (2021) Real-time soil monitoring using IoT-enabled sensors for smart farming. Sens Syst 38(6):48–55
83. Rana D, Singh A, Kumar R (2023) Monitoring root zone and soil compaction using smart robotic systems in greenhouses. J Agric Eng Automat 5(1):34–42
84. Bac CW, Hemming J, van Henten EJ (2014) Robust pixel-based classification of sweet-pepper plants using video sequences for harvesting robots. Comput Electron Agric 105:124–133. https://doi.org/10.1016/j.compag.2014.04.002

85. Shah S, Al-Sarawi S, Abbass H (2023) AI-enabled robotic systems for precision harvesting: a comprehensive survey. Artif Intell Agric 7:32–46. https://doi.org/10.1016/j.aiia.2023.02.004
86. Tang L, Zhang J, Qi H (2020) An intelligent weeding robot with machine vision and mechanical arms. Comput Electron Agric 177:105693. https://doi.org/10.1016/j.compag.2020.105693
87. Zhao X, Jin Y, Wang G (2021) Wireless energy systems for autonomous agricultural robots: a review. Renew Sust Energ Rev 135:110127. https://doi.org/10.1016/j.rser.2020.110127
88. Hemming J, Bac CW, van Tuijl BAJ, Barth R, van Henten EJ (2019) Harvesting robots for high-value crops: state-of-the-art review and challenges ahead. J Field Robot 36(4):712–730. https://doi.org/10.1002/rob.21804
89. Tsouros DC, Bibi S, Sarigiannidis PG (2019) A review on UAV-based applications for precision agriculture. Information 10(11):349. https://doi.org/10.3390/info10110349
90. Fortune Business Insights (2024) Agricultural robots market size, share & COVID-19 impact analysis, by type, application, and regional forecast, 2024–2032. Retrieved from https://www.fortunebusinessinsights.com
91. Dennison J, Kumar P, Zhao X (2025) Economic and energy implications of robotic systems in precision agriculture. Sustain Agric Rev 55:219–235
92. Selvam A, Natarajan S, Wang L (2025) AI-enabled deep learning systems for yield forecasting and crop maturity detection in vertical farms. Artif Intell Agric 9:75–86
93. Mahalingam S, Rao DS, Rani PU (2023) Integration of robotics and machine vision in controlled environment agriculture: prospects for medicinal plant production. Smart Agricultural Technology 5:100102. https://doi.org/10.1016/j.atech.2023.100102
94. Channa MA, Zhang Y, Shah A (2024) Role of robotics and AI in optimizing aquaponics for sustainable crop-fish integration. Aquac Rep 30:101584. https://doi.org/10.1016/j.aqrep.2023.101584
95. Wu Y, Chen Y, Tan J (2022) Learning from demonstration: collaborative robots in urban farming environments. J Agric Robot Automat 3(2):95–104
96. Okomoda VT, Chong CM, Hassan A (2023) Challenges and solutions in scaling aquaponics for commercial production. Aquacult Eng 101:102282. https://doi.org/10.1016/j.aquaeng.2023.102282
97. Sharma R, Thakur M, Gupta R (2024) Robotic innovations in hydroponics and medicinal plant cultivation: toward pharmaceutical-grade production systems. J Contr Environ Agric 2(1):34–47
98. Singh A, Prasad R (2024) Medicinal plant biotechnology and industrial applications. Springer India, New Delhi
99. Delaide B, Francois P, Van Havermaet R (2016) Optimization of nutrient delivery using NUTRISENSE under salinity stress conditions in hydroponic lettuce. Agric Water Manag 173:60–67
100. Han Y, Jeong SW, Cho K (2020) Impact of frequency domain reflectometry sensors on resource efficiency in hydroponic strawberry cultivation. Sensors 20(18):5283. https://doi.org/10.3390/s20185283
101. Yep B, Zheng Y (2019) Benefits of vertical farming on water use and crop yield: a comparative study. Agric Water Manag 219:272–284
102. Goddek S, Körner O (2019) Advances in aquaponics and circular agriculture. Springer, Cham
103. Purnomo CW, Susanto H, Wijaya H (2022) Nutrient recovery from aquaponic wastewater using electrodialysis and forward osmosis. J Water Process Eng 48:102814
104. Rahnama F, Nasr Esfahany M, Keshavarz Moraveji M (2021) Monitoring and controlling nutrient recycling in hydroponics using real-time sensing. Environ Technol Innov 21:101244
105. Caplan JS, Wheaton JM, Doyle MW (2019) Financial feasibility of hydroponic production in Brazil: a case study. J Agrib Dev 38(2):147–160
106. Resh HM (2013) Hydroponic food production: a definitive guidebook for the advanced home gardener and the commercial hydroponic grower, 7th edn. CRC Press
107. Kumar A, Thakur M, Singh S (2023) Enhancing Bacopa monnieri cultivation with robotics: field validation in controlled environments. Ind Crop Prod 199:116567

108. Tzachor A, Richards CE, Holt L (2021) Artificial intelligence in agriculture: challenges and opportunities for sustainable farming. Nat Food 2(8):442–447. https://doi.org/10.1038/s43016-021-00393-1
109. Bac CW, Hemming J, van Henten EJ (2014) Harvesting robots for high-value crops: state-of-the-art review and challenges ahead. J Field Robot 31(6):888–911
110. Singh A, Prasad R, Mehta P (2022) Robotics and automation for high-value crop production: a review. J Agric Eng 59(4):312–326
111. Tzachor A, Richards CE, Smidt SJ (2021) Autonomous agriculture: a strategic review of core technologies, socio-ecological implications, and future prospects. J Clean Prod 320:128591. https://doi.org/10.1016/j.jclepro.2021.128591
112. Love DC, Fry JP, Genello L et al (2015) An international survey of aquaponics practitioners. PLoS One 10(7):e0133042. https://doi.org/10.1371/journal.pone.0133042
113. Somerville C, Cohen M, Pantanella E, Stankus A, Lovatelli A (2014) Small-scale aquaponic food production: integrated fish and plant farming, FAO fisheries and aquaculture technical paper no. 589. FAO, Rome
114. Goddek S, Joyce A, Kotzen B, Burnell G (2019) Aquaponics food production systems: combined aquaculture and hydroponic production technologies for the future. Springer
115. Fernández M, Alarcón R, García J (2023) Machine vision and AI for fish locomotion monitoring in aquaponics. Aquac Rep 30:101587
116. Goddek S, Körner O (2019) A fully integrated simulation model of multi-loop aquaponics: a case study for system validation. Ecol Model 411:108817. https://doi.org/10.1016/j.ecolmodel.2019.108817
117. Rahnama M, Shamshiri RR, Saeys W (2021) Smart nutrient solution management in hydroponic systems using sensor-based automation: a review. Comput Electron Agric 189:106417. https://doi.org/10.1016/j.compag.2021.106417
118. Purnomo CW, Kurniawan SB, Imron MF, Shayya WH (2022) Nutrient recovery from hydroponic wastewater using electrodialysis-forward osmosis: a step toward circular agriculture. Environ Technol Innov 27:102488. https://doi.org/10.1016/j.eti.2022.102488
119. Carey EE, Hall BE (2017) Hydroponic capital: exploring automation to offset labor and land constraints. J Urban Agric 1(2):67–76
120. Al-Kodmany K (2018) The vertical farm: a review of developments and implications for the future of urban food production. Buildings 8(2):24
121. Kumar R, Patel H, Singh A, Mehta R (2023) Design and deployment of adaptive robotic arms for automated transplanting in vertical farming systems. J Agric Eng 60(1):45–56. https://doi.org/10.1016/j.jagreng.2023.01.006
122. Gómez-Gil J, Gonzalez-de-Santos P, Emmi L (2021) Collaborative agricultural robots for automated handling of horticultural products. Robotics 10(1):11
123. Patel N, Verma A, Saini S (2021) Challenges in robotic harvesting of medicinal and aromatic plants: a review. J Herb Sci 7(2):101–110
124. Park J, Lee H, Kim D, Cho Y (2021) Vision-based fruit detection and tracking system using faster R-CNN and centroid tracking for smart greenhouse applications. Comput Electron Agric 182:105991. https://doi.org/10.1016/j.compag.2021.105991
125. Ghorbani M, Amini A, Safa M (2022) Artificial neural network and support vector regression for nutrient deficiency diagnosis in hydroponic systems. Biosyst Eng 219:103–115. https://doi.org/10.1016/j.biosystemseng.2022.05.004
126. Liu G, Zhang H, Zhang X, Yang X (2021) Development of a robotic strawberry harvester with a soft gripper and deep learning vision. Sensors 21(12):4052. https://doi.org/10.3390/s21124052
127. Tanaka T, Yamamoto K, Nakamura Y (2020) A robotic manipulator with shape memory alloy actuators for transplantation of tissue-cultured seedlings. IEEE ASME Trans Mechatron 25(3):1294–1302. https://doi.org/10.1109/TMECH.2020.2967711
128. Al-Mohannadi M, Khan MA, Al-Thani H (2021) IoT-enabled vertical hydroponics: smart agriculture for arid environments. Agric Water Manag 255:107003. https://doi.org/10.1016/j.agwat.2021.107003

129. Rurangwa E, Goddek S, Kotzen B (2022) Simulation-based aquaponic system optimization for local climate adaptability. Aquac Rep 23:100997. https://doi.org/10.1016/j.aqrep.2022.100997
130. Al-Busaidi H, Al-Karousi H, Al-Masroori H (2023) Energy-efficient temperature control strategies in aquaponics for medicinal plant cultivation. Sustain Energy Technol Assess 56:103069. https://doi.org/10.1016/j.seta.2023.103069
131. Rodríguez JA, Morales M, Delgado A (2023) IoT-based predictive control of aquaponics for real-time crop and fish health monitoring. Comput Electron Agric 204:107493. https://doi.org/10.1016/j.compag.2023.107493
132. García M, Ortiz C, Navarro A (2021) Small-scale aquaponics as an educational and sustainable model for medicinal herb production in Latin America. Urban Agric Reg Food Syst 6(1):e20021. https://doi.org/10.1002/uar2.20021
133. Sánchez D, Pérez J, Vidal R (2022) Digital twin simulation for aquaponic farming systems: enhancing decision-making in medicinal plant cultivation. J Clean Prod 368:133047. https://doi.org/10.1016/j.jclepro.2022.133047
134. Fernández M, Alarcón R, García J (2023) YOLO-based deep learning model for fish locomotion monitoring in aquaponic environments. Aquac Rep 30:101587. https://doi.org/10.1016/j.aqrep.2023.101587

Chapter 9
Nanotechnological Interventions to Enhance Growth and Secondary Metabolite Content in Medicinal Plants

Anchal Thakur, Harpreet Kaur, Vikram Patial, Sudesh Kumar Yadav, and Amitabha Acharya

Abstract Medicinal plants have long served as vital resources in traditional medicine systems worldwide, yet their cultivation faces challenges in yield, active compound production, and environmental adaptation. The revolutionary potential of nanotechnology in overcoming these constraints is examined in this chapter. We investigated the interactions between plant systems and artificial nanomaterials, such as metal nanoparticles (NPs), carbon-based nanomaterials, and polymer nanocarriers, to improve secondary metabolite production, stress tolerance, and growth characteristics. Nanotools, including nanobiosensors, play a crucial role in the advancement of crop production by enabling the precise management and control of inputs such as fertilizers, pesticides, and herbicides. This chapter describes the

A. Thakur · H. Kaur
Nanobiology Lab, Fermentation and Phytofarming Technology, CSIR-Institute of Himalayan Bioresource Technology, Palampur, Himachal Pradesh, India

Academy of Scientific and Innovative Research (AcSIR), Ghaziabad, India

V. Patial
Academy of Scientific and Innovative Research (AcSIR), Ghaziabad, India

Pharmacology and Toxicology Laboratory, Dietetics & Nutrition Technology, CSIR-Institute of Himalayan Bioresource Technology, Palampur, Himachal Pradesh, India

S. K. Yadav (✉)
Nanobiology Lab, Fermentation and Phytofarming Technology, CSIR-Institute of Himalayan Bioresource Technology, Palampur, Himachal Pradesh, India

Academy of Scientific and Innovative Research (AcSIR), Ghaziabad, India

Pharmacology and Toxicology Laboratory, Dietetics & Nutrition Technology, CSIR-Institute of Himalayan Bioresource Technology, Palampur, Himachal Pradesh, India

A. Acharya (✉)
Nanobiology Lab, Fermentation and Phytofarming Technology, CSIR-Institute of Himalayan Bioresource Technology, Palampur, Himachal Pradesh, India

Pharmacology and Toxicology Laboratory, Dietetics & Nutrition Technology, CSIR-Institute of Himalayan Bioresource Technology, Palampur, Himachal Pradesh, India
e-mail: amitabha@ihbt.res.in

P. Kumar, A. R. Warghat (eds.), *Transforming Medicinal Plant Agriculture*, Smart Agriculture 16, https://doi.org/10.1007/978-981-95-3963-5_9

state-of-the-art applications in which nano-elicitors boost defensive mechanisms to improve the manufacture of bioactive compounds, nano-fertilizers supply nutrients with previously unheard-of precision, and nano-delivery systems maximize the growth regulator absorption. Recent field and lab research shows that varieties of medicinal plant species have significantly improved biomass accumulation, phytochemical profiles, and pharmacological effectiveness under nanoparticle treatment conditions.

Keywords Nano-elicitors · Medicinal plants · Phytoconstituents · Nano-fertilizers · In vitro propagation

9.1 Introduction

There has been a renewed interest in herbal remedies and medicinal plants in recent years. A significant proportion of the population utilizes medicinal plant items to treat ailments or preserve their health, both in developed and developing nations. The World Health Organization reports that approximately 80% of people worldwide rely on plant-based remedies for a variety of illnesses [1]. This widespread reliance directly contributes to the high demand for medicinal plants across culinary, cosmetic, and pharmaceutical sectors. Correspondingly, the expanding application of these plants in newer markets has driven a ~ 15% annual increase in demand. About 960 out of the 8000 types of medicinal plants that were collected from the wild in India are actively traded [2]. Since herbal healthcare methods are becoming more and more popular, medicinal plants have become more significant. This is because they contain bioactive chemicals that have a variety of pharmacological effects. A significant disparity now exists between the availability and demand for many therapeutic plants due to unfavorable climate conditions and lack of appropriate technologies for crop cultivation and domestication. Changes in global climate are seriously threatening both the production rate and the composition of secondary metabolites found in medicinal plant extracts [3]. The survival of endangered species is being threatened by this rising demand, though. In order to prevent the extinction of these species and to help people with phytotherapeutics using contemporary technologies, researchers are investigating several strategies. Furthermore, to guarantee the quality of medicinal herbs throughout their cultivation and processing into final herbal products, it is essential to implement rigorous regulatory oversight and precision [4]. Despite the fact that some of these issues have been addressed in the past with commendable efforts, a quick step in this direction requires the use of contemporary research methodologies and creative thinking. Researchers from the fields of biotechnology and nanotechnology are working together to develop more advance technologies, particularly in the areas of sustainable agriculture, diagnostics and medicine [5].

One such approach that has been popular in the last 10 years of research is nanotechnology, and its scope is constantly expanding. Nanotechnology is a field that encompasses a wide range of applications across nearly all scientific sectors due to the extensive research being done in this area. With new techniques and instruments to improve plant's capacity to absorb nutrients, nanotechnology holds the potential to completely transform agriculture. Molecular-level interactions between nanoparticles and living cells occur in nano-agriculture to benefit the different crops. Utilizing nanoparticles to promote plant growth and manage plant diseases is a relatively new technique [6].

9.2 Nanomaterials and Their Types

Nanomaterials (NMs) are characterized as substances that possess at least one dimension ranging from 1 to 100 nm. The term "nanomaterial" describes a substance with an amorphous or crystalline particle structure at the nanoscale. Nanomaterials comprise three-dimensional nanoscale particles, nanomembranes, nanowires, and nanoblocks of the material. Because of their unique size and structural characteristics, nanoparticles have unique qualities that are absent from many conventional materials. The unique characteristics of nanomaterials, such as their nanosize, high surface energy, and huge specific surface areas, are responsible for their different unique properties [7]. There are different categories of nanomaterials including bulk materials, discrete nanomaterials and nanoscale device materials. "Bulk materials" in nanotechnology are materials that are larger than 100 nanometers (nm) in all dimensions and have properties that are independent of size. Materials with a high surface area and quantum effects that have at least one dimension between 1 and 100 nm are the discrete nanomaterials. Discrete nanomaterials encompass three categories: 0D-materials, encompassing nanoparticles, and quantum dots; 1D materials, comprising nanowires and nanotubes; and 2D materials, exemplified by graphene nanosheets and nanofibers. Material elements that are incorporated in devices at the nanoscale are known as nano-scale devices material, predominantly as thin films. These materials exhibit the two dimensional morphology taking the form of thin sheets [8].

Despite the fact that nanotechnology has been transforming agricultural research over the last several decades, there are still relatively few studies pertaining to its role in medicinal plant research. In order to improve the status of medicinal plant research today, it is worthwhile to investigate the potential of nanobiotechnology. In light of this, the current book chapter emphasizes how nanobiotechnology has promise for launching a new chapter in the study of therapeutic plants and their sustainable application.

9.3 Plant Secondary Metabolites

Metabolites are the small, intermediate molecules produced during metabolic processes and play diverse roles in the cell. In plants, secondary metabolites are distinguished from the primary metabolites such as nucleic acids and amino acids in that they are not directly involved in the growth of plant, they play a role in defense, survival, and environmental interactions. Plants produce a number of secondary metabolites, which have a range of biological activities, making them useful for human health (Table 9.1). There are three major classes of secondary metabolites, viz. Terpenoids, alkaloids, and polyphenols [9]. Terpenoids are a diverse class of isoprenoid compounds biosynthesized by mevalonic acid pathway. They play a major role in inhibition of plant pathogens and constitute a major mediator of ecological interactions. Triterpenoid boswellic acid extracted from *Boswellia serrata* has anti-inflammatory activity [10]. Another pentacyclic triterpene, betulinic acid which is extracted from *Betula pubescens* demonstrates anticancer and antiviral activities [11]. Alkaloids are the group of compounds containing nitrogen within the heterocyclic ring and occur abundantly in flowering plants. Many alkaloids are toxic to different organisms, thereby playing a major role in plant defense. Historically,

Table 9.1 Some commercially important secondary metabolites, their plant sources and pharmacological/therapeutic activities

Plant	Secondary metabolite	Pharmacological Activity	References
Curcuma longa	Curcumin	Antibacterial, antibiofilm, and anticancer activity	[17]
Polygonum cuspidatum	Resveratrol	MAFLD, Alzheimer's disease	[18, 19]
Silybum marianum	Silibinin	Neuroprotective, Hepatoprotective, and antioxidant	[20, 21]
Agapanthus praecox	Quercetin	Anticancerous, apoptotic, antidiabetic	[22, 23]
Genista tinctoria	Genistein	antioxidant, anti-inflammatory, antibacterial, and anticancer activities	[24, 25]
Silybum marianum	Silymarin	Hepatoprotective, anti-inflammatory, and antiviral activities	[26]
Piper nigrum	Piperine	Anticancer activity including prostate cancer, hepatocellular carcinoma, lung cancer, and glioblastoma	[27]
Nigella sativa	Thymoquinone	Anti-inflammatory, antioxidant, anti-apoptotic, and anticancer activities	[28]
Taxus brevifolia	Paclitaxel	Ovarian cancer, lung cancer	[29]
Citrus paradisi	Naringenin	Anti-inflammatory, anticancer, antidiabetic	[30]
Papaver somniferum	Codeine	Analgesic, antidepressants, sedative and hypnotic properties	[31]
Atropa belladonna, Darura stramonium	Atropine	Anti cholinergic, antagonist of muscarinic acetylcholine receptors	[32]

alkaloids have been used for various medicinal purposes. Quinine and quinidine, the antimalarial drugs, are alkaloids derived from the bark of *Cinchona officinalis.* Another important alkaloids are vincristine and vinblastine which are derived from *Catharanthus roseus* and exhibit strong anticancer activities [12]. In addition to these biological activities, alkaloids obtained from a variety of plants are well known to have anti-depressant, anti-hypertensive, anti-inflammatory, antioxidant, psychotropic and hepatoprotective activity among others. Polyphenols are one of the largest groups of secondary metabolites produced by plants and consist of different classes, viz., flavonoids, tannins, phenolics, and lignans [13]. Flavonoids contribute to flower color and flavor which aids in attracting insects and subsequently promotes pollination while other polyphenols such as polyphenolic glycosides are anti-herbivores. Since polyphenols are antioxidant, they have numerous biological activities. Flavonoids are known to lower the risk for the development of atherosclerosis, while tannins have been used as astringents and adiuretic and have anti-inflammatory properties [14]. Flavones like apigenin, uteolin, and chrysin obtained from green bell pepper, thyme, parsley exhibit antimutagenic, anticarcinogenic, and anti-inflammatory properties [15]. Phenylpropanoids like rosmarinic acid (RA) is neuroprotective and anti-inflammatory derived from *Salvia rosmarinus* [16].

9.4 Nano-Elicitors

Metabolites are the by-products of metabolism that are released in reaction to outside stimuli. The plant metabolism is a tightly regulated sequence of enzyme-mediated chemical reactions that produce and utilize a broad range of molecules belonging to the categories of carbohydrates, amino acids, fatty acids, nucleotides, and polymers made from them. All of these functions are referred to as primary metabolism, and the corresponding substances that are necessary for the plants to survive are called primary metabolites. The growth and development of plants depend on primary metabolites [33]. Several substances that are part of the secondary metabolism are also produced in addition to the primary metabolites, including terpenes, steroids, anthocyanins, anthraquinones, phenols, and polyphenols. Specific species have specific tissue specialized secondary metabolites, which can only be detected at a particular stage of plant growth and development, or can only be activated during pathogen attack, stress or the depletion of nutrients. Conversely, secondary metabolites support plant growth in a variety of stressful environments. Additionally, secondary metabolites are the bioactive substances that offer considerable therapeutic advantages and are used in nutraceuticals, pharmaceuticals, and cosmetic products [34]. The yield and quality of secondary metabolites can be affected by various factors, including environmental stress, pests, diseases, and many more. The preferred method for enhancing the synthesis of these secondary metabolites is often the plant tissue culture, which uses number of strategies such as medium optimization, cell culture, elicitation, biotransformation, precursor feeding, etc. [35]. The application of diverse biocompatible nanomaterials in plant biotechnology has emerged as a crucial factor in enhancing the levels of secondary metabolites (Fig. 9.1). When plants are treated with nanoparticles, which act as elicitors, they tend to generate a higher quantity of phytoconstituents. In a research study, callus cultures of *P. vulgaris* that were treated with gold nanoparticles (AuNPs) alone and silver-gold nanoparticles (Ag-AuNPs) in a 1:3 ratio demonstrated an enhanced production of protein content, along with elevated levels of superoxide dismutase to ~0.54 and ~0.45 nM/min/mg-fresh weight (FW) and peroxidase enzyme to ~0.25 and ~0.21 nM/min/mg-FW. The greatest accumulation of phenolic content of 6.5 GAE-mg/g-dry weight (DW) was also produced by combining Ag-AuNPs with naphthalene acetic acid (NAA). The addition of Ag-AuNPs with or without NAA to the growth media resulted in the highest total flavonoid content of 7.88 and 9.5 RE-mg/g-DW [36]. The effects of several NPs, such as Nickel (Ni), Cadmium-Selenide (Cd-Se), Zinc-Silver (Zn-Ag), on the production of withanolides and the metabolism of carbohydrates have been investigated in *Withania somnifera.* The treatments with Ni and Cd-Se resulted in the most significant enhancement of photosynthesis and transpiration rates, as well as an increase in withanolide content (~2.4 and ~1.7 times, respectively) and specific carbohydrates such as glucose (~1.2 and ~1.6 times) when compared to plants treated with water [37].

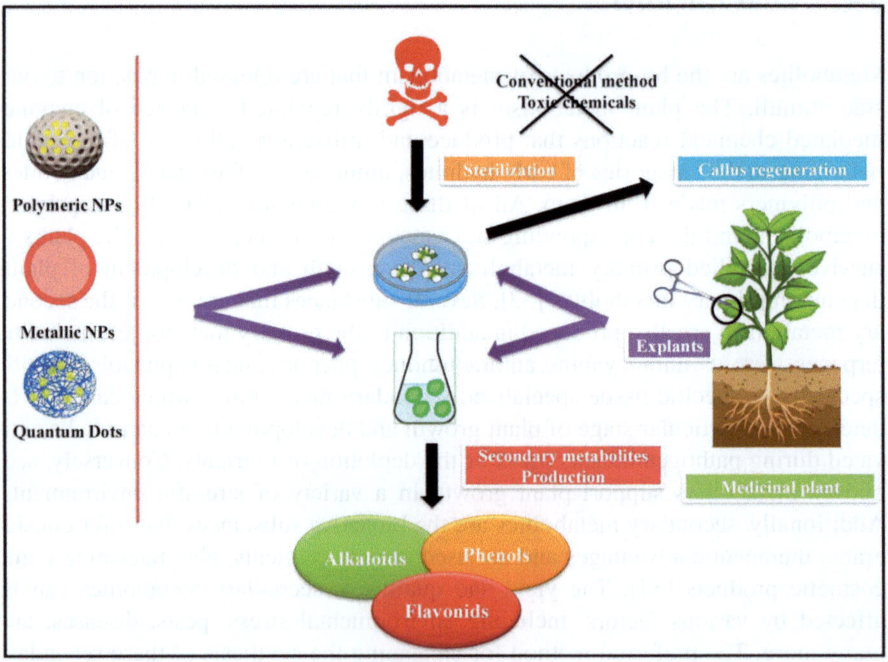

Fig. 9.1 Figure illustrating the nano-elicitation of the plant secondary metabolites under nano-sterilized conditions with enhanced growth of explants and callus regeneration (Created with Freepik)

Studies indicate that the control of transcript levels for different genes associated with the biosynthesis of secondary metabolites may be connected to the NPs-mediated modifications. The expression of a crucial gene geranyl diphosphate synthase (GPPS), in the thymoquinone production pathway, increased when Silicon dioxide (SiO_2) and Titanium dioxide (TiO_2) NPs were introduced to *Nigella sativa* plant during its early flowering phase. The expression of the GPPS gene was examined at two distinct doses; included 50 and 100 mg/L of each SiO_2 and TiO_2 nanoparticle. The induction of GPPS expression was more significant by TiO_2 NPs than by SiO_2. Additionally, compared to 50 mg/L concentration, the increase was more than the 100 mg/L treatment with nanoparticles [38].

In cell suspension cultures of *Gymnema sylvestre,* giving copper oxide NPs with plant growth medium increased the output of bioactive compounds such as flavonids, gymnemic acid II (upto ~9 folds), and phenolics [39]. Additionally, the potential of copper-based spherical nanoparticles, to improve the secondary metabolites synthesis in the *Bacopa monnieri* plant was assessed at concentrations varying from 0 to 100 mg/L. The concentration of secondary metabolites, including flavonoids, saponins, and alkaloids increased at the concentration from 5 mg/L to approximately 40 mg/L. Conversely, there was a decline in the total phenolic content after reaching 20 mg/L [40]. The treatment of *Dracocephalum* plant hairy roots with SiO_2 NPs resulted in an enhanced expression of the genes rosmarinic acid synthase (ras) and phenylalanine ammonia lyase (pal), which are responsible for the production of RA. The treatment of SiO_2 NPs at 100 mg/L concentration for 48 h, increased in RA content in hairy root culture, measuring 1016.6 μg/g-FW, which is ~8.26 more than that of the control group. Additionally, the incorporation of SiO_2 NPs led to a notable rise in anticancer flavonoids, including cirsimaritin, isokaempferide, and xanthomicrol, with increases of ~13, ~13.42, and ~10 times, respectively [41]. The RAS and PAL enzymes, which play a vital role in the biosynthesis pathway of RA, were activated when *Salvia officinalis* plants were subjected to topical treatment with AgNPs at concentrations of 50 and 100 mg/L. However, it appears that the activity of these enzymes diminished when AgNPs concentrations exceeded 100 mg/L [42].

The introduction of multi-walled carbon nanotubes (MWCNT) into the culture of *Catharanthus roseus* resulted in a ~37-fold increase in the transcript levels of the deacetyl vindoline 4-O-acetyltransferase (DAT) gene, which plays a crucial role in the biosynthesis of vindoline. Catalase and peroxidase activity was increased by two fold, alkaloids by ~1.7-fold, soluble phenols by ~23%, and phenylalanine ammonia lyase activity by an average of ~36.5% after the application of MWCNT [43].Additionally, SiO_2 NPs stimulated the expression of the Hyoscyamine 6-hydroxylase (H6H) and Putrescine N-methyltransferase (PMT) genes in various *Hyoscyamus*species at distinct concentrations. These genes are crucial to produce tropane alkaloids, which possess antispasmodic, anticholinergic, and sedative properties. In *Hyoscyamus reticulatus*, both genes showed increased expression when exposed to 100 mg/L of SiO_2 NPs. Conversely, in *Hyoscyamus pusillus*, H6H expression was induced at a concentration of 200 mg/L, while PMT exhibited a slight upregulation at both 100 and 200 mg/L concentrations [44]. In a similar manner, treatment of the Feverfew plant (*Tanacetum parthenium*) with 25 mM SiO_2 nanoparticles enhanced the expression of genes necessary for the production of β-caryophyllene and parthenolide [45]. Additionally, the application of AgNPs at a concentration of 40 mM to *Stevia rebaudiana* resulted in the upregulation of all key genes responsible for the synthesis of steviolglycosides and rebaudioside [46]. In addition to their interactions with gene networks, NPs also influence other cellular cascades to modify secondary metabolism in plants. By better understanding the mechanism of elicitation secondary metabolism in plants through NPs, we can create novel molecular pharmacological tools and methods.

9.5 Nano-Fertilizers

In order to preserve soil fertility and raise crop quality and yield, nutrient fertilization is essential. Conventional or traditional fertilizers are expensive to the producers and harmful to both the biotic and abiotic environment. As a result, people are looking for fertilizers that are not harmful to the environment, especially ones that have ahigh nutrient-use efficiency. Nanotechnology is showing promise as a substitute. Nanomaterials can be utilized in soil as delivery systems or nanostructured fertilizers to improve the effectiveness and absorption of conventional fertilizers. The gradual and efficient release of nutrients to plants facilitated by nano fertilizers diminishes environmental hazards, reduces the agitation and leaching associated with traditional fertilizers, and enhances the efficiency of fertilizer usage [47]. Research indicates that the application of magnetite nano-fertilizer notably enhances the growth attributes of *Ocimum basilicum*, resulting in an increase in total chlorophyll by ~13.7%, essential oil by approximately ~1.8%, and iron content [48]. In contrast to the conventional NPK (nitrogen, phosphorus, potassium) formulation, the application of nano-hydroxyapatite at concentrations of 500 and 1000 mg/L significantly increased the growth and production of essential oils of rosemary plants. The application of this nano-fertilizer led to increased concentrations of essential components in rosemary oil, such as endo-borneol, camphor, verbenone, eucalyptol, and alpha-pinene, when contrasted with the control group. Furthermore, the application of nano-fertilizer improved various leaf characteristics, including midvein thickness, lamina thickness, pith diameter, thickness of xylem and phloem tissue, stem diameter, and fiber thickness. This enhancement resulted in better water and nutrient transport, greater structural support, and increased resilience to environmental stress [49]. To improve the yield and of Oregano (*Origanum syriacum)* herb, the use of the nano-fertilizer lithovit, combined with a lower dosage of nitrogenous fertilizer, has demonstrated greater effectiveness compared to applying a larger amount of nitrogenous fertilizer [50]. Another investigation revealed that *Dracocephalum moldavica*, yielded greater dry mass and essential oil when treated with iron nano-fertilizer at a concentration of 1000 mg/L [51]. Furthermore, an organic nano-fertilizer at approximately 3000 mg/L has been utilized to enhance the level of anthocyanin pigment in the Domask-Rose plant, achieving a level of ~62.63 μg/g [52].

Thus, there is a lot of potential for increasing plant output and quality with nano-fertilizers. However, the lack of documentation regarding their application in plant research suggests that scientists in this field have not given them enough attention.

9.6 Nanotechnology for In Vitro Propagation

The ability of in vitro multiplication of medicinal plants to preserve consistency in secondary metabolite profiles across different batches, something that is normally challenging to control in field-grown plants, makes it of great commercial significance. In vitro culture facilitates the establishment of germplasm banks, the cultivation of medicinal plant tissues, large-scale propagation, and the prolonged preservation of plants in a suspended physiological condition [53]. However, some problems limit the success of micropropagation techniques, including poor regeneration, low germination rates, and contamination. Hence, the nanotechnology has provided the methods to get beyond these obstacles [54]. It offers gene transfer, secondary metabolites elicitation, better aseptic cultures, enhanced callus regeneration and many more (Fig. 9.2).

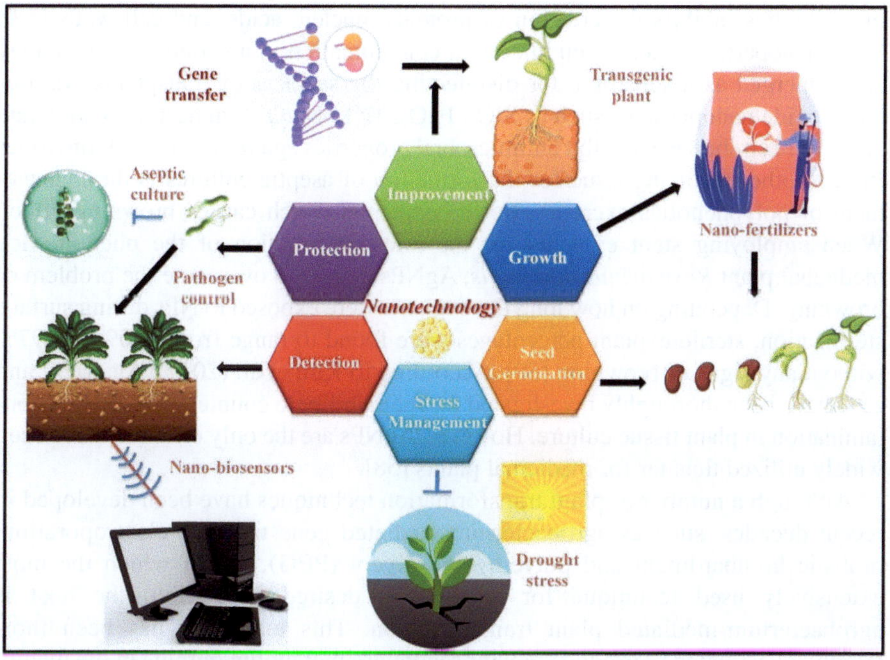

Fig. 9.2 Figure illustrating the various applications of nanoparticles in growth, crop production improvement, stress detection and management in the medicinal plants (Created with Freepik)

In the plant tissue culture field, nanoparticles have been employed for the processes of somatic embryogenesis and callus regeneration. *Prunella vulgaris* showed increased (~100%) callus proliferation in comparison to control (~95%) when exposed to different concentrations of NAA with Ag, AgAu (1:2), and AgAu (2:1) NPs. Additionally, the antioxidant activity of Ag-AuNPs (3:1) without NAA was increased by ~85%–87% [36]. The growth of callus from leaf explants of the medicinal plant *S. rebaudiana* was enhanced through the application of ZnO and CuO nanoparticles. When treated with 0.01 mg/L of ZnO, the callus had highest fresh weight (0.87 g) and for 1 mg/L CuO treatment the fresh weight and dry weight of callus was 0.51 and 0.05 g respectively [55]. In a different investigation into the plant regeneration of *O. basilicum,* copper nanoparticles markedly increased the number of plantlets (from ~4.3 to ~18.7) and the proportion of explants generated by somatic embryo (from ~15 to ~84%) [56]. These preliminary investigations indicate that NPs may have an impact on somatic embryogenesis and callus regeneration in plants. However, additional research in this area is required to determine the precise function and mechanism of action of NPs during the in vitro regeneration process of medicinal plants.

In vitro propagation of plants can lead to contamination in each subculture if explants, culture media, equipment, and other components are not properly sterilized. Conventional sterilization methods often rely on strong chemicals such as antibiotics, sodium hypochlorite, hydrogen peroxide, 70% ethanol, calcium hypochlorite, mercuric chloride and silver nitrate. Additionally, it is known that the use of antibiotics inhibits the creation of proteins, nucleic acids, and cell walls [57]. Since nanoparticles can lessen microbial contamination in a variety of plants, they have emerged as a substitute for disinfecting the surfaces of transplants. Various metal oxide nanoparticles such as SiO_2, FeO_2, TiO_2, MgO (Magnesium oxide), and Ag have been shown to exhibit antimicrobial properties against a variety of microbes [58]. Another recurring issue for the formation of aseptic cultures is the management of polyphenolics excretion by the explants, which causes browning effect. When employing stem explants for the micropropagation of the phenolic-rich medicinal plant *Rosmarinus officinalis*, AgNPs helped to overcome the problem of browning. Depending on how long the explants were exposed to NPs during surface sterilization, sterile explant percentages were found to range from ~40% to ~97% without any signs of browning [59]. According to Kim et al. (2017), Ag, TiO_2, and ZnO have been thoroughly investigated for their ability to counteract microbial contamination in plant tissue culture. However, AgNPs are the only ones that have been widely utilized thus far for medicinal plants [58].

Although a number of plant transformation techniques have been developed in recent decades, such as agrobacterium-mediated gene transfer, electroporation, biolistic bombardment and polyethylene glycol (PEG), out of which the most extensively used technique for introducing desired genes into the host is agrobacterium-mediated plant transformation. This technique has been thoroughly investigated to produce greater secondary metabolite outputs in the medicinal plants as well [60]. Nevertheless, this approach is restricted to specific species of plants and requires the refinement of techniques for plant transformation and regeneration applicable to additional species. Plant genetic transformation has recently seen the development of gene-delivery techniques based on nanotechnology. The capacity of single-walled carbon nanotubes (SWNT) to transfer genetic material directly into the chloroplast and nucleus eliminates the need for an agrobacterium-mediated approach [61, 62]. Similarly, polyethyleneimine and carbon nanotubes have been utilized to transfer genes to German chamomile because of their affordable price, advantageous gene transfer rate, and capacity to safeguard DNA [63]. The gene-delivery mechanism in plants mediated by nanomaterials is still in its infancy, and still there are numerous obstacles to its widespread use. This nano-strategy demonstrates outstanding biocompatibility, excellent transformation efficiency, sufficient preservation of external nucleic acids, and the possibility of plant regeneration.

9.7 Nanotechnology in Seed Germination Improvement

In terms of seed viability, germination, and rapid seedling multiplication, many medicinal plants encounter major challenges because of stiff seed coats, seed dormancy, and the slow growth rate of seedlings. To improve the metabolic processes that inherently trigger the initial phases of seed germination and boost development and yield, seed priming is used that is the soaking of seeds in the nanoparticle solutions [64]. The effects of many nanoparticles in different dosages on the germination of medicinal plant seeds have been assessed.

Various medicinal plants, including *Salvia mirzayanii, Carum copticum, Alyssum homolocarpum, Nigella sativa,* and *Sinapis alba,* were prompted to germinate their seeds using nano-sized TiO_2. However, the response differed based on the specific plant species and the concentration of TiO_2 used. In *N. sativa* seeds, the NP treatment with the concentration of 20 mg/L showed the greatest germination index value (GI) (~9.29). *S. alba* seeds treated with 20 mg/L had the greatest germination rate (GR) (~87%) and highest seedling vigor index of ~102.6, respectively. However, at the concentration of 10 mg/L, the seeds of *A. homolocarpum* showed the lowest mean time germination (MTG) of ~1.55 day [65].

The impact of a combination strategy that included nanopriming and inoculating plant growth promoting rhizobacteria on *Melissa officinalis* plant's physiological and metabolic characteristics was assessed. Silicon nanoparticles at different concentrations of 0, 100, and 500 mg/L were used to prime the seeds and *Pseudomonas fluorescens* and *P. putida* rhizobacteria strains were introduced into the grown seedlings. The *P. putida* strain exposure to the nano-primed (at 500 mg/L) seeds demonstrated a larger deposition of silicon nanoparticle on the root surface, whereas *P. fluorescens* application on the plants grown from 100 mg/L nano-primed seeds displayed a greater number of open stomata on the surface than the leaf than the controls. Plants infected with *P. putida,* when treated with a concentration 100 mg/L, showed the highest increase (by ~57.8%) in total dry weight when compared to the control. The yield of essential oil saw a significant enhancement through individual treatments of seed priming with nSi at concentrations of 100 and 500 mg/L, as well as through seedling inoculation with rhizobacteria. However, the most substantial increase of ~78.1% was observed when combining seed priming at 100 mg/L with inoculation of the plant using *P. putida*, in comparison to the control group [66].

Different dosages of multi-walled carbon nanotubes (MWCNTs) notably enhanced the germination rate of seeds, reduced the average germination time, and increased the lengths of both roots and stems, as well as the fresh and dry weights of these structures, in conditions of drought stress. When compared to the control, all treatments exhibited increased germination percentage (GP) at −8 bar stress, but with the 200 mg/l treatment, it exhibited the greatest GP of ~88.9%. The treatment of CNT at a concentration of 30 mg/L was shown to be ideal when there is no stress, while plants treated with 50 and 100 mg/L of CNTs exhibited improved behavior under drought stress conditions [67]. CuO nanoparticles were evaluated for their effects on physiological characteristics, seed germination, and seedling traits using *Trigonella foenum-graecum*. The highest mitotic index recorded was ~80% at a concentration of 10 mg/L, whereas it dropped to ~42.3% at 500 mg/L. The findings imply that higher concentrations of CuO NP have a negative impact on fenugreek's germination and cytophysiology [68].

9.8 Nanotechnology in Plant Stress Management

Plant growth, development, and productivity can be impeded by biotic and abiotic stressors through physiological, biochemical, morphological, and molecular alterations. Over the past 10 years, plant science has been increasingly using nanoparticles to reduce these pressures [69]. A study investigated the combined use of nanomaterials and biostimulants for the essential oil production enhancement from the *Mentha* plant during drought conditions. Under drought stress at 60% field capacity, the plants that received treatment with Kitoplus (chitosan based growth stimulant) at a concentration of 1% and iron oxide (FeO_2) NPs coated with chitosan at concentration of 10 μM exhibited the highest levels of menthone [70]. TiO_2 NPs, when utilized at appropriate concentrations, can effectively address damage resulting from water scarcity. Specifically, at 19.98 mg/L concentration, these nanoparticles have alleviated the negative effects of drought on *Verbascum sinuatum* species by enhancing the antioxidant capacity and inducing the formation of secondary metabolites [71]. Nitric oxide and selenium nanoparticle treatment of *Cichorium intybus* increased the expression of several genes, including PAL, and hydroxycinnamoyl-CoA quinate transferase (HCT1), in addition to the dehydration response element binding protein 1A (DREB1A) gene that is a key gene of stress tolerance [72]. At 1% concentration, chitosan nanoparticles also reduced drought stress in *Catharanthus roseus* by triggering an antioxidant defense cascade. By activating the genes for deacetylvindoline-4-O-acetyltransferase (DAT9), strictosidine synthase (STR), geissoschizine synthase (GS), and peroxidase 1(PRX1) the chitosan NPs in *Catharanthus roseus* enhanced the accumulation of alkaloids in drought-stressed plants [73].Nanoparticles have demonstrated efficacy in mitigating salinity stress in plants, alongside their known benefits in alleviating drought stress. In the case of *D. moldavica*, treatment with TiO_2 NPs (100 mg/L) significantly reduced the salt stress and simultaneously enhanced the levels of key essential oil components [74].

Satureja hortensis has been shown to increase its tolerance to salt when exposed to AgNPsat concentration 59.93 mg/L [75].Similarly, the ability of milk thistle seedling to withstand salt was enhanced by nanopriming them with chitosan nanoparticles [76]. The experiments performed on *Mentha piperita* in the presence of salinity stress, utilizing different concentrations of FeO_2 NPs, indicated that a concentration of 30 μM effectively alleviated the effects of salt stress [77]. Additionally, nanoparticles can help develop resistance to biotic stressors by promoting the upregulation of genes related to antioxidant defenses, their antibacterial properties, or their capacity to scavenge radicals. Findings were reported for the medicinal plant *Paeonia suffruticosa*, where the detrimental effects of salt stress were effectively reversed by nano-TiO_2 [78].

Using nanobiosensors, nanobiotechnology has also been used to detect plant diseases. Nanoparticles can be employed as pesticides to lessen the harm that viral, bacterial and fungal pathogens inflict, or they can be used as biomarkers for quick diagnosis of these pathogens [79]. A sandwich lateral flow immunoassay has been developed for the detection of potato virus Y, a virus that frequently affects the medicinal plant, *Chrysanthemum*. This assay utilizes AuNPs conjugated with antibodies, achieving the limit of detection upto 5.4 ng/mL, along with a polyvalent antigen [80]. Additionally, a colorimetric nano-biosensor leveraging the resonance of gold nanoparticles enables the detection of the tomato leaf curl virus, which infects *Acalypha indica*, without the need of other detection equipment [81]. Nanosensors have also been effectively used to identify pesticides and plant nutritional deficiencies. The surface of zinc sulfide (ZnS) quantum dots has been functionalized with cysteine hydrochloride to develop an optical nano-biosensor for the detection of herbicides [82].

9.9 Nanomaterials for Pathogen Control

The economy and plant health are severely harmed by pathogenic diseases globally. Numerous pathogens attack medicinal plant crops from the time of bloom until harvest, and then again after harvest. These pathogens significantly reduce quality, quantity of essential metabolites, oils and impact on the yield. Alternaria species, including *Alternaria brassicae, A. alternata, A. tenuissima*, and *A. tenuis*, are known to cause diseases in different medicinal plants such as leaf blight, bud rot and leaf blight. These pathogens affect hosts like *Withania somnifera, Aloe barbadensis, Chlorophytum borivilianum, Cassia angustifolia, Datura metel, Rauwolfia serpentina*, and *Ophiopogon japonicus*. Nanotechnology can improve disease diagnostics and enable the development of several novel strategies for disease suppression in the field and greenhouse. Currently, the effective nanomaterials used against these diseases include copper NPs, chitosan NPs and silver NPs, etc. [83–85]. In comparison to plants protected by fungicide, NPs of Ti and Zn reduced, the severity percentage of leaf spot disease caused by the fungus cercospora in sugar beet plant and increased the sucrose contents, particularly during severe disease stress condition. As a defense strategy against cercospora leaf spot (CLS) disease, NPs caused activation and recorded high enzyme activity values of peroxidase enzymes for up to 6 min and polyphenoloxidase enzymes for up to 4 min estimating periods than the control [86]. When compared to conventionally sized azomethines, the in vitro potential of nano-chemicals against pathogenic fungi, specifically *Sclerotium rolfsii, Rhizoctonia solani,* and *Rhizoctonia bataticola* was doubled [87]. A number of plant extracts and their phytochemical components are regarded as a good substitute for chemically synthesized antibacterial agents because of their wide range of active ingredients and minimal adverse effects. Different components were found in *Juniperus procera* fruit and leaf extracts by Gas chromatography-Mass spectrometry (GC-MS). *J. procera* leaf extracts at concentration of 100 mg/ml demonstrated ~35.83 and ~44.09% growth suppression against *Fusarium chlamydosporum,* and *Aspergillus fumigatus* respectively. However, this has been increased to ~50.55% and ~59.06% upon 50 ppm AgNPs treatment [88]. Researchers assessed the fungicidal properties of MgO NPs against soil *Phytophthora nicotianae* and *Thielaviopsis basicola* for the first time in both greenhouse and laboratory settings by performing comparison studies using macroscale MgO. Research conducted in vitro demonstrated that MgO NPs was more effective than its macroscale counterparts at inhibiting the growth of fungus, its spore germination and sporangium development. In comparison to untreated controls, 500 μg/ml treatment of MgO NPs effectively suppressed fungal invasion in the greenhouse plants, as evidenced by ~36.58 and ~42.35% decrease in tobacco black shank and black root rot disease, respectively [89]. *Xanthomonas perforans, X. gardneri, X. euvesicatoria,* and *X. vesicatoria* are the major pathogens of tomato bacterial spot disease. The over-reliance on traditional copper bactericides for decades caused the prevalence of copper-tolerant strains of Xanthomonas species, rendering the copper bactericides useless. The antibacterial activity of a copper-zinc (Cu/Zn) hybrid nanoparticle, on the copper-tolerant and sensitive strains, was examined in a study. High antibacterial activity was demonstrated by Cu/Zn hybrid nanoparticles against both copper-tolerant and sensitive strains at all the doses [90]. The antifungal activity of ZnO-NPs on *Colletotrichum* sp. was investigated in vitro. This was accomplished by monitoring the fungal growth area exposed to concentrations of 9, 12, and 15 mmol/L of ZnO-NPs. The behavior of the fungus was analyzed in comparison to the impact of the fungicide ciproconazole, which served as the positive control, as well as with a sample of the fungus that exhibited normal growth, acting as the negative control. After 6 days, the ZnO-NPs demonstrated a significant percentage of fungal growth suppression (~96%) at a concentration of 15 mmol/L, which resulted in the creation of groups of hyphal structures and the loss of certain hyphae continuity [91].

9.10 Phytonanotoxicology

Phytonanotechnology enables numerous innovative applications within the fields of agriculture and plant biotechnology. Since nanoparticles are becoming more and more common in our daily lives, environmental contact to them is inevitable. Anthropogenic releases of NPs into the environment could endanger life and health. There is growing interest in the interaction between NPs and biological processes. Consequently, NPs are adsorbed to plant surfaces, absorbed through plant apertures at the nano or micrometer scale, and moved throughout the plant body. As plants make up a significant portion of the biological system, research is being done on the effects of developing nanomaterials on plant development and how they affect morphological, physiological, and biochemical characteristics. Studies on metal, carbon NPs, nano-fertilizers, and their effects on plants, including the mechanisms of phytotoxicity and regulatory perspectives, are the major points [92]. A relatively new field of study, ecotoxicology, examines pollutants in the biosphere and how they affect its components, including humans. There is limited knowledge regarding the toxicity of nanomaterials to species that are significant to the environment. This scarcity of data hampers the quantitative risk assessment of NPs. Nevertheless, there is an increasing recognition that synthetic NPs must be evaluated for their potential environmental risks before being incorporated into products and subsequently released into the environment [93]. In the context of econanotoxicity, plants are most crucial because of their interactions with soil, water, and air. Furthermore, because animals, humans, and lower trophic level organisms devour plants, they offer a substantial chance to promote nanoparticle transfer among different species in the food web [94]. The introduction of soil fertilization based on NPs and nanotechnology into crop production appeared to be the brilliant move. It introduced minimal fertilizer use, preventing waste, improving crop nutrient absorption, boosting crop development, and increasing crop productivity while typically lowering production costs. Unfortunately, overuse of this state of the art technology has been shown to have negative impacts on phytoplankton and zooplankton, damage beneficial terrestrial and aquatic microbial populations, and pollute the environment [95].

It should be emphasized that all ecosystems contain naturally occurring NPs, including nano-sized metal oxides, to which they have evolved to adapt. Nevertheless, as the ecotoxicological understanding of synthetic nanoparticles continues to evolve, it is essential to evaluate their potential harmful effects on ecosystems. In freshwater environments, the presence of specific organic matters or natural products significantly influences the propensity of nanoparticles to aggregate in both seawater and hardwater. While the dispersion state can alter ecotoxicity, several abiotic factors, such as salinity, pH, and the presence of organic matter, must be meticulously studied as part of ecotoxicology [96].

9.11 Conclusion and Future Prospectus

In a variety of specialized fields of medical plant biotechnology, nanoparticles have demonstrated their value. Numerous applications, such as plant protection, aseptic culture establishment, genetic modification, calli differentiation, biotic stresses alleviation, secondary metabolite enhancement, bioactive component separation, somaclonal variant production, and nutrient delivery, have demonstrated their efficacy. Nevertheless, the use of NPs in medicinal plants has been limited to few, including CuO, ZnO, and Ag. Clearly, nanomaterials like graphite, carbon nanotubes, quantum dots, dendrimers, graphene, and atomic clusters hold significant potential in medicinal plant research, highlighting the need for further investigation in these domains. Research exploring the nanotechnology use to mitigate biotic stressors in medicinal plants is scarce, especially when compared to the extensive studies conducted on crop plants. This disparity presents a significant opportunity for researchers focused on medicinal plants to further investigate this field. Furthermore, the generation of reactive oxygen species (ROS), particular gene regulation and activation of antioxidant enzymes have all been connected to the well-established role of nanoparticles in enhancement of secondary metabolites; however, the underlying mechanisms are still unclear. Instead of examining the cause-and-effect relationship, in-depth research is preferred to clarify the fundamental process underlying the modification of plant's secondary metabolism. When examining how nanoparticles affect physiological parameters or secondary metabolites, researchers should also keep an eye on the biochemical and molecular alterations that occur. Also, in the fields of shoot multiplication and elongation, callus induction, rooting and organogenesis, the action mechanism of nanoparticles still not well studied. Future studies should therefore focus on the ways that nanoparticles alter the development, growth and defense mechanisms of plants by interacting with different signaling cascades. It is necessary to analyze how nanoparticles interact with cells of the plants and their constituent parts to develop a stronger theoretical knowledge that could support the safety of these particles in commercial products. Furthermore, there is immediate concern about the potential for nanoparticles to enter the food chain through herbal products. When evaluating the possible environmental impact of nanoparticles, phytotoxicity is a crucial consideration. On the surface of their roots or shoots, plants may absorb NPs, which may then be stored in various tissues or translocated. This characteristic has been the focus of most research observations and studies. At low concentrations, the harmless forms of NPs may not show any negative effects and seem to be good for both human and animal health. Most of the assessed NPs, however, induce stress and toxicity at higher concentrations, leading to the increased production of ROS that interfere with cellular metabolism.

However, there are still more promising areas of therapeutic plant biotechnology to explore. In the future period of medicinal plant nanobiotechnology, enabling technologies will be developed to safeguard medicinal plants from environmental challenges and effectively meet the needs of the herbal industry.

Acknowledgements The authors would like to thank the Director, CSIR-IHBT, for the infrastructural facilities. SKY, AA, and VP acknowledge the financial support from CSIR, India, (MLP00207) AT acknowledges DST-INSPIRE-JRF (IF240059) and HK acknowledges UGC for CSIR-UGC JRF (Award no. 23161012765).

Conflict of Interest The authors declare no competing financial interest.

Authors' Contributions Corresponding authors

Sudesh Kumar Yadav—Conceptualization, supervision, writing—original draft, review & editing, funding acquisition, investigation, project administration, review and editing.
Amitabha Acharya—Conceptualization, supervision, writing—original draft, review & editing, funding acquisition, investigation, project administration.

Authors

Anchal Thakur—Drafting the article, validation, visualization, investigation, writing—original draft, review and editing.
Harpreet Kaur—Drafting the article, validation, visualization, investigation, writing—original draft and editing.
Vikram Patial—Funding acquisition, investigation, project administration Review and editing.

References

1. Smith-Hall C, Larsen HO, Pouliot M (2012) People, plants and health: a conceptual framework for assessing changes in medicinal plant consumption. J Ethnobiol Ethnomed 8:1–11
2. Srivastava AK (2018) Significance of medicinal plants in human life. In: Synthesis of medicinal agents from plants. Elsevier, pp 1–24
3. Palit P, Mandal SC (2021) Climate change, geographical location, and other allied triggering factors modulate the standardization and characterization of traditional medicinal plants: a challenge and Prospect for Phyto-drug development. In: Evidence based validation of traditional medicines. Springer Singapore, pp 359–369
4. Halder M, Jha S (2023) The current status of population extinction and biodiversity crisis of medicinal plants. In: Medicinal Plants: Biodiversity, Biotechnology and Conservation. Springer, pp 3–38
5. Pandita D (2022) Role of Nano-biotechnology in medicinal plant production. In: Environmental challenges and medicinal plants. Springer, pp 355–384
6. Das K (2021) The healthy and sustainable growth of medicinal and aromatic plants through nanotechnology: a review. J Phytonanotech Pharmaceut Sci 1:4–9
7. Kreyling WG, Semmler-Behnke M, Chaudhry Q (2010) A complementary definition of nanomaterial. Nano Today 5:165–168
8. Ramesh KT (2009) Nanomaterials. In: Nanomaterials. Springer US, pp 1–20
9. Roaa MH (2020) A review article: the importance of the major groups of plants secondary metabolism phenols, alkaloids, and terpenes. Int J Res Appl Sci Biotechnol 7:354–358
10. Siddiqui MZ (2011) *Boswellia serrata*, a potential antiinflammatory agent: an overview. Indian J Pharm Sci 73:255
11. Rastogi S, Pandey MM, Rawat AKS (2015) Medicinal plants of the genus *Betula*—traditional uses and a phytochemical–pharmacological review. J Ethnopharmacol 159:62–83
12. Bribi N (2018) Pharmacological activity of alkaloids: a review. Asian J Bot 1:1–6
13. Garg SK, Shukla A, Choudhury S (2019) Polyphenols and flavonoids. In: Nutraceuticals in veterinary medicine. Springer, pp 187–204

14. Hossain MT et al (2021) A review study on the pharmacological effects and mechanism of action of tannins. Eur J Pharm Med Res 8:5–10
15. Mouffouk C et al (2021) Flavonols as potential antiviral drugs targeting SARS-CoV-2 proteases (3CLpro and PLpro), spike protein, RNA-dependent RNA polymerase (RdRp) and angiotensin-converting enzyme II receptor (ACE2). Eur J Pharmacol 891:173759
16. Ghasemzadeh Rahbardar M, Hosseinzadeh H (2020) Effects of rosmarinic acid on nervous system disorders: an updated review. Naunyn Schmiedeberg's Arch Pharmacol 393:1779–1795
17. Kumari A et al (2020) Green surfactant based synthesis of curcumin loaded poly lactic-co-glycolic acid nanoparticles with enhanced solubility, photo-stability and anti-biofilm activity. J Drug Deliv Sci Technol 59:101884
18. Hosseini H et al (2020) Resveratrol alleviates non-alcoholic fatty liver disease through epigenetic modification of the Nrf2 signaling pathway. Int J Biochem Cell Biol 119:105667
19. Yan Y et al (2020) Research progress on Alzheimer's disease and resveratrol. Neurochem Res 45:989–1006
20. Song X et al (2018) Estrogen receptors are involved in the neuroprotective effect of silibinin in aβ 1–42-treated rats. Neurochem Res 43:796–805
21. Haddad Y et al (2011) Antioxidant and Hepatoprotective effects of Silibinin in a rat model of nonalcoholic steatohepatitis. Evid Based Complement Alternat Med 2011:2011:647903
22. Ankit, Malviya R, Sharma A (2022) Sources, properties, and pharmacological effects of quercetin. Curr Nutr Food Sci 18:457–465
23. Anjum S et al (2022) Green synthesis of biocompatible core–shell (au–ag) and hybrid (au–ZnO and ag–ZnO) bimetallic nanoparticles and evaluation of their potential antibacterial, antidiabetic, antiglycation and anticancer activities. RSC Adv 12:23845–23859
24. Sharifi-Rad J et al (2021) Genistein: an integrative overview of its mode of action, pharmacological properties, and health benefits. Oxid Med Cell Longev 2021:3268136
25. Goh YX et al (2022) Genistein: a review on its anti-inflammatory properties. Front Pharmacol 13:820969
26. Ferraz AC et al (2021) Hepatoprotective, antioxidant, anti-inflammatory, and antiviral activities of silymarin against mayaro virus infection. Antivir Res 194:105168
27. Wiraswati HL et al (2025) Piperine: an emerging biofactor with anticancer efficacy and therapeutic potential. Biofactors 51:e2134
28. Kohandel Z et al (2021) Anti-inflammatory effects of thymoquinone and its protective effects against several diseases. Biomed Pharmacother 138:111492
29. Sharifi-Rad J et al (2021) Paclitaxel: application in modern oncology and nanomedicine-based cancer therapy. Oxid Med Cell Longev 2021:3687700
30. Singh S et al (2023) Compendium of naringenin: potential sources, analytical aspects, chemistry, nutraceutical potentials and pharmacological profile. Crit Rev Food Sci Nutr 63:8868–8899
31. Enggaard TP et al (2001) The analgesic effect of codeine as compared to imipramine in different human experimental pain models. Pain 92:277–282
32. Clark A (1926) The antagonism of acetyl choline by atropine. J Physio 61:547
33. Bocso N-S, Butnariu M (2022) The biological role of primary and secondary plants metabolites. J Nutr Food Process 5:1–7
34. Twaij BM, Hasan MN (2022) Bioactive secondary metabolites from plant sources: types, synthesis, and their therapeutic uses. Int J Plant Biol 13:4–14
35. Bhatia S et al (2015) Modern applications of plant biotechnology in pharmaceutical sciences. Academic press
36. Fazal H et al (2016) Elicitation of medicinally important antioxidant secondary metabolites with silver and gold nanoparticles in callus cultures of *Prunella vulgaris L.* Appl Biochem Biotechnol 180:1076–1092
37. Singh R et al (2019) Nanoparticles alter the withanolide biosynthesis and carbohydrate metabolism in *Withania somnifera* (Dunal). Ind Crop Prod 127:94–109
38. Kahila MMH et al (2018) Effect of nanoparticle treatment on expression of a key gene involved in thymoquinone biosynthetic pathway in *Nigella sativa* L. Nat Prod Res 32:1858–1862

39. Chung I-M et al (2019) Impact of copper oxide nanoparticles on enhancement of bioactive compounds using cell suspension cultures of *Gymnema sylvestre* (Retz.) R. Br. Appl Sci 9:2165
40. Lala S (2020) Enhancement of secondary metabolites in *Bacopa monnieri* (L.) Pennell plants treated with copper-based nanoparticles *in vivo*. IET Nanobiotechnol 14:78–85
41. Nourozi E et al (2019) Pharmaceutical important phenolic compounds overproduction and gene expression analysis in *Dracocephalum kotschyi* hairy roots elicited by SiO_2 nanoparticles. Ind Crops Prod 133:435–446
42. Moazzami Farida SH, Karamian R, Albrectsen BR (2020) Silver nanoparticle pollutants activate oxidative stress responses and rosmarinic acid accumulation in sage. Physiol Plant 170:415–432
43. Ghasempour M et al (2019) Multi-walled carbon nanotubes improved growth, anatomy, physiology, secondary metabolism, and callus performance in *Catharanthus roseus*: an in vitro study. 3 Biotech 9:404
44. Hedayati A et al (2020) Improved tropane alkaloid production and changes in gene expression in hairy root cultures of two *Hyoscyamus* species elicited by silicon dioxide nanoparticles. Plant Physiol Biochem 155:416–428
45. Khajavi M, Rahaie M, Ebrahimi A (2019) The effect of TiO2 and SiO2 nanoparticles and salinity stress on expression of genes involved in parthenolide biosynthesis in feverfew (*Tanacetum parthenium L.*). Caryologia 72:3–14
46. Ramezani M et al (2020) Effect of silver nanoparticle treatment on the expression of key genes involved in glycosides biosynthetic pathway in *Stevia rebaudiana B.* plant. Sugar Tech 22:518–527
47. Bernela M et al (2021) Nanofertilizers: applications and future prospects. In: Nanotechnology. Jenny Stanford Publishing, pp 289–332
48. Elfeky SA et al (2013) Effect of magnetite nano-fertilizer on growth and yield of *Ocimum basilicum L.* Int J Indig Med Plants 46:1286–11293
49. Elsayed AA et al (2022) Hydroxyapatite nanoparticles as novel nano-fertilizer for production of rosemary plants. Sci Hortic 295:110851
50. Hamed ES (2018) Effect of nitrogenous fertilization and spraying with nano-fertilizer on *Origanum syriacum L.* var. syriacum plants under North Sinai conditions. J Pharmacogn Phytochem 7:2902–2907
51. Yousefzadeh S, Sabaghnia N (2016) Nano-iron fertilizer effects on some plant traits of dragonhead (*Dracocephalum moldavica L.*) under different sowing densities. Acta Agri Slov 107:429–437
52. Shahsavari M, Naderi D, Gheisari MM (2019) Effects of organic nano-fertilizer and humic acid on biochemical characteristics and some essential oil of domask-rose (*Rosa damascena* mill.). Iran J Med Aromat Plants 35
53. Sidhu Y (2011) In vitro micropropagation of medicinal plants by tissue culture. Plymouth Stud Sci 4:432–449
54. Al-Qudah et al (2022) Nanotechnology applications in plant tissue culture and MolecularGenetics: a holistic approach. Curr Nanosci 18:442–464
55. Javed R et al (2018) Elicitation of secondary metabolites in callus cultures of *Stevia rebaudiana Bertoni* grown under ZnO and CuO nanoparticles stress. Sugar Tech 20:194–201
56. Ibrahim AS, Fahmy AH, Ahmed SS (2019) Copper nanoparticles elevate regeneration capacity of (*Ocimum basilicum L.*) plant via somatic embryogenesis. Plant Cell Tissue Organ Cult 136:41–50
57. Singh CR (2018) Review on problems and its remedy in plant tissue culture. Asian J Biol Sci 11:165–172
58. Kim DH, Gopal J, Sivanesan I (2017) Nanomaterials in plant tissue culture: the disclosed and undisclosed. RSC Adv 7:36492–36505
59. Nartop P (2018) Green sterilization of *Rosmarinus officinalis L.* stem surfaces with silver nanoparticles synthesized using *Rubia tinctorum L.* cell culture extracts. Iran J Sci Technol Trans A Sci 42:411–414

60. Keshavareddy G, Kumar ARV, Ramu VS (2018) Methods of plant transformation-a review. Int J Curr Microbiol Appl Sci 7:2656–2668
61. Demirer GS et al (2019) High aspect ratio nanomaterials enable delivery of functional genetic material without DNA integration in mature plants. Nat Nanotechnol 14:456–464
62. Yan Y et al (2022) Nanotechnology strategies for plant genetic engineering. Adv Mater 34:2106945
63. Demirer GS et al (2019) Carbon nanotube–mediated DNA delivery without transgene integration in intact plants. Nat Protoc 14:2954–2971
64. Rhaman MS et al (2022) Seed priming with nanoparticles: an emerging technique for improving plant growth, development, and abiotic stress tolerance. J Soil Sci Plant Nutr 22:4047–4062
65. Hatami M, Ghorbanpour M, Salehiarjomand H (2014) Nano-anatase TiO2 modulates the germination behavior and seedling vigority of some commercially important medicinal and aromatic plants. J Biol Environ Sci 8:53–59
66. Hatami M et al (2021) Silicon nanoparticle-mediated seed priming and *Pseudomonas spp.* inoculation augment growth, physiology and antioxidant metabolic status in *Melissa officinalis L.* plants. Ind Crop Prod 162:113238
67. Yousefi S, Kartoolinejad D, Naghdi R (2017) Effects of priming with multi-walled carbon nanotubes on seed physiological characteristics of Hopbush (*Dodonaeaviscosa* L.) under drought stress. Int J Environ Stud 74:528–539
68. Kavitha K et al (2022) Impact of biosynthesized CuO nanoparticles on seed germination and cyto-physiological responses of *Trigonella foenum-graecum* and *Vigna radiata*. Mater Lett 313:131756
69. Almutairi ZM (2019) Plant molecular defense mechanisms promoted by nanoparticles against environmental stresses. Int J Agric Biol 21:259–270
70. Giglou MT et al (2022) A new method in mitigation of drought stress by chitosan-coated iron oxide nanoparticles and growth stimulant in peppermint. Ind Crops Prod 187:115286
71. Karamian R, Ghasemlou F, Amiri H (2020) Physiological evaluation of drought stress tolerance and recovery in *Verbascum sinuatum* plants treated with methyl jasmonate, salicylic acid and titanium dioxide nanoparticles. Plant Biosyst 154:277–287
72. Abedi S et al (2021) Nitric oxide and selenium nanoparticles confer changes in growth, metabolism, antioxidant machinery, gene expression, and flowering in chicory (*Cichorium intybus L.*): potential benefits and risk assessment. Environ Sci Pollut Res 28:3136–3148
73. Ali EF et al (2021) A vital role of chitosan nanoparticles in improvisation the drought stress tolerance in *Catharanthus roseus* (L.) through biochemical and gene expression modulation. Plant Physiol Biochem 161:166–175
74. Gohari G et al (2020) Titanium dioxide nanoparticles (TiO2 NPs) promote growth and ameliorate salinity stress effects on essential oil profile and biochemical attributes of *Dracocephalum moldavica*. Sci Rep 10:912
75. Nejatzadeh F (2021) Effect of silver nanoparticles on salt tolerance of *Satureja hortensis* l. during in vitro and in vivo germination tests. Heliyon 7:e05981
76. Mosavikia AA et al (2020) Chitosan nanoparticle and pyridoxine seed priming improves tolerance to salinity in milk thistle seedling. Not Bot Horti Agrobo 48:221–233
77. Askary M et al (2017) Effects of iron nanoparticles on *Mentha piperita L.* under salinity stress. Biologija 63:65–75
78. Liu LI, Cao Y, Guo Q, Zhu Z (2021) Nanosized titanium dioxide seed priming enhances salt tolerance of an ornamental and medicinal plant *Paeonia Suffruticosa*. Pak J Bot 53:1167–1175
79. Khan MR, Rizvi TF, Ahamad F (2019) Application of nanomaterials in plant disease diagnosis and management. In: Nanobiotechnology applications in plant protection. Springer, pp 19–33
80. Razo SC et al (2018) How to improve sensitivity of sandwich lateral flow immunoassay for corpuscular antigens on the example of potato virus Y? Sensors (Basel) 18:3975
81. Razmi A et al (2019) Localized surface plasmon resonance biosensing of tomato yellow leaf curl virus. J Virol Methods 267:1–7

82. Kaushal M, Wani SP (2017) Nanosensors: Frontiers in precision agriculture. In: Nanotechnology. Springer Singapore, pp 279–291
83. Abdel-Aliem HA et al (2019) Antifungal efficacy of chitosan nanoparticles against phytopathogenic fungi and inhibition of zearalenone production by *Fusarium graminearum*. Comunicata Scientiae 10:338–345
84. Abdelmalek GAM, Salaheldin TA (2016) Silver nanoparticles as a potent fungicide for citrus phytopathogenic fungi. J Nanomed Res 3:00065
85. Wang M, Hou X, Wang H (2021) Genomic sequence resource of *Alternaria alternata* strain B3 causing leaf blight on *Ophiopogon japonicus*. Plant Dis 105:684–687
86. Farahat G (2018) Biosynthesis of nano zinc and using of some nanoparticles in reducing of Cercospora leaf spot disease of sugar beet in the field. Env Biodivers Soil Secur 2:103–117
87. Mondal P, Kumar R, Gogoi R (2017) Azomethine based nano-chemicals: development, in vitro and in vivo fungicidal evaluation against *Sclerotium rolfsii*, *Rhizoctonia bataticola* and *Rhizoctonia solani*. Bioorg Chem 70:153–162
88. Bakri MM et al (2020) Efficacy of *Juniperus procera* constituents with silver nanoparticles against *Aspergillus fumigatus* and *Fusarium chlamydosporum*. BioNanoSci 10:62–72
89. Chen J et al (2020) Comparative study on the fungicidal activity of metallic MgO nanoparticles and macroscale MgO against soilborne fungal phytopathogens. Front Microbiol 11:365
90. Carvalho R et al (2019) Bactericidal activity of copper-zinc hybrid nanoparticles on copper-tolerant *Xanthomonas perforans*. Sci Rep 9:20124
91. Mosquera-Sánchez LP et al (2020) Antifungal effect of zinc oxide nanoparticles (ZnO-NPs) on *Colletotrichum sp.*, causal agent of anthracnose in coffee crops. Biocatal Agric Biotechnol 25:101579
92. Shende P, Takke A (2021) Phytotoxic impact of nanomaterials for Nanosafety. In: Nanotoxicology and nanoecotoxicology, vol 2. Springer, pp 119–132
93. Oberdörster G, Oberdörster E, Oberdörster J (2005) Nanotoxicology: an emerging discipline evolving from studies of ultrafine particles. Environ Health Perspect 113:823–839
94. Jogaiah S et al (2021) Phytotoxicological effects of engineered nanoparticles: an emerging nanotoxicology. Sci Total Environ 801:149809
95. Maurer-Jones MA et al (2013) Toxicity of engineered nanoparticles in the environment. Anal Chem 85:3036–3049
96. Handy RD et al (2008) The ecotoxicology and chemistry of manufactured nanoparticles. Ecotoxicology 17:287–314

Chapter 10
Smart Hydroponics Monitoring System Using Internet of Things

Novin Chamling, Sitanshu Das, Sultan Mahamud, Premika Rai, Puja Acharya, and Sandeep Rawat

Abstract Hydroponic cultivation, a soilless plant-growing technique, has been emerged as a sustainable solution for the challenges of food insecurity, water scarcity, and urbanization. With precise control over nutrient delivery and environmental conditions, hydroponics improves crop yield, resource-use efficiency, and food quality. This chapter explores the integration of Internet of Things (IoT) technology with hydroponics systems to create smart agriculture solutions that enable real-time monitoring, automation, and data-driven decision-making. It discusses key parameters such as pH, electrical conductivity (EC), total dissolved solids (TDS), temperature, humidity, CO2 concentration, and light intensity that influence plant growth and are monitored using IoT-enabled sensors. The design and role of microcontrollers, cloud-based platforms, and automation in hydroponic monitoring are elaborated, along with case studies of server-based systems and mobile applications like ThingSpeak and Blynk. Sensor technologies, their types, advantages, and limitations are systematically reviewed. Additionally, this chapter addresses modern advancement in hydroponics, including smart greenhouse integration, open field soilless systems (OFSS), energy-efficient practices, and innovations in nutrient management. This interdisciplinary overview positions smart hydroponics as a promising model for future urban food production and climate-resilient agriculture, especially in resource-constrained regions.

Keywords Artificial intelligence · Agriculture automation · Smart agriculture · Soilless farming · Sensor technology

N. Chamling · S. Das · S. Mahamud · P. Rai · P. Acharya · S. Rawat (✉)
Sikkim Regional Centre, G.B. Pant National Institute of Himalayan Environment, Gangtok, Sikkim, India

P. Kumar, A. R. Warghat (eds.), *Transforming Medicinal Plant Agriculture*, Smart Agriculture 16, https://doi.org/10.1007/978-981-95-3963-5_10

10.1 Introduction

Hydroponic is a soilless cultivation system supplying nutrient solutions, often in combination with inert substrates such as sawdust, peat moss, rockwool, gravel, sand, pumice, perlite, vermiculite, or coco coir [1–4]. It includes either system, where plant roots are anchored in a growing medium and nourished by nutrient solution, or only nutrient solution, where roots are suspended directly in the liquid culture [4–6]. Hydroponic systems varied upon nutrients delivery system to plant roots. Some popular methods include deep water culture (DWC), nutrient film technology (NFT), wick systems, drip systems, ebb-flow setups, and aeroponics [1]. These systems support a diverse range of crops, including leafy vegetables like lettuce, kale, and basil to fruiting plants such as tomatoes, cucumbers, and strawberries [7]. Compared to traditional agriculture, hydroponics offers numerous advantages such as high crop yields, improved produce quality, optimized water and nutrient use, reduced pests and diseases incidence, and enhanced environmental control [8]. Its recent rapid growth is driven by urgent global challenges, including climate change, freshwater scarcity, decreasing arable land, and limited access to fresh produce in urban areas. This liquid culture system remains the most favored in academic and research due to its vast visibility and control over plant root systems [9]. The three principles types of hydroponic solution culture are static culture, where the nutrient solution remains stationary; flowing culture, where the nutrient solution continuously flows through the root zone; and aeroponics, where roots are suspended in air and misted periodically with nutrient solution (Fig. 10.1).

The origins of hydroponics trace back thousands of years, yet its historical context is often oversimplified or inaccurately presented in literature. The Hanging Gardens of Babylon, constructed by King Nebuchadnezzar II (604–562 BC) is the initial evidence without having much information technical details [10]. In the seventeenth century, English physician John Woodward conducted pioneering experiments cultivating plants in water, aiming to validate Van Helmont's theory that all plant matter is made of water [11]. Technical developments in hydroponic technology emerged in the nineteenth century when German scientists Julius von Sachs and Wilhelm Knop formulated nutrient solutions, laying the foundation for modern solution culture. In 1929, plant physiologists William Frederick Gericke of the University of California popularized the concept of suing solution culture not merely for research but for large-scale crop production. A landmark early application occurred in the 1930s on Wake Island, a remote refueling stop in the Pacific Ocean where fresh vegetables were grown hydroponically to serve Pan American Airways personnel and passengers [12]. This milestone demonstrated hydroponics potential for addressing food supply challenges in isolated or resource scarce environments.

The global food production system is increasingly strained by the compounded effects of climate change, depleting freshwaters resources, rapid urbanization, and continuous population growth. These stressors manifest as erratic rainfall patterns, shifting growing seasons, and reduced system reliability. These factors impacted

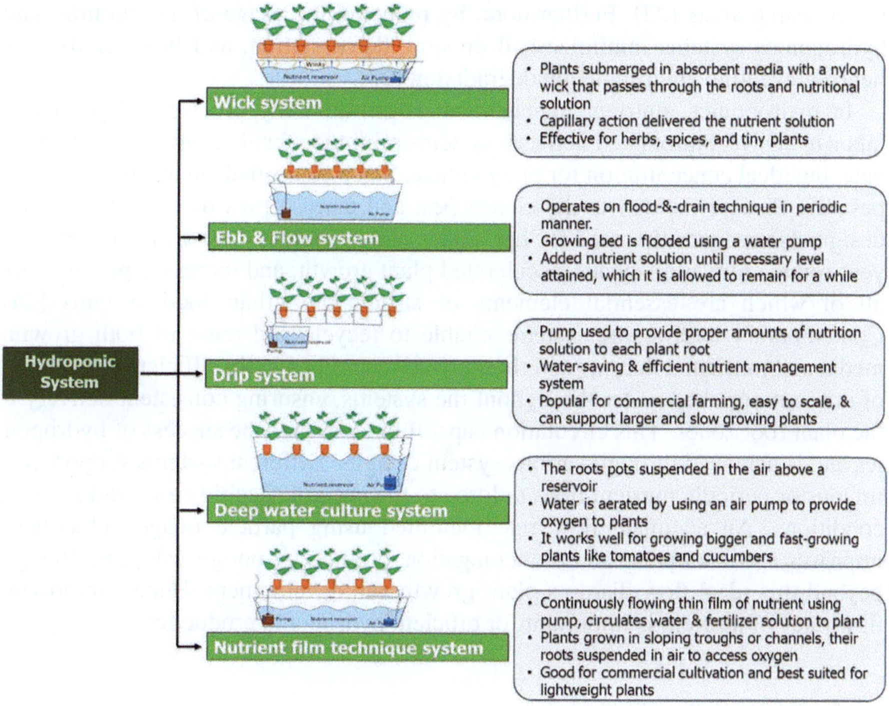

Fig. 10.1 A schematic overview of Hydroponic systems including Wick, Ebb and Flow, Drip, Deep Water Culture (DWC), and Nutrient Film Technique (NFT), highlighting their operating principles, component requirements, and crop suitability

productivity of traditionally sustained agriculture, necessitating innovative and adaptive approaches [13]. Acute shortage of water due to decreased rainfall, and excessive use of groundwater restricted the water available for irrigation and suitable habitats. Additionally, rising ocean temperatures and decrease in freshwater may disrupt phytoplankton populations, which are essential for carbon fixation [14, 15]. Concurrently, demographic change due to rapid urbanization in developing countries altered the food supply chain. Additionally, health consciousness and economic growth has altered the dietary habitat that affected food production. Thus, an evaluation of food production methods is required involving emerging technologies like precise hydroponic system [16, 17].

Hydroponics offers a viable option for urban food security under controlled environmental conditions as perfect system for rooftops, vertical farms, and urban greenhouses. It offers effective system for water utilization and precise nutrient delivery [15, 18, 19]. Particularly suited for high-value, short-cycle crops like tomatoes and leafy greens, soilless farming (SLF) systems have shown exceptional resistance to soil borne pets and diseases, decreased postharvest losses, and decreased nutrient leaching [20, 21]. By using targeted delivery and recirculation, soilless farming maximizes water application and allows for year-round food production in water scarce areas [22]. Furthermore, by reducing the usage of agricultural land, hydroponics systems minimize soil erosion, deforestation, and biodiversity loss, hence preventing environmental degradation [23].

In hydroponics, nutrient management is substantially precise and process is labor-intensive. Automated delivery systems minimize fertilizer waste, while guaranteeing ideal concentration for plant uptake. Since controlled environments reduce pest and disease incidence, the closed pest and disease pressure, the closed loop design also reduces the need for herbicides and pesticides. These systems enabled year-round cultivation cycles, accelerated plant growth, and increased productivity, all of which are essential elements of sustainable urban food systems [24]. Contemporary hydroponic systems enable to recycle and reuse of both growing media and nutrient solutions and their effectiveness lies in the efficient distribution of nutrient-enriched water throughout the systems, ensuring consistent delivery to the plant root zones. This circulation capability underpins the success of hydroponics and is achieved through various system designs. Different systems support continuous or periodic nutrient flow, tailored to the need of specific crops and growing conditions. An adequate flow rate identified using particle image velocimetry improves nutrient uptake and root elongation, promoting root growth [25]. Though, beyond this ideal flow disrupts plant growth and development. Hence, improving flow rates of nutrient is important for efficient hydroponic production.

10.2 Automation and IoT Integration in Hydroponics

Automation plays a pivotal role in enhancing the efficiency, precision, and sustainability of modern agriculture, particularly within resource-constrained urban environments. Hydroponic systems can optimize the use of water, nutrients, and energy while maximizing crop output and minimizing labor input by incorporating automated technologies. This is particularly advantageous in areas with a high labor cost and land resources. Automation increases production and opens up new agri-tech business in precise consulting, equipment maintenance, and system design. These issues highlight the significance of creating inclusive digital literacy and reskilling programs. Adaption to automation is still hampered by cost, particularly for urban and smallholder farmers. In order to encourage wider adoption, financial support mechanisms such as cooperatives, subsidies, and regulatory incentives are crucial. Food systems can become more robust and inclusive when automation is used fairly [26, 27]. Because automation allows precise control over environmental and nutrient conditions, hydroponics greatly improves crop quality and system reliability. Climate regulation systems, automated nutrient dosing, and the robotic planting and harvesting are a few examples of the kinds of technologies that will reduce the range of variability and the need for human resources. Through the use of artificial intelligence (AI) and machine learning, also predictive maintenance, yield forecasting, and early system failure detection become some more [28, 29].

The Internet of Things (IoT) is the integration of intertwined sensors, devices, decision support systems and platforms that can exchange data without direct human interaction. In an automatic hydroponics system, IoT can facilitate control of the entire farm utilizing complete data from the virtual environment, thereby promoting data-driven farming. Smart farming and IoT have stretched the boundaries of hydroponics as a new and highly productive method of growing food to address the needs of a rapidly growing population, ever-changing weather, and soil deterioration [30]. It is quite possible that autonomous field machinery, nutrient sensors, drone-based surveillance, precision nutrient management, greenhouse climate control as well as real-time data recording in specific crop management in agriculture, and hydroponics can be the potential applications of IoT. Wireless sensors and cloud-based systems enable real-time monitoring and field management, the generation of yield maps, and the utilization of data for the decision-making process with the aim of improving crop quality and reducing costs [31].

A hardware becomes most important component of future farming [32]. Hydroponic farming, which is the method of no soil but nutrient-rich water, is the one that the Internet of Things' precision and IoT's automation are meant for [33]. As there is a growing concern about water scarcity, food security, and soil degradation, a global level the integrated hydronic systems of IoT become a real and more environmental alternative of the standard agricultural approach to conventional agriculture [33]. Linking sensors, controllers, and cloud platforms, IoT systems create a connected infrastructure that can be used to monitor and control the quality and health of the hydroponics farm such as pH, electrical conductivity (EC), total dissolved solids (TDS), dissolved oxygen, nutrient concentration, temperature, light and humidity. The data collected through the sensors and other software tools has a great advantage in the decision-making process, as it enables quick, easy and real-time control and adjustment of the plant development conditions that suit the most [34]. These systems will be expanded on agriculture through the means of AI and data analytics, which will provide benefits like early identifies the presence of diseases, growth predicts and automatically alert for actions [35, 36]. The various developing Smartphone application software and online platforms are the new remote control center for basic operation modules of a hydroponics system such as nutrient circulation, lighting, and ventilation system. Real-time, user-friendly management, and operation of hydroponic systems make it easier to control and process the amounts of nutrients, light, and fresh air [36]. The speed and convenience of these monitoring and controlling tools during the tests lead to a more effective farming platform. Hydroponic is a grow system effortless and yet complicated in nature. This system can ensure a consistent and effective supply of nutrients to the plants if in a shorter pH range of 5.5–6.5, and with the handheld pH sensor</pH meter> < pH = “x-p” > being used as a spectrometer for the detection of the nutrient at the required intensity [35]. Besides, Electrical Conductivity (EC) used to express the concentration of dissolved ions, suggesting presence of mineral salts.

ESP32-based systems adjust nutrient inputs, when TDS deviates from optimal ranges [37]. Ideal temperature and humidity within range of 50–70% are monitored using automated heaters, fans and coolers [38]. Light Intensity and photoperiod of systems are monitored through IoT controlled LED lights programmed to ensure plant growth, by optimizing photosynthetic activity throughout crop development stages. Carbon Dioxide (CO_2) is monitored and adjusted to maintain 800–1200 ppm during the light period, enhancing photosynthetic efficiency [39]. The water level and flow rate of systems are ensured nutrient delivery uninterrupted and prevents system failures in DWC and NFT setups. Advanced monitoring parameters include dissolved oxygen and flow rate sensors which manages aeration and circulation via pumps and valves for root zone health. Plant health monitoring by IoT systems integrate imaging, biosensors, and algorithms to detect diseases and stress indicators early [40]. Atmospheric pressure and air quality contribute to refined environmental control for sensitive crops.

By continuously analyzing environmental and nutrient parameters, IoT-integrated hydroponics systems enable automated responses through actuators that optimize water, energy, and input use. These systems not only reduce labor requirements but also improve productivity, resource efficiency, and crop quality. As machine learning algorithms become more integrated, smart hydroponics is evolving into a data driven, predictive cultivation systems that is scalable, climate resilient and sustainable.

Hydroponics relies heavily on precise environmental and nutrient regulation to support optimal plant growth. Key factors such as pH (potential of hydrogen), total dissolved solids (TDS), electrical conductivity (EC), temperature and light intensity require continuous monitoring and control. The integration of sensor technology with IoT platforms has made it possible to automate and manage these variables in real time, thereby improving system sustainability and crop productivity [41]. Maintaining the appropriate pH range of the nutrient solution is crucial, as it directly affects nutrient solubility and plant uptake efficiency. Most hydroponic crops thrive within a pH range of 5.5–6.5, although specific optimal values may vary by species. Deviation outside this range can hinder nutrient absorption, leading to deficiencies and suboptimal plant performance. Several type of pH sensors are used in hydroponic systems such as glass electrode sensors operate on the potential difference between a references electrode and a pH sensitive glass membrane. These glass sensors are widely adopted due to their accuracy (pH $\pm$ 0.10) and are suitable for both commercial and small-scale applications [42, 43]. Ion-Sensitive Filed-Effect Transistor (ISFET) sensors offer faster response times and greater durability, making them ideal for integration into automated and remote monitoring setups. However, consistent calibration, storage, and maintenance are essential to ensure accurate performance [44]. Optical pH sensors utilize the optical properties (e.g., absorbance, fluorescence) of a pH-sensitive dye and require no direct electrical contact with the nutrient solution. Their non-invasive nature makes them substantially suitable for real time monitoring in dynamic hydroponic environments [45]. Total Dissolved Solids (TDS) sensors estimate the concentration of dissolved salts in the nutrient solution, expressed in parts per million (ppm). These sensors are vital for assessing nutrient solution strength and supporting crop-specific fertilization strategies. For example, lettuce typically requires TDS levels of 560–840 ppm, whereas tomato plants perform best within the range of 1400–3500 ppm [46]. When interfaced with microcontrollers like Arduino UNO, sensors such as SEN0244 can provide real time TDS data with an error margin as low as 2.8% [47, 48]. Advanced calibration techniques, including Sectioned-Polynomial Regression (SecPR) have been developed to improve TDS sensor accuracy up to 91%.

Electrical Conductivity (EC) sensors measure the ionic concertation of the nutrient solutions, offering direct insight into nutrient availability. Maintaining EC within appropriate limits is essential: high EC values may induce osmotic stress and nutrient toxicity, while low values may signal deficiencies [49]. These sensors are often embedded into IoT-enabled hydroponic systems to automate nutrient dosing and maintain optimal plant health [50]. Beyond nutrient parameters, monitoring plant growth and environmental conditions is essential for predictive management and yield optimization. Important indicators include morpho-physiological characters [51]. These parameters provide real-time feedback for adjusting growing conditions and optimizing system performance. Temperature plays a key significant role in nutrient solubility and metabolic activity. Maintaining water temperatures between 20 and 24 °C and ambient air temperatures between 18 and 27 °C helps prevent nutrient precipitation and supports enzymatic function [52]. Relative humidity must be controlled within 40%–80% to promote transpiration and reduce disease risk [53]. Light quality and intensity are equally critical. Advanced lighting systems now allow for precision control of photosynthetically active radiation (PAR), spectral quality, photoperiod, and intensity. For instance, lettuce has been shown to thrive under photosynthetic photon flux densities (PPFD) of 150–250 μmol m^{-2} s^{-1} in controlled artificial light environments. Light conditions not only influence biomass but also impact the nutritional and sensory qualities of crops [54–56].

10.3 Server Design

Now-a days, various servers are available to integrate for processing of inputs of parameters and making precise decision in automation of Hydroponic system (Table 10.1). The most commonly-used microcontroller server ESP32 is used in smart hydroponics for its built-in Wi-Fi capabilities, low power consumption, and multi-sensor compatibility (Fig. 10.2). It enables simultaneous real-time monitoring of pH, TDS, temperature, humidity, water level, and light intensity. This data is transmitted to cloud platforms for automated control of pumps, fans, lights, and alerts [37]. Blynk is a cloud-based IoT interface designed for ease of use in smart agriculture systems. The key features consist of such real-time dashboards with customizable widgets for sensor monitoring, remote control of actuators such as irrigation and lighting systems, alerts and notifications for parameters deviations, data logging and visualization for performance analysis and optimization, integration with platforms like Blynk provides a seamless ecosystem where growers regardless of technical expertise can operate hydroponic systems efficiently using smartphones or web applications [57].

Table 10.1 Comparison of IoT Platforms for Smart Hydroponics Applications

Platform	Real-Time Monitoring	Dashboard	Alerts & Notifications	Data Storage	Ease of Use	Ideal For	Free Tier	Reference
ThingSpeak	Yes	Basic	Limited via MATLAB	Up to 3 M data entries	Easy	Education, DIY	Yes	[58]
Blynk	Yes	Customizable	Push/email alerts	Limited in free tier	Very easy	Hobbyists, prototyping	Yes	[59]
Adafruit IO	Yes	Simple widgets	Triggers available	30 days history	Easy	Makers, small projects	Yes	[60]
Arduino IoT cloud	Yes	Native dashboards	Email alerts	Cloud-based	Moderate	Arduino users	Yes	[61]
Ubidots	Yes	Pro dashboards	Advanced alerts	Cloud database	Moderate	Startups, research	Limited	[62]
AWS IoT Core	Yes	(use with AWS Grafana or custom)	Advanced via lambda	Scalable cloud DB	Complex	Enterprise	Free tier	[63]
Azure IoT hub	Yes	(use with power BI or azure monitor)	Smart alerts	Scalable storage	Complex	Enterprise, research	Yes	[64]
Node-RED	Yes	Custom UI possible	With plugins	Local/cloud	Moderate	Custom, open-source setups	Open-source	[65]

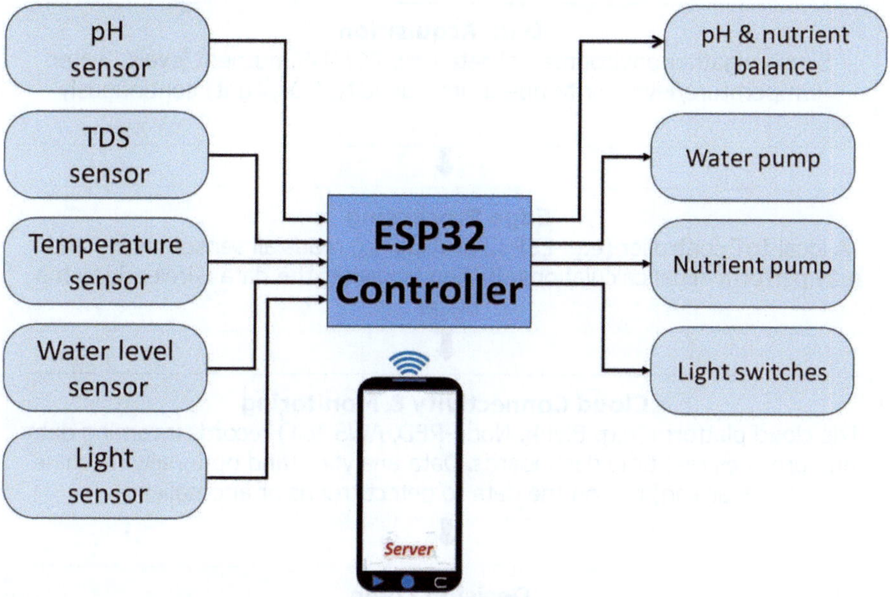

Fig. 10.2 A system block flowchart representing interphase of input signals of variable parameters to EPS32 after processing of data in microprocessor server signal forwarded for action

On the other hand, Thingspeak served as platform for accessing, evaluating, and displaying the sensor data. Using the HTTP and MQTT protocols, this free IoT application and API is used to store and retrieve data from objects. It makes it possible to create sensor logs that facilitate data tracking, aggregation, analysis, and display. The sensor's codes and programming were uploaded as AIA files to the Thingspeak platform. We configured the system to sound a buzzer and send an email alert in the event of any parameter irregularity or threshold. The suggested system processes sensor data using an Atmega328P microcontroller before sending it to the data bank on ThingSpeak. For monitoring, our system employs a two-way system communication approach. Communication between the environment and the sensors is the initial stage. The microcontroller unit and the ThinkSpeak IoT-based data telemetry form the second communication channel. Through an app, the user receives updates on the state of the hydroponic farm from the Thinspeak repository. In the event of any abnormality, the user receives an email and a buzzer sound at the farm site a brief workflow of IoT based hydroponic control for real time monitoring and control has been given in Fig. 10.3. In a closed-loop automated IoT-enabled hydroponic system, that collects sensor data, analyzes it, and adjusts pH, nutrients, EC, water flow, and environmental factors through a feedback loop. When conditions deviate, corrective actions are triggered, ensuring optimal plant growth, high yield, and efficient resource use (Fig. 10.4).

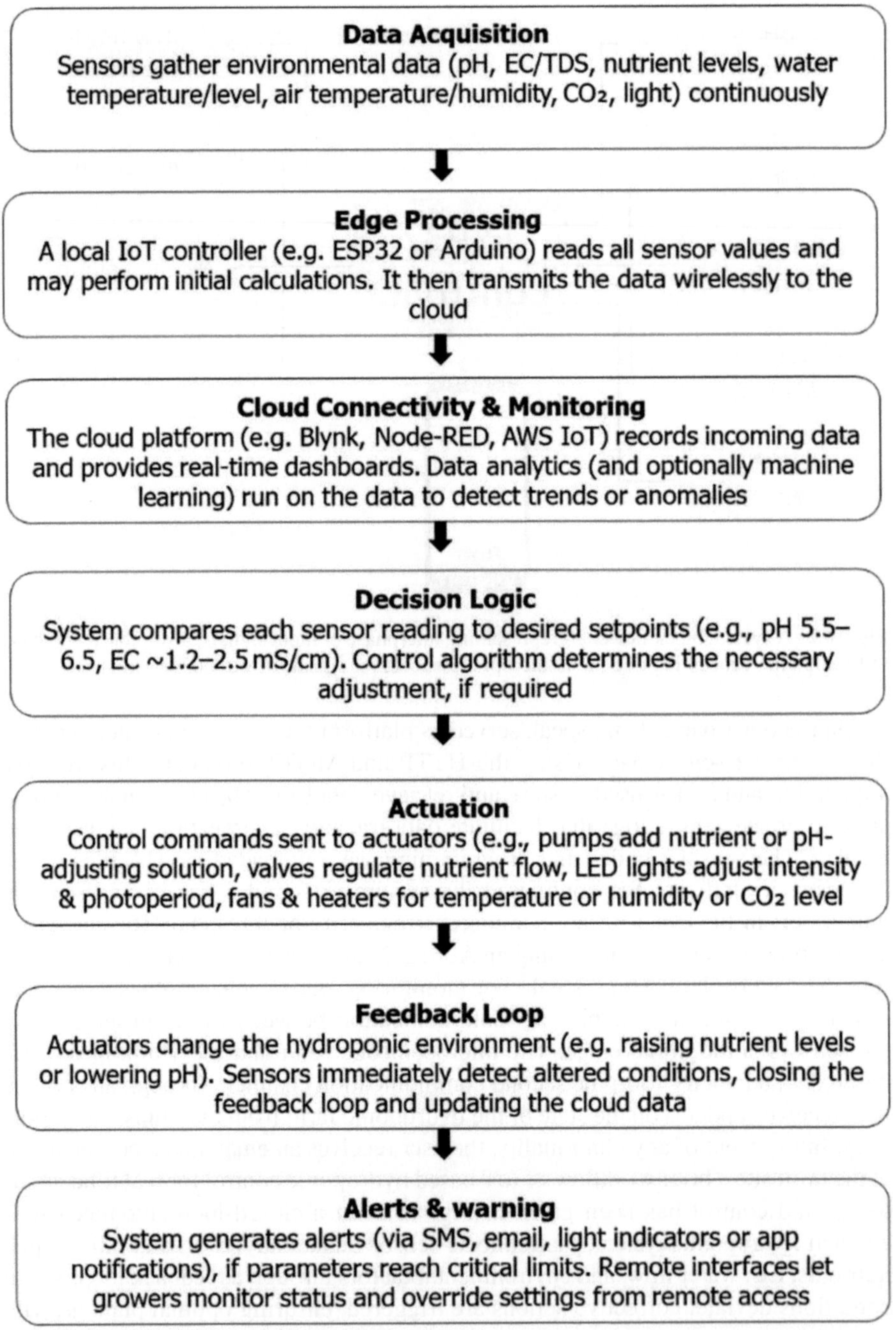

Fig. 10.3 IoT-based hydroponic control system workflow showing sensor data flow, decision-making, actitation, and feedback for real-time environmental management

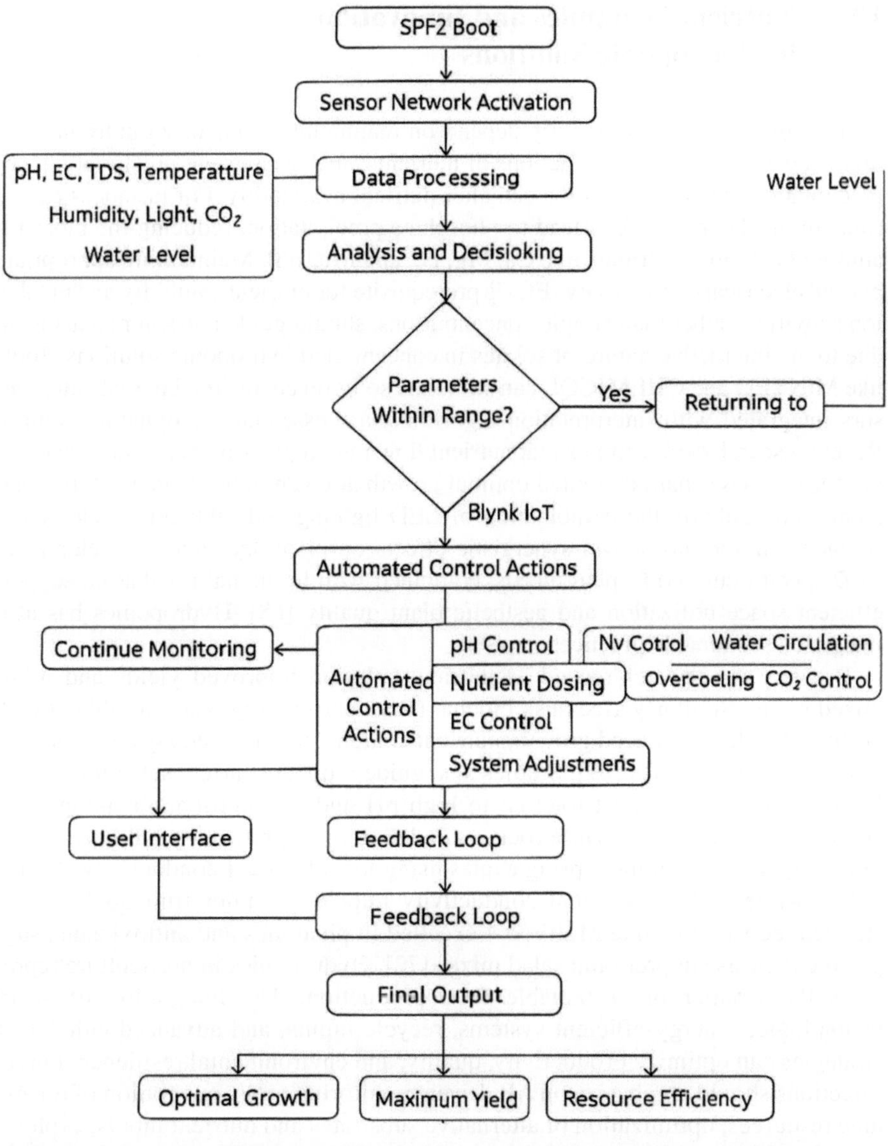

Fig. 10.4 Iot-Based Automation Flow in Hydroponic Cultivation. (a) Basic sensor data workflow using ThingSpeak for monitoring and alerts. (b) Detailed automated control system integration sensors, Blynk IoT, feedback loops, and decision-making for optimized growth, yield, and efficiency

10.4 Nutrient Dynamics and Innovation in Hydroponic Solutions

Success of soilless systems fully depend on maintaining a balanced nutrient solution. However, complex interactions of nutrients, such as nutrient precipitation, co-precipitation, and complexation can alter nutrient availability. For instances excess calcium or pH above 7 can lead to phosphate precipitation, reducing the bioavailability of essential nutrients like Ca, Mg, Fe, and Mn [66]. Maintaining appropriate pH and electrical conductivity (EC) is prerequisite for nutrient solubility and uptake. Ion activities, rather than simple concentrations, should guide nutrient management due to the interactive nature of solutes in concentrated hydroponic solutions. Tools like MINTEQ and CHEMICQL can stimulate solution chemistry, but real-time sensors integrated with interpretation algorithms are essential for dynamic control. Recent research demonstrates that nutrient flow rate impacts plant performance; for instances, Swiss chard exhibited optimal growth at 6 L/min flow rate [67]. In ornamental horticulture, the combination of LED lighting and gibberellic acid (GA_3) supplementation has shown synergistic effects on plant and flower development. LED spectra tailored to plant needs, combined with hormonal regulation, support efficient space utilization and aesthetic plant quality [68]. Hydroponics has also integrated sustainable practices.

Brewery wastewater reused as nutrient solution improved yields and minimized waste. Similarly, rice husk biochar (RB), particularly when combined with perlite (PL), has enhanced growth, nutrient content, and reduced algal contamination in leafy vegetables [69]. Studies also guide crop and variety selection. Green basil and mint were most tolerant to high pH and low nutrients, making them suitable for aquaponics, while rocket salad suffered significant yield losses [70]. For strawberries, summer spring cycles using low electrical conductivity yielded better, while higher electrical conductivity improved winter fruit quality [71]. Red lettuce varieties like Multired 4 excelled in phenolics and antioxidants, suggesting their use in premium salad mixes [72]. Hydroponics in horticulture represents the frontier of sustainable food production. The integration of smart technologies, energy-efficient systems, recycle inputs, and advanced cultivation strategies can optimize productivity, quality, and environmental resilience. Future directions should emphasize on AI- driven monitoring tools, integration of renewable resources, optimization of alternative substrates and nutrient inputs, exploration of resilient crop varieties for controlled environments. These systems will not only mitigate the effects of climate change but also enhance global food and nutrition security.

In recent years, the design of sanitation technologies has been linked to agricultural systems through hydroponic system but it required precise monitoring. Essential nutrients (nitrogen and phosphorus) present in municipal wastewater can be efficiently utilized for crop production. Hydroponic systems can efficiently recover nutrients from municipal wastewater without utilizing land to contribute toward food security [73, 74]. Besides nutrients, this untreated water contain pollutants, pathogens and organic matter, which discharged directly into the environment (80% of global waste) leading to contaminated water bodies putting the risk of waterborne diseases [75]. Additionally, pollution of heavy metal in untreated waste water also causes threat to human health. Hydroponic technology gained significant interest for their ability to phytoremediation property. Few crops such as spinach and many other have ability to accumulate high concentration of heavy metals. *Helianthus annuus* has ability to absorb multiple toxic metals like As, Cr, Cd, Ni, etc. and a significant increase of the As and Cd enrichment phenomenon, while *Iris sibirica* has good phytoremediation potential for Pb. Hence, further research is required to develop efficient phytoremediation system through algal and fast-growing plant. In this regard multiple stage hydroponics using algae or aquatic weeds might be efficient technology using precise environment control using IoT to removing pollutants from wastewater.

10.5 Application of Hydroponics in Modern Horticulture

Hydroponics plays a vital role in modern agriculture by offering highly efficient soilless cultivation techniques that optimizes crop growth through precise control of nutrients. With rapid urbanization, water scarcity, climate unpredictability, and shrinking arable land, hydroponics offers a controlled environment farming solution with efficient use of resources. Agriculture currently occupies 11% of the Earth's land and consumes 70% all freshwater, contributing significantly to greenhouse gas (GHG) emissions [76]. As an alternative, hydroponic systems not only reduce environmental burden but also align with circular economy principles through efficient resource utilization and waste minimization [77, 78]. Hydroponic systems commonly used in horticulture include Nutrient Film technique (NFT), Deep Water Culture (DWC), Ebb and Flow, Aeroponics, and Substrate-Based Drip irrigation. Each method is chosen based on the crop type, growth cycle, and infrastructure. NFT and DWC are preferred for leafy greens and herbs, while substrate-based systems are ideal for fruiting vegetables like tomatoes, cucumbers, and peppers, and strawberries. Greenhouses offer an ideal environment for hydroponics, allowing for year-round cultivation. Innovations like dye-sensitized solar cell (DSSC) roofs have enabled energy-efficient greenhouses, improving fruit quality and reducing electrical consumption. Spectral reflectance tools and machine learning models have further supported automation in fertigation and nutrient stress detection, achieving classifications accuracies up to 91% in fruit crops. In urban areas, hydroponics provides a scalable solution for local food production. Integrated with the Internet of Things (IoT), smart hydroponic systems allow for real-time monitoring and automated adjustments of key parameters such as pH, temperature, light, humidity, and nutrient concentration [78, 79]. Sensors and microcontrollers optimize water and nutrient delivery, improve crop yield and quality, and reduce manual labor. Hydroponics is considered a critical solution to meet the projected 110% increase in food demand by 2050 [21]. Saffron cultivation, traditionally confined to specific regions like Kashmir, is now being piloted under smart hydroponic setups using nutrient film technique (NFT) systems, solar energy, and advanced sensor arrays [79]. These systems not only improve resource-sue efficiency but also allow year-round cultivation of high-value crops in urban environments. An innovative adaptation of hydroponic principles to tree fruit crops is the Open Filed Soilless Systems (OFSS). Originally developed for citrus, OFSS integrates drip irrigation, coconut coir substrates bags, and fertigation management in field conditions. Unlike greenhouse, hydroponics, OFSS enables cultivation of perennial fruit crops like persimmon, peach, and mandarin orange with improved water use efficiency (WUE) and nutrient use efficiency (NUE) [80]. OFSS systems use substrate bags partially buried or placed on the soil surface. These bags, filled with high-retention organic media like coconut coir, serve as targeted zones for nutrient delivery and root growth. OFSS has demonstrated yield and fruit quality enhancement in field trails under both full and deficit irrigation conditions. Particularly, permission and peach tress exhibited increased mean fruit weight and water use efficiency when grown in substrate enhanced conditions. Smart hydroponics and OFSS systems manage ideal range is 5.5–6.5 with light sensor to optimized to photosynthetic and photoperiodic needs, temperature with 21–27 °C, and CO_2 and humidity to optimized enhanced transpiration and photosynthetic efficiency (Fig. 10.5).

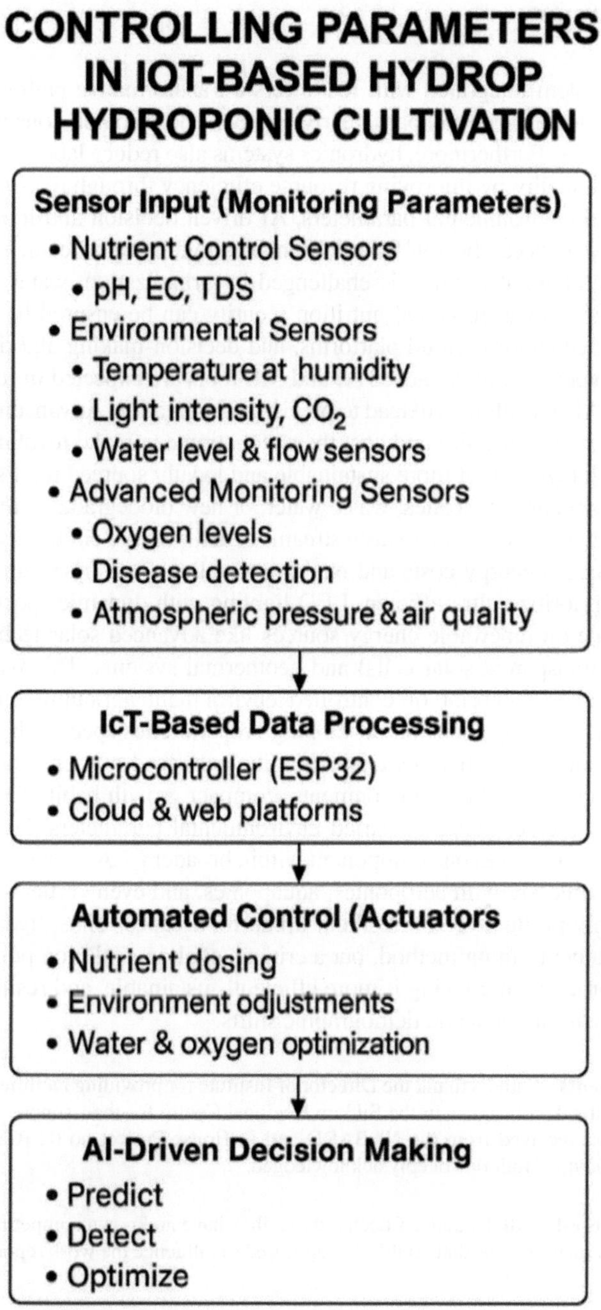

Fig. 10.5 A schematic representation of controlling parameters in an IoT integrated hydroponic system, showing sensor inputs, ICT-based data processing, automated control through actuators, and AI- driven decision-making for efficient monitoring management

10.6 Summary

Hydroponic system integrated with IoT offers a transformative pathway for sustainable and precision agriculture, particularly in urban and resource-constrained environmental conditions. Furthermore, hydronics systems also reduce labor cost and enhance crop yield and quality by improving resource efficiency through real-time monitoring of nutrient and environmental parameters, AI driven decision and input automation. Future food systems can benefit from IoT-enabled hydroponics resilience, data-driven approach as traditional framing is challenged by urbanization, water shortages, and climate change. Global food and nutrition security can be ensured by smart agriculture if sensor technology, cloud platforms, and decision-making algorithms continue to advance. Integration of advanced AI and ML tools are expected in future and more sophisticated AI algorithms will lead to precise autonomation. Advancement in nutrient specific Sensor (ion specific) and cost their reduction can make revolution in modern hydroponics. Integration of more sustainable and locally sourced inert substrates such as advanced biochar composites, waste water, or new biodegradable materials can be potentially integrated with other waste streams to achieve true circularity and minimize external inputs. As energy costs and environmental concerns rise, future hydroponic systems will prioritize ultra-efficient LED lighting with dynamic spectral tuning and greater reliance on renewable energy sources like advanced solar technologies (e.g., DSSC roofs, transparent solar cells) and geothermal systems. This will significantly reduce the carbon footprint of controlled environment agriculture. Plant breeding efforts will increasingly focus on developing crop varieties specifically optimized for hydroponic conditions. This includes traits like enhanced nutrient uptake efficiency, disease resistance in soilless environments, compact growth habits for vertical farming, and improved resilience to varied environmental parameters. Hydroponics will increasingly be viewed as one component within broader CEA ecosystems, potentially integrating seamlessly with aeroponics, aquaponics, and even vertical forest concepts to create highly productive and resilient urban foodscapes. Thus, hydroponics is not just an alternative farming method, but a critical, evolving solution poised to redefine global food production, making it more efficient, sustainable, and resilient in the face of ongoing environmental and demographic shifts.

Acknowledgments Authors thank the Director of Institute for providing facilities and encouragement. We also thank colleagues in the Sikkim Regional Centre for their support and help. Partial financial support received from the NABARD and In-House Project no 02 (CSED) of Institute (MoEF&CC, Govt. of India) is deeply acknowledged.

Conflict of Interest All the authors declared that they have no known competing financial interests or personal relationships that could have appeared to influence the work reported in this paper.

References

1. Lee S, Lee J (2015) Beneficial bacteria and fungi in hydroponic systems: types and characteristics of hydroponic food production methods. Sci Hortic 195:206–215. https://doi.org/10.1016/j.scienta.2015.09.011

2. Niu G, Masabni J (2022) Hydroponics. Plant factory basics, applications and advances. Academic Press, Elsevier, Amsterdam, pp 153–166. https://doi.org/10.1016/B978-0-323-85152-7.00023-9
3. Sharma N, Acharya S, Kumar K, Singh N, Chaurasia OP (2018) Hydroponics as an advanced technique for vegetable production: an overview. J Soil Water Conserv 17:364–371. https://doi.org/10.5958/2455-7145.2018.00056.5
4. Walters KJ, Behe BK, Currey CJ, Lopez RG (2020) Historical, current, and future perspectives for controlled environment hydroponic food crop production in the United States. HortSci 55:758–767. https://doi.org/10.21273/HORTSCI14901-20
5. Gomez C, Currey CJ, Dickson RW, Kim HJ, Hernández R, Sabeh NC, Raudales RE, Brumfield RG, Laury-Shaw A, Wilke AK, Lopez RG (2019) Controlled environment food production for urban agriculture. HortScienc 54:448–1458. https://doi.org/10.21273/HORTSCI14073-19
6. Caron J, Zheng Y (2021) Glossary of terms and basic characteristics to be reported in scientific publications on growing media. In: II International Symposium on Growing Media, Soilless Cultivation, and Compost Utilization in Horticulture 1317, pp 55–64. https://doi.org/10.17660/ActaHortic.2021.1317.7
7. Kozai T (2019) Towards sustainable plant factories with artificial lighting (PFALs) for achieving SDGs. Int J Agric Biol Eng 12:28–37. https://doi.org/10.25165/j.ijabe.20191205.5177
8. Lages Barbosa G, Almeida Gadelha FD, Kublik N, Proctor A, Reichelm L, Weissinger E, Wohlleb GM, Halden RU (2015) Comparison of land, water, and energy requirements of lettuce grown using hydroponic vs. conventional agricultural methods. Int J Environ Res Public Health 12:6879–6891
9. Gericke WF (1937) Hydroponics-crop production in liquid culture media. Science 85:177–178. 10.1126/science.85.2198.177
10. Resh HM (1990) Hydroponic home food gardens. Woodbridge Press, Santa Barbara
11. Hershey DR (1991) Digging deeper into Helmont's famous willow tree experiment. Am Biol Teach 53(8):458–460. https://doi.org/10.2307/4449369
12. Anonymous (1940) What's become of hydroponics? Bus Week:38–39
13. Yan H, Acquah SJ, Zhang J, Wang G, Zhang C, Darko RO (2021) Overview of modelling techniques for greenhouse microclimate environment and evapotranspiration. Int J Agric Biol Eng 16:1–8. https://doi.org/10.25165/j.ijabe.20211406.3948
14. Huang S, Yan H, Zhang C, Wang G, Acquah SJ, Yu J, Li L, Ma J, Darko RO (2020) Modeling evapotranspiration for cucumber plants based on the Shuttleworth-Wallace model in a Venlo-type greenhouse. Agric Water Manag 228:105861. https://doi.org/10.1016/j.agwat.2019.105861
15. Shoaib A, Rahman S, Shalabi MM, Alshayeb FI, Shatnawi ZN (2022) Climate resilience and environmental sustainability: how to integrate dynamic dimensions of water security modeling. Agriculture 12:303. https://doi.org/10.3390/agriculture12020303
16. Lakhiar IA, Yan H, Zhang J, Wang G, Deng S, Bao R, Zhang C, Syed TN, Wang B, Zhou R Wang X (2024) Plastic pollution in agriculture as a threat to food security, the ecosystem, and the environment: an overview. Agronomy 14: 548. https://doi.org/10.3390/agronomy14030548
17. Lam HM, Remais J, Fung MC, Xu L, Sun SSM (2013) Food supply and food safety issues in China. Lancet 381:2044–2053. https://doi.org/10.1016/S0140-6736(13)60776-X
18. Luo B, Dou X (2024) Climate change, agricultural transformation and climate smart agriculture development in China. Heliyon 10:21. https://doi.org/10.1016/j.heliyon.2024.e40008
19. Mohamed TMK, Gao J, Tunio M (2022) Development and experiment of the intelligent control system for rhizosphere temperature of aeroponic lettuce via the internet of things. Int J Agric Biol Eng 15:225–233. https://doi.org/10.25165/j.ijabe.20221503.6530
20. Regmi A, Rueda-Kunz D, Liu H, Trevino J, Kathi S, Simpson C (2024) Comparing resource use efficiencies in hydroponic and aeroponic production systems. Technol Horticult 4:12988. https://doi.org/10.48130/tihort-0024-0002
21. Sousa RD, Bragança L, da Silva MV, Oliveira RS (2024) Challenges and solutions for sustainable food systems: the potential of home hydroponics. Sustainability 16:817. https://doi.org/10.3390/su16020817
22. Lakhiar IA, Gao J, Syed TN, Chandio FA, Tunio MH, Ahmad F, Solangi KA (2020) Overview of the aeroponic agriculture–an emerging technology for global food security. Int J Agric Biolog Eng 13:5156. https://doi.org/10.25165/j.ijabe.20201301.5156

23. Al Meselman (2024) Hydroponics: the future of sustainable farming. In: Hydroponics: the future of sustainable farming. Springer, New York, pp 101–122. https://doi.org/10.1007/978-1-0716-3993-1_6
24. Gruda N, Gianquinto G, Tüzel Y, Savvas D (2016) Soilless culture. In: Encyclopedia of soil sciences, 3rd edn. CRC Press/Taylor & Francis Group, Boca Raton, pp 533–537
25. Baiyin B, Tagawa K, Yamada M, Wang X, Yamada S, Yamamoto S, Ibaraki Y (2021) Effect of the flow rate on plant growth and flow visualization of nutrient solution in hydroponics. Horticulture 7:225. https://doi.org/10.3390/horticulturae7080225
26. Dietze V, Feindt PH (2023) Innovation systems for controlled-environment food production in urban contexts: a dynamic case study analysis of combined plant, fish and insect production in Berlin. Int J Agric Sustain 21:2166230. https://doi.org/10.1080/14735903.2023.2166230
27. Fussy A, Papenbrock J (2022) An overview of soil and soilless cultivation techniques—chances, challenges and the neglected question of sustainability. Plants 11(9):1153. https://doi.org/10.3390/plants11091153
28. Nair CS, Manoharan R, Nishanth D, Subramanian R, Neumann E, Jaleel A (2025) Recent advancements in aquaponics with special emphasis on its sustainability. J World Aquacult Soc 56:13116. https://doi.org/10.1111/jwas.13116
29. Channa AA, Munir K, Hansen M, Tariq MF (2024) Optimisation of small-scale aquaponics systems using artificial intelligence and the IoT: current status, challenges, and opportunities. Encyclopedia 4:313–336. https://doi.org/10.3390/encyclopedia4010023
30. Khan S, Singh R, Khan S, Ngah AH (2023) Unearthing the barriers of Internet of Things adoption in food supply chain: a developing country perspective. Green Technol Sustain 1:100023. https://doi.org/10.1016/j.grets.2023.100023
31. Sinha BB, Dhanalakshmi R (2022) Recent advancements and challenges of Internet of Things in smart agriculture: a survey. Futur Gener Comput Syst 126:169–184. https://doi.org/10.1016/j.future.2021.08.006
32. Friha O, Ferrag MA, Shu L, Maglaras L, Wang X (2021) Internet of things for the future of smart agriculture: a comprehensive survey of emerging technologies. IEEE/CAA J Automat Sinica 8:718–752. https://doi.org/10.1109/JAS.2021.1003925
33. Rahman MW, Islam R, Hasan A, Bithi NI, Hasan MM, Rahman MM (2022) Intelligent waste management system using deep learning with IoT. J King Saud Univ-Comput Inform Sci 34:2072–2087. https://doi.org/10.1016/j.jksuci.2020.08.016
34. Suresh V, Logasundari T, Sravani VS, Ali M, Srinivasan S (2024) IOT based automated indoor hydroponic farming system. E3S Web Conf 547:02002
35. Jain S, Kaur M (2024) Design and implementation of an IoT-based automated EC and pH control system in an NFT-based hydroponic farm. Eng Technol Appl Sci Res 14:13078–13081. https://doi.org/10.48084/etasr.6393
36. Ramakrishnam Raju SVS, Dappuri B, Ravi Kiran Varma P, Yachamaneni M, Verghese DMG, Mishra MK (2022) Design and implementation of smart hydroponics farming using IoT-based AI controller with mobile application system. J Nanomater 2022:4435591. https://doi.org/10.1155/2022/4435591
37. Sneineh AA, Shabaneh AA (2023) Design of a smart hydroponics monitoring system using an ESP32 microcontroller and the Internet of Things. MethodsX 11:102401. https://doi.org/10.1016/j.mex.2023.102401
38. Nikolov NV, Atanasov AZ, Evstatiev BI, Vladut VN, Biris SS (2023) Design of a small-scale hydroponic system for indoor farming of leafy vegetables. Agriculture (Basel) 13:1191. https://doi.org/10.3390/agriculture13061191
39. Li Yong Wei, Ding Ying LY, Li DaoLiang DY, Miao Zheng LD, MZ (2018) Automatic carbon dioxide enrichment strategies in the greenhouse: a review. Biosyst Eng 171:101–119. https://doi.org/10.1016/j.biosystemseng.2018.04.018
40. Prema K, Belinda CM (2019) Smart farming: IoT based plant leaf disease detection and prediction using deep neural network with image processing. Int J Innov Technol Explor Eng 8:3081–3083
41. Kiktev N, Lendiel T, Vasilenkov V, Kapralyuk O, Hutsol T, Glowacki S, Kuboń M, Kowalczyk Z (2021) Automated microclimate regulation in agricultural facilities using the air curtain system. Sensors 21:8182. https://doi.org/10.3390/s21248182

42. Bugbee B (2003) Nutrient management in recirculate hydroponic culture. South Pacific Soil Cult Conf-SPSCC 648:99–112. https://doi.org/10.17660/ActaHortic.2004.648.12
43. Barus S, Dhamma PYA (2024) Development of hydroponic application based on web and internet of things for the community to monitor pH and total dissolved solids. Int J Res Commun Services 5:130–137
44. Riedel V, Hinck S, Peiter E, Ruckelshausen A (2024) Concept and realisation of ISFET-based measurement modules for infield soil nutrient analysis and hydroponic systems. Electronics 13:2449. https://doi.org/10.3390/electronics13132449
45. Borisov S, Wolfbeis M, OS (2008) Optical biosensors. Chem Rev 108:423–461. https://doi.org/10.1021/cr068105t
46. Resh HM (2012) Plant culture. Hydroponic food production: a definitive guidebook for the advanced home gardener and the commercial hydroponic grower. CRC Press, New York, pp 387–474
47. Hidayatullah M, Sofyan S, Topan PA, Andriani T, Nurhairunnisah N (2022) Monitoring system of water quality on hydroponic planting media using total dissolved solid (TDS) sensor based Arduino Uno R3. J Fis Uni And 14:108–115
48. Suseno JE, Munandar MF, Priyono AS (2020) The control system for the nutrition concentration of hydroponic using web server. J Phys Conf Ser 1524:012068. https://doi.org/10.1088/1742-6596/1524/1/012068
49. Resh HM (2022) Hydroponic food production: a definitive guidebook for the advanced home gardener and the commercial hydroponic grower. CRC press, Boca Raton. https://doi.org/10.1201/9781003133254
50. Savvas D, Gruda N (2018) Application of soilless culture technologies in the modern greenhouse industry-a review. Eur J Hortic Sci 83:280–293. https://doi.org/10.17660/eJHS.2018/83.5.2
51. Resh HM (2013) Hydroponic food production: a definitive guidebook for the advanced home gardener and the commercial hydroponic grower, 7th edn. CRC Press, Boca Raton
52. Abbasi R, Yanes AR, Villanuera EM, Ahmad R (2021) Real-time implementation of digital twin for robot-based production line. Proc Conf Learn Fact (CLF). https://doi.org/10.2139/ssrn.3860500
53. Piovene C, Orsini F, Bosi S, Sanoubar R, Bregola V, Dinelli G, Gianquinto G (2015) Optimal red: blue ratio in led lighting for nutraceutical indoor horticulture. Sci Hortic 193:202–208. https://doi.org/10.1016/j.scienta.2015.07.015
54. Boros I, Székely F, Balázs G, Csambalik L, Sipos LL (2023) Effects of LED lighting environments on lettuce (*Lactuca sativa* L.) in PFAL systems-a review. Sci Hortic 321112351. https://doi.org/10.1016/j.scienta.2023.112351
55. Carotti L, Pistillo A, Zauli I, Meneghello D, Martin M, Pennisi G, Gianquinto G, Orsini F (2023) Improving water use efficiency in vertical farming: effects of growing systems, far-red radiation and planting density on lettuce cultivation. Agric Water Manag 285:108365. https://doi.org/10.1016/j.agwat.2023.108365
56. Naznin MT, Lefsrud M, Gravel V, Azad MOK (2019) Blue light added with red LEDs enhance growth characteristics, pigments content, and antioxidant capacity in lettuce, spinach, kale, basil, and sweet pepper in a controlled environment. Plants 8:93. https://doi.org/10.3390/agriculture13061191
57. Hostalrich D, Pelegri-Sebastia J, Sogorb T, Pellicer V (2022) Intelligent management of hydroponic systems based on IoT for agrifood processes. J Sens 2022:9247965. https://doi.org/10.1155/2022/9247965
58. Pasha S (2016) ThingSpeak based sensing and monitoring system for IoT with Matlab analysis. Int J New Technol Res 2:19–23
59. Azimi FH, Aw SR, Syafie LS, Aris RSNAR, Fikri KI, Shafie WRW (2020) IOT monitoring in NFT hydroponic system using Blynk-an android platform. Int J Syner Eng Technol 1:1
60. Sawlani K, Patole D, Kelkar V, Pange M, Kardile K (2024) Self-adaptive control system for hydroponics. Technologies for energy, agriculture, and healthcare. CRC Press, pp 472–486
61. Arduino (2020). Bring the power of easy connectivity into everyday life and business with the Arduino IoT Cloud. [online] Available at: https://blog.arduino.cc/2020/08/21/experience-the-power-of-easy-connectivity-into-everyday-life-and-business-with-the-arduino-iot-cloud/[Accessed 27 June 2025]

62. Rakshitha M, Shwetha HL, Roopa TR, Prasad AS (2018) Automation of hydroponics system using android application and ubidots platform. Int J Eng Res Technol, NCESC Conf Proc:1–4
63. Wittig A, Wittig M (2023) Amazon Web Services in action: an in-depth guide to AWS. Simon and Schuster
64. Rosca CM, Stancu A, Popescu M (2025) The impact of cloud versus local infrastructure on automatic IoT-driven hydroponic systems. Appl Sci 15:2076–3417. https://doi.org/10.3390/app15074016
65. De Side GN, Putra GMD, Setiawati DA (2025) IoT (internet of things) microclimate monitoring system using node-red platform at plant factory. IOP Conf Ser Earth Environ Sci, IOP Publishing 1441:012012. https://doi.org/10.1088/1755-1315/1441/1/012012
66. Sambo P, Nicoletto C, Giro A, Pii Y, Valentinuzzi F, Mimmo T, Lugli P, Orzes G, Mazzetto F, Astolfi S, Terzano R (2019) Hydroponic solutions for soilless production systems: issues and opportunities in a smart agriculture perspective. Front Plant Sci 10:923. https://doi.org/10.3389/fpls.2019.00923
67. Sawada M, Sumi Y, Mimura M (2021) Measuring desiccation-induced tensile stress during cracking process. Soils Found 61:915–928. https://doi.org/10.1016/j.sandf.2021.03.006
68. Mills-Ibibofori T, Dunn B, Maness N, Payton M (2019) Use of diatomaceous earth as a silica supple on potted ornamentals. Horticulture 5:21. https://doi.org/10.3390/horticulturae5010021
69. Awad Y, Lee M, Ahmed SE, MBM Vu, Farooq NT, Kim M, Ok IS, YS (2017) Biochar, a potential hydroponic growth substrate, enhances the nutritional status and growth of leafy vegetables. J Clean Prod 156:581–588. https://doi.org/10.1016/j.jclepro.2017.04.070
70. Fimbres-Acedo Y, Traversari E, Cacini S, Costamagna S, Ginepro G, Massa MD (2022) Testing the effect of high pH and low nutrient concentration on four leafy vegetables in hydroponics. Agronomy 13:41. https://doi.org/10.3390/agronomy13010041
71. Caruso G, Villari G, Melchionna G, Conti S (2011) Effects of cultural cycles and nutrient solutions on plant growth, yield and fruit quality of alpine strawberry (*Fragaria vesca* L.) grown in hydroponics. Sci Hortic 129:479–485. https://doi.org/10.1016/j.scienta.2011.04.020
72. Mampholo BM, Maboko MM, Soundy P, Sivakumar D (2016) Phytochemicals and overall quality of leafy lettuce (*Lactuca sativa* L.) varieties grown in closed hydroponic system. J Food Qual 39:805–815. https://doi.org/10.1111/jfq.12234
73. Giri L, Hussain M, Angmo JC, Mustafa G, Singh B, Bahukhnadi A, Pradhan R, Kumar R, Mukherjee S, Bhatt ID, Nautiyal S (2025) Enhancing tomato (Solanum lycopersicum) yield and nutrition quality through hydroponic cultivation with treated wastewater. Food Chem 463:141079. https://doi.org/10.1016/j.foodchem.2024.141079
74. Magwaza ST, Magwaza LS, Odindo AO, Mditshwa A (2020) Hydroponic technology as decentralised system for domestic wastewater treatment and vegetable production in urban agriculture: a review. Sci Total Environ 698:134154. https://doi.org/10.1016/j.scitotenv.2019.134154
75. Rajan M, Karunanidhi D, Jaya J, Preethi B, Subramani T, Aravinthasamy P (2024) A comprehensive review of human health hazards exposure due to groundwater contamination: a global perspective. Physics and Chemistry of the Earth, Parts A/B/C 135:103637. https://doi.org/10.1016/j.pce.2024.103637
76. Spiertz JHJ (2009) Nitrogen, sustainable agriculture and food security: a review. In: Lichtfouse E, Navarrete M, Debaeke P, Véronique S, Alberola C (eds) Sustainable agriculture. Springer, Dordrecht. https://doi.org/10.1007/978-90-481-2666-8_39
77. Boneta A, Rufí-Salís M, Ercilla-Montserrat M et al (2019) Agronomic and environmental assessment of a polyculture rooftop soilless urban home garden in a Mediterranean city. Front Plant Sci 10:341. https://doi.org/10.3389/fpls.2019.00341
78. Khan S, Purohit A, Vadsaria N (2020) Hydroponics: current and future state of the art in farming. J Plant Nutr 44:1515–1538. https://doi.org/10.1080/01904167.2020.1860217
79. Kour K, Gupta D, Gupta K, Dhiman G, Juneja S, Viriyasitavat W, Islam MA (2022) Smart-hydroponic-based framework for saffron cultivation: a precision smart agriculture perspective. Sustainability 14:1120. https://doi.org/10.3390/su14031120
80. Rubio-Asensio JS, Parra M, Intrigliolo DS (2020) Open field hydroponics in fruit crops: developments and challenges. In: Srivastava AK, Hu C (Eds) Fruit crops, Elsevier. Amsterdam, pp. 419–430. https://doi.org/10.1016/B978-0-12-818732-6.00030-7

Chapter 11
AI and Smart Farming: Transforming the Future of Agriculture

Kritika Sharma and Vishal Acharya

Abstract An effective answer to modern agriculture dilemmas like insufficient resources, climate change, and population growth is the implementation of artificial intelligence (AI) into smart agriculture. The emphasis of this chapter is on how AI technologies are utilized in hydroponics, which is an advanced non-soil based farming system with high resource use efficiency and controlled growth conditions. Specifically, it explores the use of computer vision, machine learning, statistical forecasting, and the Internet of Things (IoT) to optimize key plant cultivation parameters, such as maximizing growth, predicting yields, and managing resources. This chapter highlights how AI can contribute to sustainable agriculture through automated harvesting, plant health monitoring, and precise nutrient delivery. It also examines the benefits of drone technology for smallholder farmers, emphasizing the importance of developing appropriate, affordable, user-friendly, and accessible solutions tailored to their specific needs. This chapter concludes by suggesting that AI implementation can lead to multi-purpose farming technologies capable of meeting food demands while minimizing environmental impact. It also underscores the need for collaboration among stakeholders and the development of ethical frameworks.

Keywords Artificial intelligence · Hydroponics · Smart farming · Soil less farming system · Computer vision · Predictive analysis · Machine learning · Sustainable agriculture

---

K. Sharma
Artificial Intelligence for Computational Biology (AICoB) Lab, Biotechnology Division, CSIR-Institute of Himalayan Bioresource Technology, Palampur, Himachal Pradesh, India

V. Acharya (✉)
Artificial Intelligence for Computational Biology (AICoB) Lab, Biotechnology Division, CSIR-Institute of Himalayan Bioresource Technology, Palampur, Himachal Pradesh, India

Academy of Scientific and Innovation Research (AcSIR), Ghaziabad, India
e-mail: vishal@ihbt.res.in

P. Kumar, A. R. Warghat (eds.), *Transforming Medicinal Plant Agriculture*, Smart Agriculture 16, https://doi.org/10.1007/978-981-95-3963-5_11

11.1 Introduction to Smart Agriculture

The core nutritional needs that support people worldwide are provided by traditional agriculture, which has long been a cornerstone of human society [1]. However, a number of significant factors have combined to create unprecedented challenges for this long-standing practice: resource scarcity, particularly with regard to freshwater and arable land; climate change, which is changing weather patterns and impacting crop viability [2]; and a rapidly growing global population, which is projected to exceed nearly 10 billion by 2050, which will increase the demand for food [1]. These serious problems highlight the dire need for creative answers that can support sustainable farming methods and guarantee food security [2].

In consideration of this, artificial intelligence (AI) is currently beginning to revolutionize the agriculture industry [3]. The term AI refers to a variety of technologies, including as robots, data analytics, and machine learning, all of which have the capacity to analyze large, complicated information, automate time-consuming tasks, and maximize resource use [1, 3]. Farmers and other agricultural stakeholders can improve output, lessen the adverse consequences of environmental stressors, and make better decisions by utilizing AI [3, 4].

This chapter will especially look at how AI is being used in the field of agriculture, with an emphasis on how it can work in tandem with hydroponics. The cutting-edge soil-less growth method known as hydroponics enables plants to be grown in high in nutrients water solutions, thereby providing notable benefits with regard to of both resource effectiveness and the capacity to establish regulated growing conditions [5]. This technique allows for year-round production, regardless of seasonal limitations, while also conserving water—up to 90% compared with conventional soil-based cultivation farming [5, 6].

We will investigate how AI may transform hydroponic systems by improving accuracy in tracking and controlling plant growth conditions. AI systems, for example, are able to evaluate real-time information from sensors that monitor environmental variables including moisture and temperature as well as variables like pH levels and nutrient concentrations [4].

Furthermore, by accurately forecasting the required quantities of both nutrients and water needed at various growth stages, AI can maximize resource consumption, avoiding waste and cutting expenses [6]. Farmers may protect crop health and yield by using predictive analytics to foresee possible problems like outbreaks of pests or nutrient deficits before they become serious [5]. In the end, incorporating AI into hydroponic cultivation systems has the potential to greatly increase agricultural outputs and support a more robust and sustainable food supply. As we learn more about this subject, we will emphasize case studies and instances that show how AI technologies have been successfully applied in hydroponics, demonstrating how these developments can open the door to a more environmentally conscious agricultural future [7].

11.2 Basics of Artificial Intelligence in Agriculture

AI primarily circles around the growth of autonomous machines that are capable of understanding their immediate environment, deductive reasoning by means of intricate information, gaining knowledge from statistics, and executing initiatives to accomplish particular goals [8].

AI is a potent tool in the agricultural industry that allows machines to analyze large datasets, including sensor readings, environmental data, and historical records.

This allows for data-driven decision-making and automates tasks that have historically relied on human labor and intuition. Because AI improves production, sustainability, and efficiency, its application in agriculture is especially revolutionary [8]. Several key AI technologies are especially pertinent to the agricultural landscape:

11.2.1 Machine Learning (ML)

A well-known branch of AI called machine learning (ML) enables machines to derive information from data without any explicit programming. ML algorithms may find complex patterns, predict future results, and optimize growth conditions by looking at past statistics that include weather patterns, crop yields, pest outbreaks, soil conditions (or, in the case of hydroponic agriculture, nutrient solution compositions), and weather patterns [9, 10].

The following specialized ML techniques are especially pertinent to hydroponics as shown in Fig. 11.1.

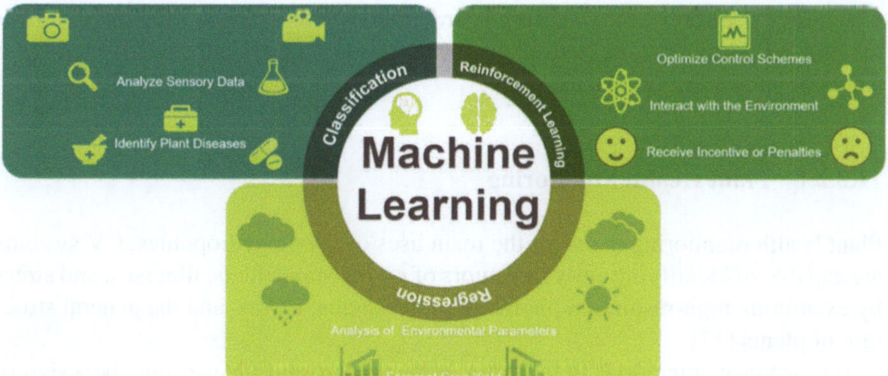

Fig. 11.1 Machine learning techniques implementation

11.2.1.1 Regression

By examining the connection between several environmental parameters and the final agricultural production, this statistical method is frequently used in the field of agriculture to forecast crop yields [11]. This approach entails developing a computational framework that illustrates how adjustments to particular parameters—like temperature, moisture, nutrient levels, and light intensity—affect crop yields [12].

11.2.1.2 Classification

This is an important technique in the agricultural sector that uses visual or sensory data analysis to identify plant diseases [14]. By using sophisticated algorithms to divide data into discrete groupings, this method enables early disease detection, safeguarding crop productivity and health [13].

11.2.1.3 Reinforcement Learning

This is an advanced AI technique that works especially well for optimizing control schemes in dynamically growing environments. This approach is founded on the idea that an agent (in this particular instance, the hydroponic setup) evolves by making decisions by interacting with its surroundings along with getting reinforcement in the form of incentives or penalties [15].

11.2.2 Computer Vision (CV)

CV technologies facilitate computer systems with the potential to “see” and comprehend visual details obtained through images and videos [16]. This cutting-edge technology is essential to hydroponic gardening, where it can be used for a number of vital purposes that improve plant management and system effectiveness as a whole as shown in Fig. 11.2:

11.2.2.1 Plant Health Monitoring

Plant health monitoring is one of the main uses of CV in hydroponics. CV systems are capable of identifying early indicators of nutritional deficits, illnesses, and stress by examining high-resolution photos of root systems, leaves, and the general structure of plants [17].

For instance, alterations in the color, texture, or form of leaves may be a sign of infections or certain nutrient deficiencies [18]. CV can categorize these visual clues

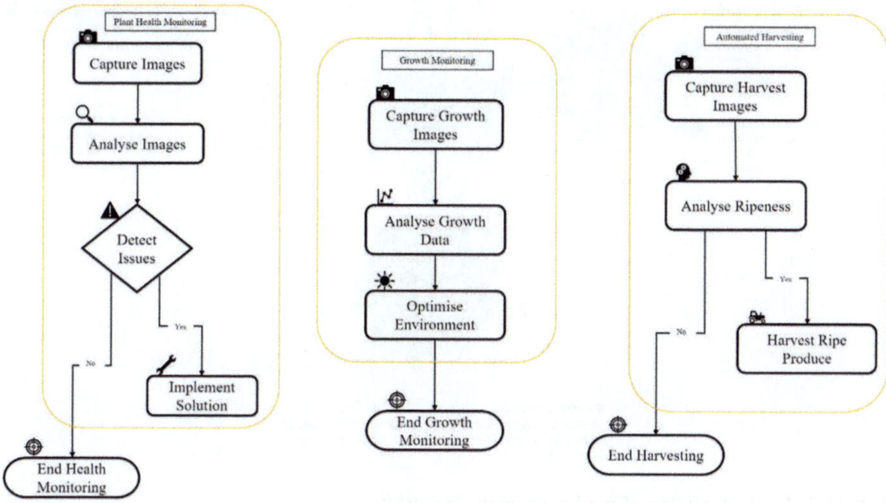

Fig. 11.2 Computer Vision (CV) implementation flowchart for hydroponic systems

using ML algorithms and image processing techniques, enabling prompt remedial action [18].

11.2.2.2 Growth Monitoring

Growth monitoring is another important way that CV is used in hydroponics. This entails using routine imaging to continuously monitor the growth and development of plants over time [19]. CV systems can examine variations regarding plant dimensions, surface area of leaves, and total biomass by taking pictures at various stages of growth.

To increase productivity, environmental factors like temperature, humidity, and light intensity can be optimized with the use of these data [19, 20]. For example, a CV system can make changes to the growing circumstances to give plants the best possible setting for growth if it notices that they are growing more slowly than anticipated.

11.2.2.3 Automated Harvesting

In hydroponic systems, CV is also essential for automating the harvesting process. Systems can recognize ripe produce by using CV algorithms, which take into account visual traits including hue, dimension, and shape [21]. This feature reduces waste and guarantees that produce is collected at its best quality by enabling autonomous harvesting systems to precisely and effectively gather crops at the appropriate moment.

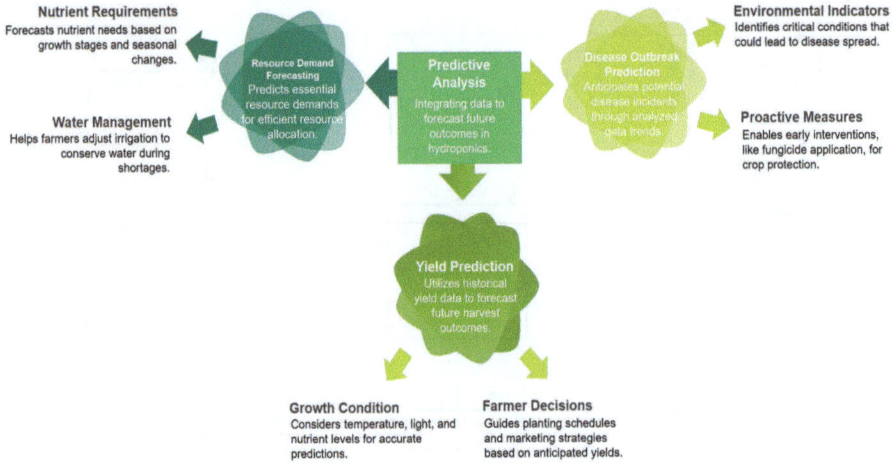

Fig. 11.3 Implementation of predictive analysis workflow

A robotic arm with CV capabilities, for instance, may scan plant rows, detect ripe tomatoes, and harvest them only, abandoning fruits that are unripe on the vine [22]. This improves overall efficiency and lowers labor expenses while streamlining the harvesting process [21, 22].

11.2.3 Predictive Analytics

To predict future occurrences, predictive analytics combines real-time sensor readings, historical data, and machine learning algorithms. This feature is especially helpful in hydroponics, where exact control over environment and materials is necessary to maximize sustainability and output [23]. Predictive analytics has several important uses in hydroponic systems also shown in Fig. 11.3:

11.2.3.1 Yield Prediction

Yield prediction is one of the main uses of predictive analytics in hydroponics. Predictive models anticipate future harvest yields by comparing historical yield data with present growing circumstances, such as temperature, light exposure, and nutrient levels [24]. Farmer decision-making on planting schedules, inventory management, and production cycle planning are all made possible by this information [25].

For example, farmers might plan for higher demand and modify their marketing methods if a predictive model predicts a higher-than-average output for a specific crop given the current circumstances.

11.2.3.2 Resource Demand Forecasting

Forecasting resource demand also heavily relies on predictive analytics. By reviewing trends in significant information along with real-time sensor measurements, predictive models can foresee the required number of crucial supplies such as nourishment, water, and energy [23, 26].

With the use of this ability, farmers can allocate resources as efficiently and effectively as possible. For instance, farmers can proactively modify their irrigation plans to avoid water shortages and save waste if a model forecasts a surge in water demand as a result of rising temperatures [27].

11.2.3.3 Disease Outbreak Prediction

Predicting disease outbreaks is a noteworthy use of predictive analytics in hydroponics [28]. Predictive models can find early warning indicators of possible disease outbreaks by examining past statistics on the condition of plants, environmental factors, and disease occurrences [29]. For example, the model might notify farmers to take precautions when similar conditions occur if certain environmental factors, like high levels of moisture and low light, are linked to past disease outbreaks [28, 30].

11.2.4 Internet of Things (IoT)

Despite not being directly categorized as AI, IoT is crucial to the agriculture industry because it provides the vital data needed for AI algorithms to operate efficiently [31]. Sensors are incorporated into hydroponic systems to gather information on a number of variables, such as light intensity, temperature, humidity, fertilizer concentrations, and pH levels [32]. In order to optimize growing conditions and guide decision-making, this data is sent to a central system and examined by AI algorithms [33]. In conclusion, a major development in agricultural operations is the incorporation of AI technologies like ML, CV, predictive analytics, and the IoT into hydroponic systems. In addition to increasing hydroponic farming's effectiveness and production, these technologies help create a more resilient and sustainable food system that can handle the problems brought on by environmental change and the world's expanding population [34].

As shown in Fig. 11.4, the working of IoT based techniques in agriculture involves the use of several sensors which collect various readings and send them to a server through a gateway which connects the server to the data base and incoming information from the sensors, here a junction is created between the farmer or field owner and the controller of the systems and sensors along with the server and a database which collects real-time data and stores it to improve functionality in the long run.

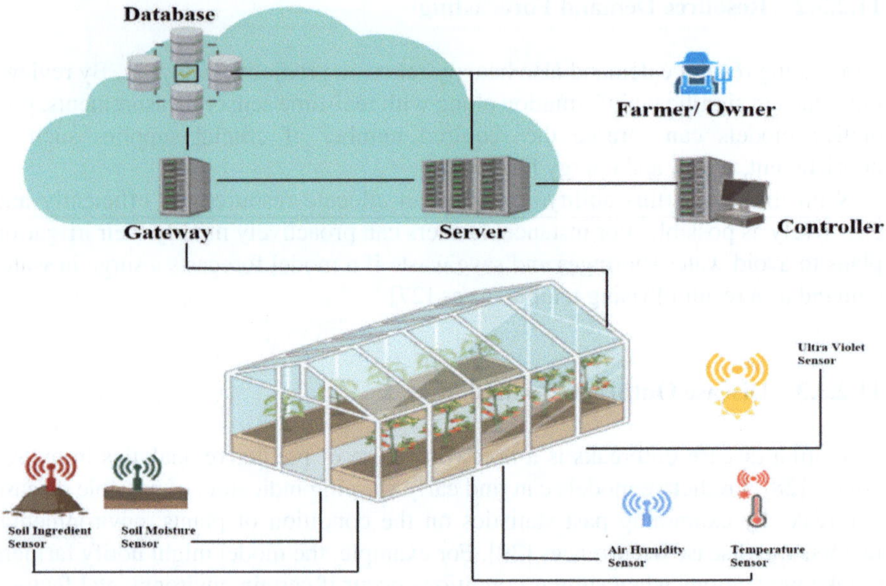

Fig. 11.4 IoT application for smart agriculture

11.3 Hydroponic Farming and Its Methods

The cutting-edge agricultural technique known as hydroponics develops many crops in aqueous solutions rich with nutrients, doing away with any requirement for soil [35]. It enables managing several important ecological aspects such as pH, temperature and nutrient content, greatly maximizing plant growth and increasing crop yields. Because they use much less water than normal farming, hydroponic systems work well in areas without much water [36]. Hydroponics can considerably improve a number of food security concerns and strengthen many ecologically sound approaches to farming [37].

Hydroponic farming encompasses several techniques, each with unique benefits and challenges. Here is a breakdown of some popular methods:

11.3.1 Nutrient Film Technique (NFT)

NFT systems are ideal for plants that grow quickly and have shallow roots. These systems keep nutrient solution moving over the roots all the time as shown in Fig. 11.5a [38].

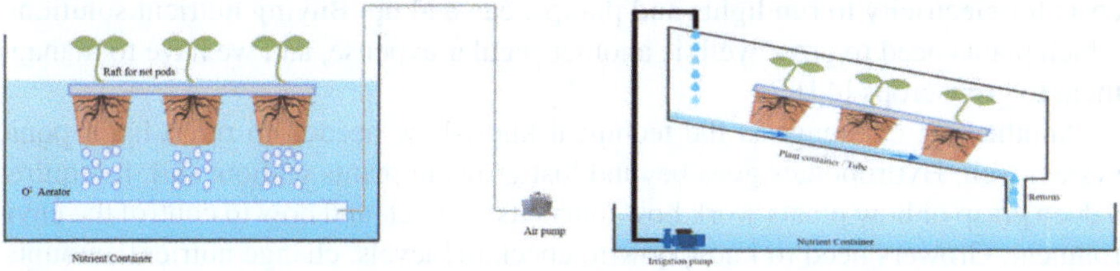

Fig. 11.5 (**a**) Deep water culture; (**b**) Nutrient film technique

11.3.2 Deep Water Culture (DWC)

In DWC systems, plant roots dangle in a nutrient solution aerated for oxygen. Many hobbyists like this setup because it is simple and easy to maintain as shown in Fig. 11.5b [39].

11.3.3 Aeroponics

Roots are usually kept suspended in the air and are missed with a solution that contains nutrients for macro aeration. It is great for city farming, but it also calls for careful measures in regard to monitoring humidity and nutrients [40].

11.3.4 Substrate-Based Systems

The sustenance systems utilize materials such as coconut husks or rockwool to preserve and promote plant growth while maintaining proper moisture and air circulation. It helps beginners and seasoned gardeners alike. Regular testing of pH and EC is necessary to encourage healthy growth of the plants [41].

11.4 Challenges Faced in Smart Farming (Hydroponic Systems)

Setting up hydroponics has a big impact on making farming more sustainable, but it comes with its own set of problems that need solving to get the most out of it [42]. The biggest issue is the high cost to start a hydroponic system. We need to spend a lot of money on special gear like pumps, grow lights, nutrient tanks, and often a controlled space such as a greenhouse. These costs can make it hard for small farmers or those in poorer areas with less money to get started [43]. Also, the ongoing costs for electricity to run lights and pumps, can add up. Buying nutrient solutions, which plants need to grow well, is another regular expense, and we have to manage them to grow crops [44].

Another big challenge is the technical know-how needed to run a hydroponic system well. Hydroponics goes beyond just growing plants without soil; it requires a deep grasp of how plants work how nutrients interact, and how to control the environment. Growers need to know how to check pH levels, change nutrient amounts, and fix problems like too much algae or not enough nutrients. This can be hard for people who have not had proper training or cannot get the right information, which might make it tough for them to grow as much as they could. Keeping everything balanced in the hydroponic system means always being on the lookout and ready to solve problems before they get big [44].

For example, power cuts can stop the system from working, which stresses plants or even kills them. When equipment breaks down, like pumps failing, it can affect crop health. While diseases do not spread as much in hydroponic systems as they do in regular farming, they can still spread in a hydroponic setup if not dealt with fast enough. Because plants grow close together, one infection can put a whole crop at risk. So, it is crucial to have backup plans and strong monitoring systems to reduce these risks [45].

Finally, whereas hydroponics is usually more efficient with water than traditional farming systems it still requires careful management the water applied is of too much consequence [42, 44]. Growers will have to remove impurities that could hinder plant health. That will certainly help us remember that we will need a grow room and filters to succeed in indoor gardening; yes, most people start at it without any commercial equipment [44]. Losing the nutrients from used nutrient solutions is a problem. The concentration of these solutions may be higher than others and if they are let into the environment. Ultimately, responsible disposal or recycling systems are crucial in order to minimize the environmental impact of this new type of farming [45].

In short, if the hurdles posed by hydroponics are to be tackled, then it becomes crucial that every aspect of these be addressed from the start. But with education, technology and sustainable farming systems, farmers can build their capacity to solve problems such as these in ways that contribute to a more flexible, food infrastructure for all involved [44, 45].

11.5 Artificial Intelligence Implementations in Smart Farming

AI is enabling more accurate, effective, and sustainable farming methods, which is transforming the agricultural industry. AI-driven solutions are revolutionizing the whole agricultural value chain, from automating difficult operations to optimizing resource utilization. Important uses include:

11.5.1 Precision Agriculture

Precision agriculture is a revolutionary method of farming that uses cutting-edge technology to control crops at the smallest level, thereby maximizing yields and optimizing resources [46, 47].

This accuracy is made possible by AI, which improves the effectiveness of operations and permits decision-making based on information [48]. The following are some essential elements of precision agriculture:

11.5.1.1 Variable Rate Technology

Variable Rate Technology (VRT) analyses data from detectors, drones, as well as satellites, among other sources, using AI-powered systems [49]. By using this technology to produce intricate maps of agricultural areas, producers may spot differences in the nutrient levels, soil characteristics, and general crop health [49, 50].

11.5.1.2 Real-Time Crop Monitoring

AI enhances real-time crop monitoring through the integration of data from various sources, including:

Unmanned Aerial Vehicles (UAVs) and Drones Drones that are outfitted with multispectral and hyperspectral imaging systems are able to take high-resolution pictures of crops [51].

Satellite Imagery By offering a more comprehensive view of vast agricultural regions, satellite technology helps farmers track crop development, evaluate water stress, and track changes in land usage over time. For resource management and strategic planning, this all-encompassing perspective is crucial [52].

Ground Sensors These sensors gather local information on temperature, nutrient levels, soil moisture, and other environmental factors. Making educated decisions on pest control, fertilizer, and irrigation requires knowledge of this information [53].

11.5.1.3 Predictive Analytics

Another essential element of precision agriculture is predictive analytics, which empowers farmers to take pre-emptive measures based on data analysis. AI can predict future events by looking at past data on market trends, pest outbreaks, weather patterns, and soil conditions [52].

For instance, by combining past data on insect behaviors with weather projections, AI can forecast the probability of a pest outbreak. Farmers may protect their crops and maximize harvests by using pre-emptive steps like applying insecticides or modifying planting schedules thanks to this forethought [52].

11.5.2 Autonomous Farming

AI is driving the development of autonomous systems that can perform tasks with minimal human intervention:

11.5.2.1 Robotic Systems

Several agricultural chores, including the planting process, picking, weeding, and pruning, are automated through the use of robotic systems [53].

To ensure ideal timing and minimize plant damage, robotic harvesters, for example, can be taught to recognize ripe produce and harvest it selectively [54].

11.5.2.2 Autonomous Vehicles

In agricultural environments, autonomous vehicles are essential for improving operational efficiency [55]. Without requiring human assistance, these vehicles can move equipment, materials, and plants across fields or inside greenhouses [55].

11.5.3 AI-Powered Decision Support Systems

AI provides farmers with powerful tools to make informed decisions:

11.5.3.1 Expert Systems

These are complex knowledge-based programs created to simulate human experts' decision-making in particular fields [56]. Expert systems can help farmers and other agricultural professionals make well-informed decisions that improve sustainability and productivity by utilizing an enormous repository containing agricultural understanding and sophisticated algorithms [57].

Crop Management Expert systems are able to examine a number of variables that impact crop growth, including crop rotation techniques, soil composition, and climatic conditions [58]. The system may suggest ideal planting times, appropriate crop types, and best procedures for fertilization and irrigation by incorporating this data [59]. For example, based on past yield statistics and current weather projections, an expert system may recommend particular crops that do well in a farmer's region's climate.

Pest Control By examining information on pest numbers, evolution, and environmental factors, these systems can offer recommendations for efficient pest management techniques [60]. An intelligent system might, for instance, use historical trends and real-time monitoring data to notify farmers of an approaching pest outbreak, enabling prompt and efficient responses [61, 62].

Resource Optimization Sustainable agriculture depends on effective resource management [63]. Expert systems are able to evaluate information on energy use, nutrition levels, and water usage and offer suggestions for the best possible resource allocation [64]. These devices assist farmers in cutting waste and operating expenses by recommending accurate hydration time frames, nutritional applications, and energy-efficient techniques [63, 64].

11.5.3.2 Decision Support Tools

Advanced AI-driven systems known as "decision support tools" are made to help farmers and agricultural management make well-informed decisions by evaluating data from various sources [65]. These resources compile data on a range of agricultural techniques and offer practical suggestions for the best times to plant, when to water, how much fertilizer to use, and other crucial management techniques [66]. Decision support technologies improve agricultural operations' efficacy and efficiency by utilizing data analysis and ML algorithms in a simple work design as shown in Fig. 11.6.

Data Integration Various data sources, such as crop growth models, satellite imagery, soil moisture readings, and weather forecasts, are combined via decision support tools [66].

Ideal Planting Times Decision support technologies can suggest the ideal times to plant particular crops by examining past weather trends, soil characteristics, and crop growth cycles [67]. To ensure ideal germination and growth, a tool might, for example, recommend planting maize when soil temperatures hit a specific threshold [66].

Management of Pests and Diseases By examining historical data and environmental variables, decision support technologies can also help identify possible hazards from pests and diseases. These tools allow farmers to take preventive measures to safeguard their crops by offering early alerts and suggestions for pest control techniques [68].

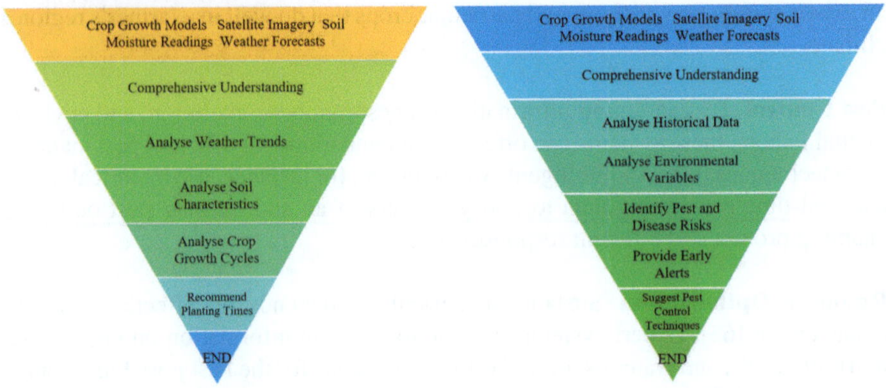

Fig. 11.6 Decision support tools working methodology

11.5.4 Sustainable Agriculture

Through better crop resilience, resource management, and environmental impact reduction, AI is essential to the advancement of sustainable farming methods. AI supports sustainable agriculture in the following important areas:

11.5.4.1 Optimization of Water

By optimizing irrigation plans based on real-time data, AI technology can greatly improve agricultural water management [69]. AI systems can make accurate suggestions for irrigation amount and timing by evaluating weather patterns, soil moisture levels, and the unique water requirements of various crops [71].

This focused strategy guarantees that crops receive the appropriate amount of water at the appropriate time while lowering total water usage and waste. In order to avoid over-irrigation and save precious water resources, an AI-driven irrigation system, for instance, can dynamically modify watering schedules in response to changing rainfall forecasts [71].

11.5.4.2 Climate-Resilient Crop Development

By examining enormous genetic information, AI plays a key role in hastening the creation of crop types that are tolerant to climate change [71]. AI can assist breeders in choosing and creating crops that are more resilient to the difficulties brought on by climate change by detecting genetic features linked to resistance to heat, drought, pests, and diseases [70].

In addition to improving food security, this skill helps farmers adjust to shifting environmental conditions [71]. To expedite the process of breeding and shorten the time needed to create new, hardier crop varieties, AI algorithms, for example, can examine the genetic makeup of current crop varieties to identify features that confer resilience [70].

11.5.4.3 Enhancing Resilience

Identifying Resilient Traits AI can help identify key traits that enhance resilience, such as root depth, leaf morphology, and water-use efficiency [72]. By focusing on these traits, breeders can develop cultivars that are better equipped to withstand extreme weather events, such as droughts, floods, and temperature fluctuations [73].

Accelerated Breeding Programs The integration of AI in breeding programs can significantly accelerate the development of resilient crop varieties [74]. By streamlining the selection process and reducing the time required for traditional breeding methods, AI enables faster deployment of climate-resilient cultivars in both field and soilless systems [75].

Phenotyping Technologies AI-driven phenotyping technologies, such as remote sensing and image analysis, can assess plant traits in real time [76]. These technologies provide valuable insights into plant health and stress responses, allowing for the identification of resilient varieties based on their performance under varying environmental conditions [77].

11.5.5 Enhancing Integrated Pest Management (IPM) Through AI

Integrated Pest Management (IPM) is a sustainable approach to controlling pests that combines various strategies to reduce damage while protecting human health and the environment [78]. The introduction of artificial intelligence (AI) is revolutionizing IPM by improving its effectiveness and accuracy [79]. Here is a closer look at how AI is enhancing IPM practices:

11.5.5.1 Automated Pest Detection and Identification

Computer Vision AI technologies, such as cameras and sensors, capture high-resolution images of crops, allowing for the automated detection and identification of pests [80]. This early detection helps prevent significant crop damage [81].

Machine Learning AI algorithms can learn to recognize specific pest species based on their unique visual characteristics, streamlining the identification process and reducing the need for manual inspections [82].

Real-Time Monitoring Continuous surveillance by AI systems provides immediate alerts about pest presence and population changes, enabling farmers to act quickly to prevent outbreaks [83].

11.5.5.2 Predictive Modeling for Pest Outbreaks

Data Analysis AI can analyze large datasets, including weather patterns and historical pest data, to predict potential pest outbreaks in specific areas and times [84].

Risk Assessment By evaluating various risk factors, AI helps farmers identify the most vulnerable areas, allowing for targeted pest management strategies [84, 85].

Optimized Timing AI determines the best times for applying pest control measures, enhancing their effectiveness while reducing environmental impact [85].

11.5.5.3 Precision Pest Control Strategies

Targeted Application AI systems enable precise application of pesticides or biological agents only where pests are detected, minimizing overall pesticide use and protecting non-target organisms [86].

Variable-Rate Technology AI allows for adjusting application rates based on specific field conditions, improving resource efficiency and reducing costs [86].

Site-Specific Management By integrating data from various sources, AI creates customized pest management plans tailored to the unique conditions of each field [87].

11.5.5.4 Smart Traps and Real-Time Monitoring

Remote Monitoring AI-powered traps can monitor pest populations remotely, providing real-time data and reducing the need for manual inspections [88].

Automated Alerts AI systems can send alerts when pest populations exceed critical levels, enabling timely interventions [88].

Data-Driven Decisions Information from smart traps, combined with other data sources, offers a comprehensive view of pest pressures, guiding strategic management decisions [88].

11.5.5.5 Autonomous Pest Management Devices

Robotic Systems AI-equipped robots can navigate fields, identify pest issues, and perform targeted actions like spraying herbicides or releasing beneficial insects, enhancing efficiency and reducing labor reliance [89].

Precision Application These devices deliver pest control measures accurately, minimizing environmental impact and operational costs [88, 89].

Continuous Monitoring Robotic systems provide ongoing field surveillance, offering real-time insights into pest presence and crop health for proactive management [77].

11.5.5.6 Benefits of AI-Enhanced Integrated Pest Management

Reduced Pesticide Use Targeted applications lead to lower pesticide usage and decreased environmental risks [90].

Improved Pest Control Early detection and timely responses significantly reduce pest outbreaks and crop damage [90].

Increased Efficiency Automation and real-time monitoring lessen the need for manual labor, streamlining pest management operations [91].

Enhanced Sustainability AI-driven IPM supports sustainable practices by minimizing environmental impacts and protecting biodiversity [92].

Data-Driven Decision-Making AI provides farmers with valuable insights, empowering informed pest management decisions [92].

In summary, integrating AI into Integrated Pest Management not only boosts the precision and effectiveness of pest control but also promotes sustainable agricultural practices, ultimately strengthening food systems against evolving pest challenges.

11.6 AI Farming Solutions for Small Farmers

Even though AI has the potential to completely transform the agriculture industry, small-scale farmers, who are essential to maintaining global food security, have a difficult time embracing AI [93]. In order to improve the productivity and resilience of smallholder farmers, this section will examine how AI can be applied to their unique requirements and limitations [94].

11.6.1 Leveraging IoT for Data Collection and Transmission

Robust data gathering is the cornerstone of AI-driven smart farming since it is necessary for optimizing agricultural practices and making well-informed judgments [95]. The deployment of several sensor networks that collect real-time data on vital aspects influencing agricultural productivity and health is made possible by the IoT, which is essential to this process. The overall process is depicted in Fig. 11.7 as how the data transmission from sensors can be done, providing usefulness. These are the main IoT elements used in smart farming:

11.6.1.1 Sensor Systems

Installing dependable and reasonably priced sensor networks is essential for gathering data in real time on a range of agricultural characteristics [96]. These sensors offer insightful information that can improve resource management and decision-making. Important kinds of sensors include:

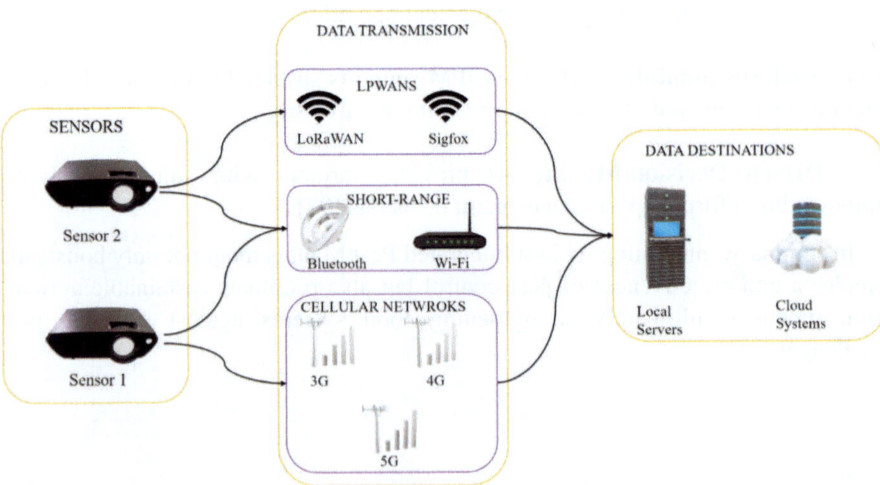

Fig. 11.7 Data transmission and connectivity process

Soil Moisture Sensors By determining the amount of water in the soil, these sensors help farmers better control irrigation. Farmers can improve watering schedules and guarantee that crops are getting the proper quantity of water while reducing waste by using real-time data on moisture in the soil levels [97].

Climate and Humidity Sensors Understanding microclimates and determining plant stress depend on the measurement of air and soil temperatures as well as humidity levels. Farmers can use this information to make well-informed decisions on crop management techniques, like whether to fertilize or irrigate [98].

Weather Stations These observatories gather information on a range of meteorological factors, such as sun radiation, wind speed, and rainfall. Farmers are able to predict fluctuations in weather patterns and modify their activities accordingly thanks to this knowledge, which is crucial for forecasting the weather and irrigation planning [99].

Water Quality Sensors These devices measure variables including pH, dissolved oxygen levels, and electrical conductivity (EC) in aquaculture or hydroponic systems. The health of aquatic life and plants depends on maintaining ideal water quality, and these sensors offer real-time monitoring to make sure that conditions stay ideal [100].

11.6.1.2 Data Transmission

Data must be sent to local servers or cloud-based systems for analysis after it has been gathered [101]. This data transmission can be facilitated by a number of technologies, guaranteeing that information is available for decision-making:

Cellular Networks (3G, 4G, and 5G) These networks are ideal for sending massive volumes of data from distant detectors to centralized systems because of their broad coverage and comparatively high capacity. Farmers can obtain data in real time from any location by connecting to cellular networks, which improves their capacity to react to shifting situations [102].

Low-Power Wide Area Networks (LPWANs) Long range wide area network (LoRaWAN) and Sigfox are two examples of technologies that offer long-distance communication with minimal power usage [103]. These networks enable sensors to send data over long distances without requiring substantial power sources, making them perfect for isolated locations with inadequate infrastructure [104].

Bluetooth and Wi-Fi These technologies can be used to link sensors in a small greenhouse or farm for short-range communication [104]. These technologies facilitate local monitoring and control by enabling fast data transfer between devices [110].

11.6.2 Addressing Specific Challenges Faced by Small Farmers

A careful strategy that takes into account their particular difficulties is required for small farmers to successfully use AI technologies. Global food security depends on smallholder farmers, but they frequently encounter major obstacles when attempting to implement cutting-edge technologies. The following are some of the main issues and possible fixes:

11.6.2.1 Restricted Infrastructure and Technology Access

Many small farmers work in areas with limited access to energy, dependable internet connectivity, and cutting-edge equipment. The deployment of AI technologies may be hampered by this lack of infrastructure.

Solution Farmers may benefit from useful information and tools by creating AI-powered smartphone applications that operate offline or with limited bandwidth access. Without requiring continuous internet access, these apps can provide features like crop monitoring tools, pest control guidance, and weather forecasts. Developers can enable small farmers to efficiently employ AI by making sure these solutions are accessible, easy to use, and flexible enough to accommodate local conditions.

11.6.2.2 Quality and Scarcity of Data

Due to a lack of resources for gathering fresh data and old data, small farms frequently struggle with data scarcity. The efficacy of AI models, which often need large datasets for training, may be hampered by this restriction.

Solution Models can be trained with little data by utilizing AI techniques like transfer learning along with small data learning. Transfer learning enhances accuracy even with sparse data by enabling models trained on bigger sets of data to be modified for particular tasks in smaller environments. Furthermore, by combining information from several farms, crowdsourcing data might assist overcome data shortages and facilitate the creation of more reliable AI models that take into account a range of farming circumstances.

11.6.2.3 Digital Literacy and the Skill Gap

It may be difficult for many small farmers to successfully adopt and use new technology due to their lack of technical expertise and digital literacy.

Solution Farmers can be prepared to handle new technologies by putting in place AI-driven education initiatives that employ straightforward language and real-world examples. Through hosting of workshops, online courses, or in person training, these programs can be brought closer to the farmers. In order to ensure that farmers feel comfortable utilizing AI tools to improve their practices, local agricultural extension organizations can be quite helpful in offering continuous training and assistance.

11.6.2.4 Accessibility and Affordability

Due of their frequently limited resources, small farmers may find the expense of applying AI solutions to be a major obstacle.

Solution Improving affordability requires the development of inexpensive hardware and software. Using open-source technologies can lower expenses and give farmers access to strong tools without having to pay high license fees. Additionally, small farmers can invest in AI solutions without experiencing severe financial burden by investigating cutting-edge finance approaches like micro financing or cooperative purchasing agreements.

11.6.3 Developing Effective Strategies for Technological Integration

Careful planning and execution are necessary for the effective incorporation of machine learning into agricultural processes in order to guarantee that the solutions are practical, affordable, and long-lasting. These are some crucial factors to take into account in order to accomplish this integration, especially for small-scale farmers:

11.6.3.1 Interfaces That Are Easy to Use

For novice farmers with little technical expertise, AI-powered equipment must have straightforward, user-friendly interfaces. Developers may improve users' experiences by creating user interfaces with usability as their top priority.

Solution Usability can be greatly increased by including indigenous languages and aids with imagery like infographics and icons. For example, an app that offers pest control guidance can display pictures of crops and pests to assist farmers in rapidly recognizing problems and comprehending suggested courses of action. This method not only increases user confidence but also renders the technology more approachable.

11.6.3.2 Solutions That Are Scalable and Adaptable

AI systems must to be customized to small-scale farmers' unique requirements, taking into consideration variables like farm size, crop variety, regional climate, and resource availability.

Solution By letting farmers choose the aspects that are most pertinent to their operations, customizable solutions make sure that the technology takes into account their particular difficulties. Scalability is also crucial; farmers should be able to progressively embrace more sophisticated features as their demands change and they grow more accustomed to the technology. This adaptability promotes sustained use of AI techniques.

11.6.3.3 Participatory and Community-Based Methods

Involving farmers and local communities in the development and application of AI technologies is essential to guaranteeing their uptake and sustainability.

Solution Farmers can adopt technology more quickly and develop a sense of ownership by forming farmer cooperatives and encouraging knowledge exchange. Solutions can be more closely matched to farmers' requirements and preferences by including them in the development process. Feedback sessions and community workshops can also aid in improving the technology by drawing on practical experiences.

11.6.3.4 Combining with Current Methods

AI systems must to be made to easily mesh with current agricultural methods and regional expertise.

Solution This integration makes the transition easier and helps farmers comprehend the advantages of the new technology. For instance, an AI tool that makes irrigation suggestions ought to take into account currently used conventional irrigation techniques, enabling farmers to apply AI insights without radically changing their ways. Farmers are more willing to use AI if it is shown how, it can improve current practices.

11.6.3.5 Local Collaborations and Assistance

Working together with regional research institutes, NGOs, and agricultural extension agencies can offer important resources and assistance for putting AI solutions into practice.

Solution These collaborations can help with financial access, training program facilitation, and continuing technical assistance. Local groups can assist in bridging the gap between end users and technology suppliers because they frequently have existing contacts with farmers. Applications of AI can be implemented more successfully and sustainably by utilizing local resources and skills.

11.7 Benefits of AI for Small-Scale Farmers

There are many benefits for small-scale farmers when AI is successfully incorporated into farming methods. These farmers can see notable gains in a number of areas of their operations by utilizing AI technologies, which will ultimately improve their standard of living and help ensure the security of the world's food supply. The main advantages consist of:

11.7.1 Increased Productivity and Yields

AI can enhance crop management techniques and maximize resource use, resulting in increased yields and productivity. Farmers may minimize crop losses and increase yield by using data-driven insights to guide their decisions about planting times, succession planting, and pest control.

11.7.2 Reduced Costs

Small-scale farmers can reduce input costs related to water use, pesticides, and fertilizers by utilizing AI technologies. Farmers may apply inputs more effectively and make sure supplies are being utilized only where and when needed by using precision agriculture techniques. Further increasing total resource efficiency, automation of some processes can result in lower labor expenses.

11.7.3 Enhanced Sustainability

By lowering dependency on synthetic fertilizers and enhancing resource management, AI adoption can support ecologically friendly farming methods. AI-powered technologies can make it easier to monitor pest populations, water use, and soil health, empowering farmers to use environmentally friendly and productive sustainable methods.

11.7.4 Improved Market Access

Small-scale farmers can connect with potential customers and strengthen their negotiating position by using AI to increase their access to market information. Farmers may make more informed choices about the best time and place to sell their goods by using AI-driven platforms to obtain information into pricing, demand, and market trends. Better prices and more stable incomes may result from this expanded market access.

11.7.5 Increased Resilience to Climate Change

AI tools can help small-scale farmers lessen the effects of global warming on their livelihoods and adjust to shifting weather patterns. AI can assist farmers in making proactive decisions that increase their adaptability to climate-related risks like droughts, floods, and changing growing seasons by offering real-time meteorological information, statistical modeling, and climate-smart agricultural practices.

11.8 Revolutionizing Agriculture: Exploring the Challenges and Opportunities of AI-Powered Smart Farming

Although AI has the potential to revolutionize agriculture, there are obstacles in the way of its effective application. In order to fully utilize AI in agriculture, these issues must be handled sensibly and successfully. However, new developments in technology offer promising prospects for other breakthroughs that can improve sustainability and farming methods.

11.8.1 Challenges to Address

11.8.1.1 Data Security and Privacy

Privacy and security issues are brought up by the gathering and analysis of enormous volumes of agricultural data. Farmers may be reluctant to divulge private information about their business for fear of abuse or illegal access. Gaining farmers' trust requires implementing strong data protection procedures and open data usage guidelines.

11.8.1.2 Integration with Current Systems

A lot of small-scale farmers may be reluctant to embrace new technology since they depend on conventional farming methods. To guarantee a seamless transition, significant preparation and training are needed when integrating AI technologies with current procedures and systems. By offering assistance and training, farmers may better grasp the advantages of AI and how it might enhance their existing practices.

11.8.1.3 Cost of Implementation

For many farmers, especially smallholders with limited resources, the upfront costs associated with AI technologies might be a major deterrent. To promote broad adoption, AI solutions must be made accessible and reasonably priced. This financial load can be lessened with the use of creative financing techniques like cooperative purchasing or microloans.

11.8.1.4 Technical Expertise

A certain amount of specialized expertise and digital literacy are frequently necessary for the effective application of AI technologies. It is possible that many farmers lack the expertise needed to use AI products efficiently. By offering specialized training courses and materials, farmers can be better equipped to use AI and improve their farming methods.

11.8.2 Opportunities for Advancement

11.8.2.1 Precision Agricultural

The application of precision agricultural techniques, which maximize resource utilization and enhance crop management, is made possible by the development of AI technologies. Farmers may improve sustainability and production by using data-driven insights to inform their decisions.

11.8.2.2 Climate-Smart Agriculture

AI has the potential to be a key component in creating climate-smart farming methods that assist farmers in adjusting to shifting environmental circumstances. AI can offer robust varieties of crops and sustainable agricultural methods by examining soil data and weather trends.

11.8.2.3 Improved Market Access

By offering real-time data on demand, pricing, and market trends, emerging AI-driven systems can help farmers have better access to markets. Farmers are better equipped to decide which locations to sell their goods because to this access, which eventually increases their revenue and stability.

11.8.2.4 Cooperation and Information Exchange

Farmers, scientists, and technology companies work together more when AI is incorporated into agriculture. Stakeholders can collaborate to create creative solutions that tackle shared problems and improve farming methods by exchanging information and resources.

11.9 Summary

A revolutionary technical frontier, the intersection of agricultural science with AI gives previously unheard-of chances to tackle the intricate problems confronting the world's food systems. We can create savvy, sustainable agricultural systems that can meet the rising demand for food worldwide while protecting the resources of our planet by meticulously considering computational, natural, and human factors—from maximizing resource use and reducing environmental impact to providing farmers with data-driven insights. A coordinated and cooperative effort is necessary to fully utilize AI in agriculture. Persistent interdisciplinary research that encourages cooperation between social scientists, engineers, agricultural scientists, and AI researchers must be our top priority. This partnership is essential for creating novel and practical solutions, concentrating on strong AI algorithms, combining various data sources, and comprehending the intricate relationships found in agricultural ecosystems.

In order to encourage innovation and research, infrastructure development, including rural broadband access, along with educational initiatives involving farmers and agricultural professionals, governments, businesses, and research institutions must make strategic investments. AI-powered agricultural technology can be deployed more quickly through collaborative technological development via open-source projects, information exchange platforms, and public-private collaborations. For solutions to be applicable, easily available, and user-friendly, cooperation among technology developers, agricultural companies, and farmer communities is essential.

Furthermore, in order to ensure an accountable and equitable use of AI in agriculture, it is imperative that clear ethical principles and regulatory frameworks be established for its development and implementation. These frameworks ought to cover topics like algorithmic bias, data privacy, and the possible effects on agricultural labor.

In order to promote adoption and establish public trust, it is crucial to educate farmers, consumers, and politicians about the advantages and potential of AI in agriculture. Education and awareness are equally critical.

More food production is not the only goal of agriculture's future but it also needs to be done more sustainably, smartly, and efficiently. It is about equipping farmers with the skills and information they need to prosper in a world that is changing. The goal is to preserve our planet's valuable resources while providing food security for the growing global population. We can create a more robust, just and resilient agricultural system that benefits farmers and consumers alike by incorporating AI and its revolutionary applications, ensuring a wealthy and nutrition-secure future for future generations.

Acknowledgements We are thankful to HiCHiCoB Centre supported by Department of Biotechnology, Government of India, iHUB &HCI Foundation Grants (GAP-0338; GAP0332) and MLP-201 under CSIR-Network Institutional project.

References

1. Holzinger A, Fister I Jr, Fister I, Kaul HP, Asseng S (2024) Human-centered AI in smart farming: towards agriculture 5.0. IEEE Access
2. Snowdon RJ, Wittkop B, Chen TW, Stahl A (2021) Crop adaptation to climate change as a consequence of long-term breeding. Theor Appl Genet 134(6):1613–1623

3. Mana AA, Allouhi A, Hamrani A, Rahman S, el Jamaoui I, Jayachandran K (2024) Sustainable AI-based production agriculture: exploring AI applications and implications in agricultural practices. Smart Agricultural Technology:100416
4. Altayeb, J.M., Eleyan, H., Wishah, N.D., Elmahmoum, A.E., Khalil, A.J., Abu-Nasser, B.S. and Abu-Naser, S.S., 2024. AI-driven innovations in agriculture: transforming farming practices and outcomes
5. Barrett GE, Alexander PD, Robinson JS, Bragg NC (2016) Achieving environmentally sustainable growing media for soilless plant cultivation systems–a review. Sci Hortic 212:220–234
6. Fussy A, Papenbrock J (2022) An overview of soil and soilless cultivation techniques—chances, challenges and the neglected question of sustainability. Plants 11(9):1153
7. Liakos KG, Busato P, Moshou D, Pearson S, Bochtis D (2018) Machine learning in agriculture: a review. Sensors 18(8):2674
8. Benos L, Tagarakis AC, Dolias G, Berruto R, Kateris D, Bochtis D (2021) Machine learning in agriculture: a comprehensive updated review. Sensors 21(11):3758
9. Jagtap ST, Phasinam K, Kassanuk T, Jha SS, Ghosh T, Thakar CM (2022) Towards application of various machine learning techniques in agriculture. Mater Today Proc 51:793–797
10. Altieri MA, Nicholls CI (2017) The adaptation and mitigation potential of traditional agriculture in a changing climate. Clim Chang 140:33–45
11. Sharma P, Dadheech P, Aneja N, Aneja S (2023) Predicting agriculture yields based on machine learning using regression and deep learning. IEEE Access 11:111255
12. Monteleone B, Borzí I, Bonaccorso B, Martina M (2023) Quantifying crop vulnerability to weather-related extreme events and climate change through vulnerability curves. Nat Hazards 116(3):2761–2796
13. Fraser ED, Campbell M (2019) Agriculture 5.0: reconciling production with planetary health. One Earth 1(3):278–280
14. Lesk C, Rowhani P, Ramankutty N (2016) Influence of extreme weather disasters on global crop production. Nature 529(7584):84–87
15. Elavarasan D, Vincent PD (2020) Crop yield prediction using deep reinforcement learning model for sustainable agrarian applications. IEEE access 8:86886–86901
16. Patrício DI, Rieder R (2018) Computer vision and artificial intelligence in precision agriculture for grain crops: a systematic review. Comput Electron Agric 153:69–81
17. Ding W, Abdel-Basset M, Alrashdi I, Hawash H (2024) Next generation of computer vision for plant disease monitoring in precision agriculture: a contemporary survey, taxonomy, experiments, and future direction. Inf Sci 665:120338
18. Tian H, Wang T, Liu Y, Qiao X, Li Y (2020) Computer vision technology in agricultural automation—a review. Inform Process Agric 7(1):1–19
19. Ghazal S, Munir A, Qureshi WS (2024) Computer vision in smart agriculture and precision farming: techniques and applications. Artif Intellig Agric 13:64–83
20. Luo J, Li B, Leung C (2023) A survey of computer vision technologies in urban and controlled-environment agriculture. ACM Comput Surv 56(5):1–39
21. Dhanush G, Khatri N, Kumar S, Shukla PK (2023) A comprehensive review of machine vision systems and artificial intelligence algorithms for the detection and harvesting of agricultural produce. Scientific African 21:e01798
22. Benavides M, Cantón-Garbín M, Sánchez-Molina JA, Rodríguez F (2020) Automatic tomato and peduncle location system based on computer vision for use in robotized harvesting. Appl Sci 10(17):5887
23. Gupta DSK, Malik S (2022) Application of predictive analytics in agriculture. TTIDMKD 2(4):1–5
24. Nagini S, Kanth TR, Kiranmayee BV (2016, December) Agriculture yield prediction using predictive analytic techniques. In: 2016 2nd international conference on contemporary computing and informatics (IC3I). IEEE, pp 783–788

25. Sharma K, Sharma C, Sharma S, Asenso E (2022) Broadening the research pathways in smart agriculture: predictive analysis using semiautomatic information modeling. J Sensor 2022(1):54. Kerr A, Keane P (1997) Prediction of disease outbreaks. In: Brown JF, Ogle HJ (eds) Plant pathogens and plant diseases, pp 299–313
26. Prabhu BB, Dakshayini M (2017, November) Demand-prediction model for forecasting AGRI-needs of the society. In: 2017 International Conference on Inventive Computing and Informatics (ICICI). IEEE, pp 430–435
27. Oliveira Pezente ADSOP (2018) Predictive demand models in the food and agriculture sectors: an analysis of the current models and results of a novel approach using machine learning techniques with retail scanner data (Doctoral dissertation, Massachusetts Institute of Technology)
28. Gent DH, Mahaffee WF, McRoberts N, Pfender WF (2013) The use and role of predictive systems in disease management. Annu Rev Phytopathol 51(1):267–289
29. Mehra LK, Cowger C, Gross K, Ojiambo PS (2016) Predicting pre-planting risk of Stagonospora nodorum blotch in winter wheat using machine learning models. Front Plant Sci 7:390
30. Kim WS, Lee WS, Kim YJ (2020) A review of the applications of the internet of things (IoT) for agricultural automation. J Biosyst Eng 45:385–400
31. Tzounis A, Katsoulas N, Bartzanas T, Kittas C (2017) Internet of things in agriculture, recent advances and future challenges. Biosyst Eng 164:31–48
32. Shenoy J, Pingle Y (2016, March) IOT in agriculture. In: 2016 3rd International Conference on Computing for Sustainable Global Development (INDIACom). IEEE, pp 1456–1458
33. Quy VK, Hau NV, Anh DV, Quy NM, Ban NT, Lanza S, Randazzo G, Muzirafuti A (2022) IoT-enabled smart agriculture: architecture, applications, and challenges. Appl Sci 12(7):3396
34. Jan S, Rashid Z, Ahngar TA, Iqbal S, Naikoo MA, Majeed S, Bhat TA, Gul R, Nazir I (2020) Hydroponics–a review. Int J Curr Microbiol App Sci 9(8):1779–1787
35. Velazquez-Gonzalez RS, Garcia-Garcia AL, Ventura-Zapata E, Barceinas-Sanchez JDO, Sosa-Savedra JC (2022) A review on hydroponics and the technologies associated for medium-and small-scale operations. Agriculture 12(5):646
36. Rufí-Salís M, Calvo MJ, Petit-Boix A, Villalba G, Gabarrell X (2020) Exploring nutrient recovery from hydroponics in urban agriculture: an environmental assessment. Resour Conserv Recycl 155:104683
37. Wibisono V, Kristyawan Y (2021) An efficient technique for automation of the NFT (Nutrient Film Technique) hydroponic system using Arduino. Int J Artif Intell Robot (IJAIR) 3(1):44–49
38. Nursyahid A, Setyawan TA, Sa'diyah K, Wardihani ED, Helmy H, Hasan A (2021, March) Analysis of Deep Water Culture (DWC) hydroponic nutrient solution level control systems. In: IOP conference series: materials science and engineering, vol 1108, No. 1. IOP Publishing, p 012032
39. Garzón J, Montes L, Garzón J, Lampropoulos G (2023) Systematic review of technology in aeroponics: introducing the technology adoption and integration in sustainable agriculture model. Agronomy 13(10):2517
40. Patil ST, Kadam US, Mane MS, Mahale DM, Dhekale JS (2020) Hydroponic growth media (substrate): a review. Int Res J Pure Appl Chem 21(23):106–113
41. Ekaputri RZ, Jumiarni D, Usman B, Karyadi B, Putra EP (2021) The potential of hydroponics as a solution for food security and economics resilience: issues and challenges of smart-agriculture. Engagement: Jurnal Pengabdian Kepada Masyarakat 5(2):493–505
42. Armas KL, Lorenzo EG, Cruz CD (2023) Financial viability of business models for engineered vertical hydroponics systems for sustainable onion production in The Philippines. J Appl Eng Technol Sci (JAETS) 4(2):864–872
43. Souza V, Gimenes RMT, de Almeida MG, Farinha MUS, Bernardo LVM, Ruviaro CF (2023) Economic feasibility of adopting a hydroponics system on substrate in small rural properties. Clean Techn Environ Policy 25(8):2761–2775

44. Khatri L, Kunwar A, Bist DR (2024) Hydroponics: advantages and challenges in soilless farming. Big Data Agric (BDA) 6:81–88
45. Dusadeerungsikul PO, Nof SY (2024) Precision agriculture with AI-based responsive monitoring algorithm. Int J Prod Econ 271:109204
46. Akintuyi OB (2024) Adaptive AI in precision agriculture: a review: investigating the use of self-learning algorithms in optimizing farm operations based on real-time data. Res J Multidiscip Stud 7(02):016–030
47. Anurag D, Roy S, Bandyopadhyay S (2008, May) Agro-sense: precision agriculture using sensor-based wireless mesh networks. In: 2008 first Itu-t kaleidoscope academic conference-innovations in ngn: future network and services. IEEE, pp 383–388
48. Sawyer JE (1994) Concepts of variable rate technology with considerations for fertilizer application. J Prod Agric 7(2):195–201
49. Šarauskis E, Kazlauskas M, Naujokienė V, Bručienė I, Steponavičius D, Romaneckas K, Jasinskas A (2022) Variable rate seeding in precision agriculture: recent advances and future perspectives. Agriculture 12(2):305
50. Sharma K, Shivandu SK (2024) Integrating artificial intelligence and internet of things (IoT) for enhanced crop monitoring and management in precision agriculture. Sensors International:100292
51. Brook A, De Micco V, Battipaglia G, Erbaggio A, Ludeno G, Catapano I, Bonfante A (2020) A smart multiple spatial and temporal resolution system to support precision agriculture from satellite images: proof of concept on Aglianico vineyard. Remote Sens Environ 240:111679
52. Sishodia RP, Ray RL, Singh SK (2020) Applications of remote sensing in precision agriculture: a review. Remote Sens 12(19):3136
53. Eaton R, Katupitiya J, Siew KW, Howarth B (2010) Autonomous farming: modelling and control of agricultural machinery in a unified framework. Int J Intell Syst Technol Appl 8(1–4):444–457
54. Gackstetter D, von Bloh M, Hannus V, Meyer ST, Weisser W, Luksch C, Asseng S (2023) Autonomous field management–an enabler of sustainable future in agriculture. Agric Syst 206:103607
55. Pota H, Eaton R, Katupitiya J, Pathirana SD (2007, August) Agricultural robotics: a streamlined approach to realization of autonomous farming. In: 2007 international conference on industrial and information systems. IEEE, pp 85–90
56. Elbasi E, Mostafa N, Zaki C, AlArnaout Z, Topcu AE, Saker L (2024) Optimizing agricultural data analysis techniques through ai-powered decision-making processes. Appl Sci 14(17):8018
57. Zhai Z, Martínez JF, Beltran V, Martínez NL (2020) Decision support systems for agriculture 4.0: survey and challenges. Comput Electron Agric 170:105256
58. Jeffrey L, Bommu R (2024) Innovative AI solutions for agriculture: enhancing crop management and yield. Int J Adv Eng Technol Innov 1(3):203–221
59. Shankar P, Werner N, Selinger S, Janssen O (2020, September) Artificial intelligence driven crop protection optimization for sustainable agriculture. In: 2020 IEEE/ITU international conference on Artificial Intelligence for Good (AI4G). IEEE, pp 1–6
60. Panwar LC (2024) Artificial intelligence in farming: advancing crop management, pest control, and sustainable practices. J Punjab Acad Sci 24:86–92
61. Rahman SM, Ravi G (2022) Role of artificial intelligence in pest management. Curr Topic Agric Sci 7:64–81
62. Raman RK, Kumar A, Sarkar S, Yadav AK, Mukherjee A, Meena RS, Kumar U, Singh DK, Das S, Kumar R, Babu S (2024) Reconnoitering precision agriculture and resource management: a comprehensive review from an extension standpoint on artificial intelligence and machine learning. Indian Res J Exten Educ 24(1):108–123
63. Titirmare S, Margal PB, Gupta S, Kumar D (2024) AI-powered predictive analytics for crop yield optimization. In: Agriculture 4.0. CRC Press, Boca Raton, pp 89–110

64. Mishra H, Mishra D (2024) AI for data-driven decision-making in smart agriculture: from field to farm management. In: Artificial intelligence techniques in smart agriculture. Springer, Singapore, pp 173–193
65. Suneetha K (2023) Machine learning and artificial intelligence in agricultural decision support systems. Digital agriculture, p 36
66. Javaid M, Haleem A, Khan IH, Suman R (2023) Understanding the potential applications of artificial intelligence in agriculture sector. Adv Agrochem 2(1):15–30
67. Selvaraj MG, Vergara A, Ruiz H, Safari N, Elayabalan S, Ocimati W, Blomme G (2019) AI-powered banana diseases and pest detection. Plant Methods 15:1–11
68. Bhagat PR, Naz F, Magda R (2022) Artificial intelligence solutions enabling sustainable agriculture: a bibliometric analysis. PLoS One 17(6):e0268989
69. Alreshidi E (2019) Smart sustainable agriculture (SSA) solution underpinned by internet of things (IoT) and artificial intelligence (AI). arXiv preprint arXiv:1906.03106
70. Ashoka P, Devi BR, Sharma N, Behera M, Gautam A, Jha A, Sinha G (2024) Artificial intelligence in water management for sustainable farming: a review. J Sci Res Report 30(6):511–525
71. Usigbe MJ, Asem-Hiablie S, Uyeh DD, Iyiola O, Park T, Mallipeddi R (2024) Enhancing resilience in agricultural production systems with AI-based technologies. Environ Dev Sustain 26(9):21955–21983
72. Jung J, Maeda M, Chang A, Bhandari M, Ashapure A, Landivar-Bowles J (2021) The potential of remote sensing and artificial intelligence as tools to improve the resilience of agriculture production systems. Curr Opin Biotechnol 70:15–22
73. Rai KK (2022) Integrating speed breeding with artificial intelligence for developing climate-smart crops. Mol Biol Rep 49(12):11385–11402
74. Khan MHU, Wang S, Wang J, Ahmar S, Saeed S, Khan SU, Xu X, Chen H, Bhat JA, Feng X (2022) Applications of artificial intelligence in climate-resilient smart-crop breeding. Int J Mol Sci 23(19):11156
75. Sheikh M, Iqra F, Ambreen H, Pravin KA, Ikra M, Chung YS (2024) Integrating artificial intelligence and high-throughput phenotyping for crop improvement. J Integr Agric 23(6):1787–1802
76. Nabwire S, Suh HK, Kim MS, Baek I, Cho BK (2021) Application of artificial intelligence in phenomics. Sensors 21(13):4363
77. Kariyanna B, Sowjanya M (2024) Unravelling the use of artificial intelligence in management of insect pests. Smart Agric Technol 8:100517
78. Zhou W, Arcot Y, Medina RF, Bernal J, Cisneros-Zevallos L, Akbulut ME (2024) Integrated pest management: an update on the sustainability approach to crop protection. ACS omega 9(40):41130–41147
79. Li C, Wang M (2024) Pest and disease management in agricultural production with artificial intelligence: innovative applications and development trends. Adv Resour Res 4(3):381–401
80. Karar ME, Alsunaydi F, Albusaymi S, Alotaibi S (2021) A new mobile application of agricultural pests recognition using deep learning in cloud computing system. Alex Eng J 60(5):4423–4432
81. Kasinathan T, Singaraju D, Uyyala SR (2021) Insect classification and detection in field crops using modern machine learning techniques. Inform Process Agric 8(3):446–457
82. Ahmad I, Pothuganti K (2020, September) Smart field monitoring using ToxTrac: a cyber-physical system approach in agriculture. In: 2020 international conference on smart electronics and communication (ICOSEC). IEEE, pp 723–727
83. Klassen W, Vreysen MJB (2021) Area-wide integrated pest management and the sterile insect technique. In: Sterile insect technique. CRC Press, Boca Raton, pp 75–112
84. Hendrichs J, Vreysen MJB, Enkerlin WR, Cayol JP (2021) Strategic options in using sterile insects for area-wide integrated pest management. In: Sterile insect technique. CRC Press, Boca Raton, pp 841–884
85. Iost Filho FH, Heldens WB, Kong Z, de Lange ES (2020) Drones: innovative technology for use in precision pest management. J Econ Entomol 113(1):1–25

86. Karlsson Green K, Stenberg JA, Lankinen Å (2020) Making sense of Integrated Pest Management (IPM) in the light of evolution. Evol Appl 13(8):1791–1805
87. Preti M, Moretti C, Scarton G, Giannotta G, Angeli S (2021) Developing a smart trap prototype equipped with camera for tortricid pests remote monitoring. Bull Insectol 74(1)
88. Lima MCF, de Almeida Leandro MED, Valero C, Coronel LCP, Bazzo COG (2020) Automatic detection and monitoring of insect pests—a review. Agriculture 10(5):161
89. Preti M, Verheggen F, Angeli S (2021) Insect pest monitoring with camera-equipped traps: strengths and limitations. J Pest Sci 94(2):203–217
90. Sharma K, Shivandu SK (2024) Integrating artificial intelligence and internet of things (IoT) for enhanced crop monitoring and management in precision agriculture. Sensor Int:100292
91. Deguine JP, Aubertot JN, Flor RJ, Lescourret F, Wyckhuys KA, Ratnadass A (2021) Integrated pest management: good intentions, hard realities. A review. Agron Sustain Dev 41(3):38
92. Ngulube P (2025) Leveraging information and communication technologies for sustainable agriculture and environmental protection among smallholder farmers in tropical Africa. Discover Environ 3(1):1–17
93. Shafik W, Tufail A, Apong RAAHM, De Silva LC (2024) Internet of things for smart agricultural practices. In: Internet of things applications and technology. Auerbach Publications, pp 190–217
94. Liang C, Shah T (2023) IoT in agriculture: the future of precision monitoring and data-driven farming. Eigenpub Rev Sci Technol 7(1):85–104
95. Yin H, Cao Y, Marelli B, Zeng X, Mason AJ, Cao C (2021) Soil sensors and plant wearables for smart and precision agriculture. Adv Mater 33(20):2007764
96. Pramanik M, Khanna M, Singh M, Singh DK, Sudhishri S, Bhatia A, Ranjan R (2022) Automation of soil moisture sensor-based basin irrigation system. Smart Agric Technol 2:100032
97. Ma Z, Fei T, Zhang T (2023) An overview: sensors for low humidity detection. Sensors Actuators B Chem 376:133039
98. Ioannou K, Karampatzakis D, Amanatidis P, Aggelopoulos V, Karmiris I (2021) Low-cost automatic weather stations in the internet of things. Information 12(4):146
99. Yang H, Kong J, Hu H, Du Y, Gao M, Chen F (2022) A review of remote sensing for water quality retrieval: progress and challenges. Remote Sens 14(8):1770
100. Dey K, Shekhawat U (2021) Blockchain for sustainable e-agriculture: literature review, architecture for data management, and implications. J Clean Prod 316:128254
101. Avşar E, Mowla MN (2022) Wireless communication protocols in smart agriculture: a review on applications, challenges and future trends. Ad Hoc Netw 136:102982
102. Kuntke F, Sinn M, Reuter C (2021, August) Reliable data transmission using low power wide area networks (LPWAN) for agricultural applications. In: Proceedings of the 16th international conference on availability, reliability and security, pp 1–9
103. Liya ML, Arjun D (2020, October) A survey of LPWAN technology in agricultural field. In: 2020 fourth international conference on I-SMAC (IoT in social, Mobile, analytics and cloud) (I-SMAC). IEEE, pp 313–317
104. Balakrishna K, Rakshith N, Shashank R, Nithish Y, Praveen KM (2023) Implementation of LoRa and Bluetooth technology in farming application with performance analysis. Int J Comput Artif Intell 4(1):45–50

Chapter 12
Integrating IoT and Bioinformatics for Nutrient Monitoring in Hydroponic Systems

Tatenda Justice Gunda, Nedhi Jasrotia, Ali Baba Eshawu, Marbi Ete, Hena Dhar, and Karnika Thakur

Abstract Hydroponics offers a promising future for sustainable agriculture, but its full potential hinges on our ability to monitor and manage nutrient dynamics with precision and biological insight. The integration of Internet of Things (IoT) technologies with bioinformatics tools to create intelligent, data-driven nutrient management systems offers new smart farming technologies. There have been technological advances in development of environmental sensors measuring variables such as pH, electrical conductivity, and ion concentrations that can be paired with transcriptomic analysis to reveal how plants internally respond to nutrient availability. IoT technologies can be coupled with sensors and as such allowing the transmitting of data via cloud connected platforms for real-time analysis. By treating the hydroponic plants as a biological sensor, there is a shift from reactive management to proactive, plant-informed decision-making. This chapter discusses the design of integrated systems, data synchronization challenges, and methodological strategies for linking sensor data with gene expression profiles. As agriculture moves toward digital transformation, this work advocates for systems that are not only automated but also biologically aware offering a blueprint for the next generation of smart, sustainable hydroponic farming.

Keywords IoT-based nutrient monitoring · Hydroponic systems · Transcriptomics · Bioinformatics · Precision agriculture · Smart farming technologies

T. J. Gunda · N. Jasrotia · A. B. Eshawu · M. Ete · H. Dhar (✉) · K. Thakur (✉)
School of Biotechnology, Faculty of Applied Sciences and Biotechnology, Shoolini University, Solan, Himachal Pradesh, India

P. Kumar, A. R. Warghat (eds.), *Transforming Medicinal Plant Agriculture*, Smart Agriculture 16, https://doi.org/10.1007/978-981-95-3963-5_12

12.1 Introduction

The escalating global demand for food, coupled with shrinking arable land and the intensifying impact of climate change, calls for innovative, resource-efficient farming systems. Among these, hydroponics a soilless cultivation method where plants are grown in nutrient-rich water has emerged as a promising solution [1]. By delivering nutrients directly to the root zone, hydroponics not only accelerates plant growth but also optimizes water and land use, making it especially valuable in urban, water-scarce, or climate-vulnerable regions [2, 3].

However, the true potential of hydroponics lies beyond its physical structure. Modern hydroponic systems are increasingly being transformed by smart agriculture technologies, particularly the Internet of Things (IoT) and bioinformatics. These technologies are redefining how data is captured, analyzed, and used to guide crop management decisions in real time. IoT-enabled sensors continuously monitor environmental and nutrient parameters, while bioinformatics processes this data often alongside genomics and phenomics information to model plant responses and optimize nutrient delivery strategies [4, 5].

The convergence of IoT and bioinformatics represents a paradigm shift in controlled-environment agriculture. It enables the development of intelligent, adaptive farming systems capable of self-regulation, predictive modeling, and precision resource management. These digital-physical integrations, often described as components of Agriculture 5.0, aim to create sustainable, high-yield cultivation platforms that are both data-driven and biologically informed [6, 7]. This chapter explores the integration of IoT and bioinformatics in hydroponic systems. It begins with a discussion on precise nutrient monitoring and the role of smart sensors, followed by advancements in automation, data analytics, and predictive modeling. Finally, it illustrates how bioinformatics tools from machine learning to multi-omics are enhancing the decision-making capacity of hydroponic farms, ushering in a new era of precision agriculture.

12.2 IoT in Hydroponic Nutrient Monitoring

Hydroponics is a soilless cultivation system where plants are grown in nutrient-rich water solutions. In such systems, precise monitoring and control of environmental and nutrient parameters are critical, as plants rely entirely on the solution for all essential inputs. Traditionally, monitoring these parameters required manual sampling and laboratory analysis, which limited responsiveness and scalability [8, 9]. In contrast, IoT-based systems offer the advantages over traditional monitoring (Table 12.1). The Internet of Things (IoT) has transformed this landscape by enabling real-time, automated monitoring and management of hydroponic systems, thereby increasing efficiency, reducing labor, and improving crop yield and quality [10].

Table 12.1 Comparison of IoT-based monitoring with traditional monitoring

Parameter	Traditional monitoring	IoT-based monitoring	References
Frequency	Periodic (manual)	Continuous (real-time)	[11]
Labor demand	High	Low	[12]
Data storage	Manual logs	Digital/cloud-based	[13]
Decision-making	Delayed	Instantaneous/autonomous	[14]
Scalability	Limited	High	[15]

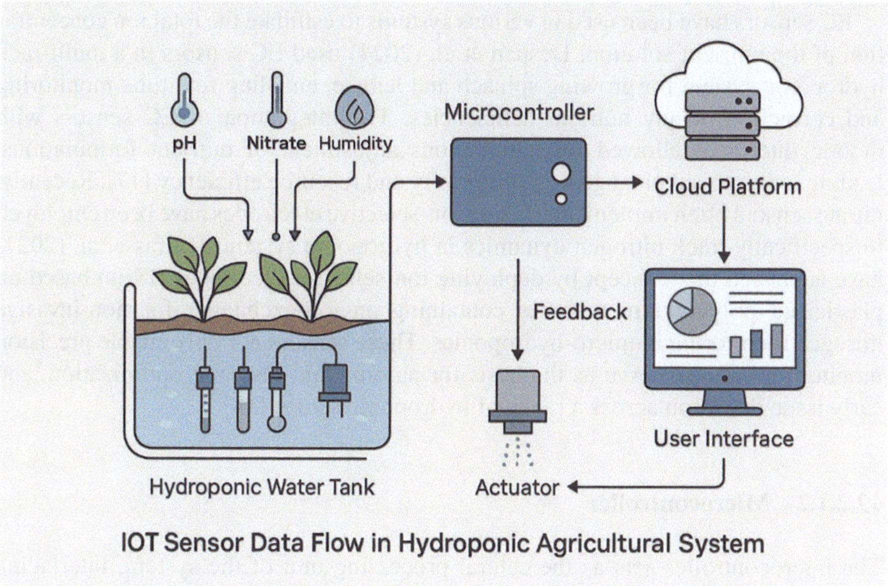

Fig. 12.1 Schematic representation of IoT data flow in a hydroponic system. Sensor inputs (e.g., pH, EC, nitrate, humidity) are processed by a microcontroller and transmitted to a cloud platform for visualization and decision-making. Actuators respond based on feedback to maintain optimal nutrient and environmental conditions

12.2.1 Key IoT Components in Hydroponics

An IoT-enabled hydroponic system typically consists of four core components: sensors, microcontrollers, data acquisition platforms, and feedback-actuation mechanisms. Together, they create an integrated, self-regulating framework for continuous monitoring and dynamic nutrient management (Fig. 12.1).

12.2.1.1 Sensors

In IoT-enabled hydroponic systems, sensors form the foundational layer by enabling continuous environmental monitoring and guiding nutrient decision-making. The most deployed sensors include those for pH, electrical conductivity (EC), temperature, humidity, and specific ion concentrations, each selected for their role in regulating plant growth and nutrient uptake. These devices have been widely validated across a range of hydroponic crops and system designs. pH sensors, essential for monitoring the acidity or alkalinity of the nutrient solution, directly influences nutrient solubility and availability. In the study conducted by Putra et al. (2022), pH sensors were integrated into a fuzzy logic control system in a Nutrient Film Technique (NFT) hydroponic setup for growing leafy greens. The system maintained optimal pH levels by triggering peristaltic pumps based on fuzzy rule evaluations, thereby improving stability and reducing nutrient imbalances [16].

EC sensors have been used in various systems to estimate the total ion concentration of the nutrient solution. Deshan et al. (2024) used EC sensors in a multi-rack hydroponic system for growing spinach and lettuce, enabling real-time monitoring and correction of any nutrient deficiencies. The integration of EC sensors with mobile interfaces allowed for autonomous adjustment of nutrient formulations, leading to improved plant growth uniformity and resource efficiency [17]. Recently, nitrate sensors often implemented using ion-selective electrodes have been employed to specifically track nitrogen dynamics in hydroponic systems. Miras et al. (2021) have advanced this concept by deploying ion-selective electrodes (ISEs) based on plasticized polyvinyl membranes containing an ion-exchanger for non-invasive nitrogen monitoring in micro-hydroponics. These sensors not only enable precision monitoring but also serve as the basis for automation, resource optimization, and early issue detection across a range of hydroponic setups [18].

12.2.1.2 Microcontroller

The microcontroller acts as the central processing unit of the system, interfacing between sensors and actuators. It receives raw input from sensors, processes it locally (often via edge computing), and transmits it to cloud platforms for storage and advanced analytics [19, 20]. Popular microcontrollers like Arduino and Raspberry Pi are often used due to their flexibility and low cost. They are also responsible for executing control logic (e.g., adjusting nutrient dosing) based on predefined thresholds or machine learning models.

12.2.1.3 Real-Time Data Acquisition and Management

Data collected from sensors is transmitted to cloud-based platforms where it is stored, visualized, and analyzed through user-friendly dashboards or mobile apps [16, 21]. Algorithms help filter out noise from raw data, especially in fluctuating environments. The incorporation of edge computing allows initial data processing to be performed at the sensor or controller level, reducing latency and bandwidth demands while enabling real-time responsiveness [22].

12.2.1.4 Feedback and Actuation

The final layer involves actuators such as peristaltic pumps, solenoid valves, grow lights, and fans that respond to sensor data in real time. This feedback loop enables automated adjustment of pH, nutrient concentrations, irrigation cycles, and environmental conditions. For example, if nitrate levels fall below a set threshold, the system can trigger nutrient injection to restore balance without human intervention. This closed-loop control ensures a stable environment and consistent crop performance.

12.2.2 Notable IoT Platforms for Hydroponic Integration

The effective implementation of IoT-based nutrient monitoring systems in hydroponics hinges not only on robust hardware like sensors and microcontrollers but also on suitable software platforms. These platforms facilitate real-time data acquisition, remote monitoring, visualization, and automation of cultivation processes. Among the most widely adopted platforms for hydroponic IoT systems are Blynk, Node-RED, and ThingsBoard each offering unique features in terms of flexibility, scalability, and ease of integration [23, 24].

12.2.2.1 Blynk

Blynk is a web- and mobile-based IoT platform that is easy to use, cost-effective, and supports microcontrollers like Arduino, ESP8266, and Raspberry Pi. Blynk enables users to construct customized dashboards using an easy drag-and-drop system for real-time device visualization and control. This is best applicable to smallholder farmers, hobbyists, and scientists with limited resource capabilities [25–27]. In hydroponics, Blynk has been extensively used in measuring and controlling parameters such as pH, EC, temperature, humidity, and water level. For example, researchers have constructed an automated hydroponic system employing Arduino Uno and sensors with the Blynk app for real-time monitoring of the environment and nutrients [28]. Another study utilized Blynk in an NFT system for remotely controlling light and nutrient supply based on environmental feedback [29]. Moreover, Blynk also supports actuator control like triggering water pumps or nutrient dosing systems based on sensor readings. For example, a smart hydroponic monitoring system utilized NodeMCU and Arduino to regulate irrigation and lighting remotely, demonstrating the capability of Blynk platform for automation and precision farming [23]. In general, Blynk is a user-friendly solution for entry-level and experimental hydroponic applications, offering easy and cost-effective control in real time [3].

12.2.2.2 Node-RED

Node-RED, a web services, API, and device integration flow-based web browser programming tool from IBM, allows users to create logic-based applications through a visually based interface where data flow is built through linking modular nodes by interlinking "nodes" [23, 25]. Node-RED is an appropriate middlewares platform for dealing with hydroponics sensor data processing, rule-based automation, and database or cloud services interfaces such as AWS IoT or Microsoft Azure [23, 30, 31]. For instance, users can program Node-RED to monitor EC or pH level reading, apply data filter algorithms, and activate actuators (like nutrient pumps) when threshold is crossed. Its modular architecture allows seamless integration with Message Queuing Telemetry Transport (MQTT) protocols and other IoT frameworks, hence making it ideal for real-time decision-making in smart agriculture. It can also be combined with machine learning algorithms to use adaptive control methods in hydroponic systems [32].

12.2.2.3 ThingsBoard

ThingsBoard is a full-featured enterprise-level IoT platform with real-time telemetry visualization, device management, rule-based automation, as well as storage for historical data. It also features strong support for a variety of communication protocols such as MQTT, CoAP, and HTTP and is thus best suited for cloud as well as on-site deployments [33, 34]. ThingsBoard's multi-tenancy nature makes it even more useful for business or hydroponic large-scale operations, which can monitor different greenhouses or farms centrally. Its high-powered and sophisticated rule engine can be used to automate the dosing of fertilizing, send alerts when environmental conditions are outside pre-specified norms, and analyze historical trends to allow predictive maintenance and crop optimization [35]. Additionally, ThingsBoard supports other platforms for integration. For instance, Node-RED can be employed for preprocessing data from the sensors prior to sending it to ThingsBoard for centralized monitoring and control [23]. This multi-level approach encourages scalability and responsiveness of the system with less intricate handling of data.

Each of the three IoT platforms, Blynk, Node-RED, and ThingsBoard has its strengths. Blynk is strongest with small-scale system minimalism and low-cost deployment, Node-RED is strongest with a generic development environment for bespoke logic, and ThingsBoard offers enterprise-class capabilities for commercial hydroponic companies and large-scale initiatives. The final platform to be utilized would depend on technological sophistication, deployment scale, and knowledge of the user [36]. For most applications, integration of the two systems can improve system performance and allow more intelligent, more sustainable hydroponic cultivation.

12.2.3 Case Studies of IoT-Enabled Hydroponic Nutrient Monitoring

Increasing numbers of reports demonstrate effective use of IoT in hydroponic technology, highlighting both technological feasibility as well as agricultural impact. The following case studies show implementation in the field at different scales and environments:

- **Fuzzy Logic Control in NFT System (Indonesia)**

  Indonesian researchers designed an IoT-based Mamdani Fuzzy Inference System to regulate pH and Total Dissolved Solids (TDS) in a Nutrient Film Technique (NFT) hydroponic tomato crop growing system. Arduino microcontrollers, EC and pH sensors, and actuators were used by the system for automatic controlling of nutritional solution supply. EC error and input volume of nutrients were processed using fuzzy logic algorithms to dynamically adjust the concentrations of nutrients. The 1:1440 scaled model to mimic the whole tomato development process provisioned for change over time because of nutrient intake. Input in real time and rapid response of the system facilitated three iterations of control stabilizing without over-correction or nutrient loss. Aside from technological innovation, the system was also resource efficient in terms of resource savings in water conservation and fertilizer application and providing a sustainable cost-reduction method to controlled-environment agriculture [24, 29].

- **Multi-Rack Monitoring System (Cibodas Village, Indonesia)**

  In another example from Cibodas Village is a modular monitoring system built with ESP32 microcontrollers and Firebase for real-time data processing. The platform allowed for the simultaneous monitoring of several hydroponic racks, each customized to different crop requirements. A mobile interface enabled farmers to remotely monitor and manage important variables like pH and nutrient flow. The system claimed over 98% measurement accuracy, considerably increasing production and input efficiency through precise, rack-level nutrition management [21].

- **IoT in Aeroponic Greenhouses (China)**

  A high-tech aeroponic greenhouse in China connected sensors for pH, EC, temperature, and humidity to a cloud-based IoT platform. This allowed for real-time data analysis and autonomous, feedback-driven control of the nutrient misting system. The strategy increased water and nutrient use efficiency while reducing manual intervention. The scalability and automating feature of the system rendered it to be a suitable model for high-tech greenhouse businesses and inner-city agriculture [37].

Collectively, these case studies illustrate the versatility of IoT architecture in hydroponic systems NFT, rack module, and aeroponic types and how they can improve accuracy, resource optimization, and sustainability of controlled cultivation systems.

12.2.4 Challenges and Considerations

Although IoT in hydroponics is promising, it has several disadvantages. The major problem lies in sensor reliability due to drift, fouling, and calibration error that may adversely affect data accuracy and lead to ineffectiveness in nutrient management. Poor internet connectivity and electricity supply in most rural or resource-scarce environments deter real-time monitoring and automation. Furthermore, the data generated in IoT systems needs to be managed, stored, preprocessed, and interpreted meaningfully efficiently. This requires not just strong system design but also inter-disciplinary skills in computer-embedded systems, sensor integration, software development, data analysis, and agronomy. Shortages in manpower of qualified human resources in these disciplines could restrict the scalability and sustainability of IoT solutions. Reducing these issues calls for the development of cost-effective, affordable, and reliable systems tailored to different operating environments.

12.3 Bioinformatics Tools for Nutrient Response Analysis

In hydroponics, where environmental parameters are controlled with accuracy, understanding the molecular mechanisms governing nutrient uptake and response becomes crucial to ensure that cultivation is high-yielding and resource-saving. Bioinformatics becomes central to this in enabling one to analyze and interpret response at the transcriptomic as well as genetic level [38]. While Internet of Things (IoT) technologies track nutrient concentrations and environmental conditions in real time, bioinformatics deciphers how plants react at a molecular level to those conditions bridging the gap between external stimuli and internal biological responses [34].

12.3.1 Nutrient Response Modeling in Hydroponics

In a hydroponic setup, plants rely entirely on nutrient solutions for their growth and development. These systems eliminate soil as a buffer, making plants highly sensitive to nutrient imbalances. Bioinformatics helps decode how plants perceive and respond to these nutrient inputs at the gene level (Fig. 12.2). Transcriptomic analyses, particularly RNA sequencing (RNA-Seq), have revealed key nutrient-responsive genes such as NRT1.1, GLN1.2, PHT1, and IRT1, which are involved in the uptake and internal transport of nitrogen, phosphorus, and iron respectively [22]. Understanding the regulation of such genes helps in designing nutrient regimes that go beyond general recommendations and instead cater to the plant's real-time physiological needs.

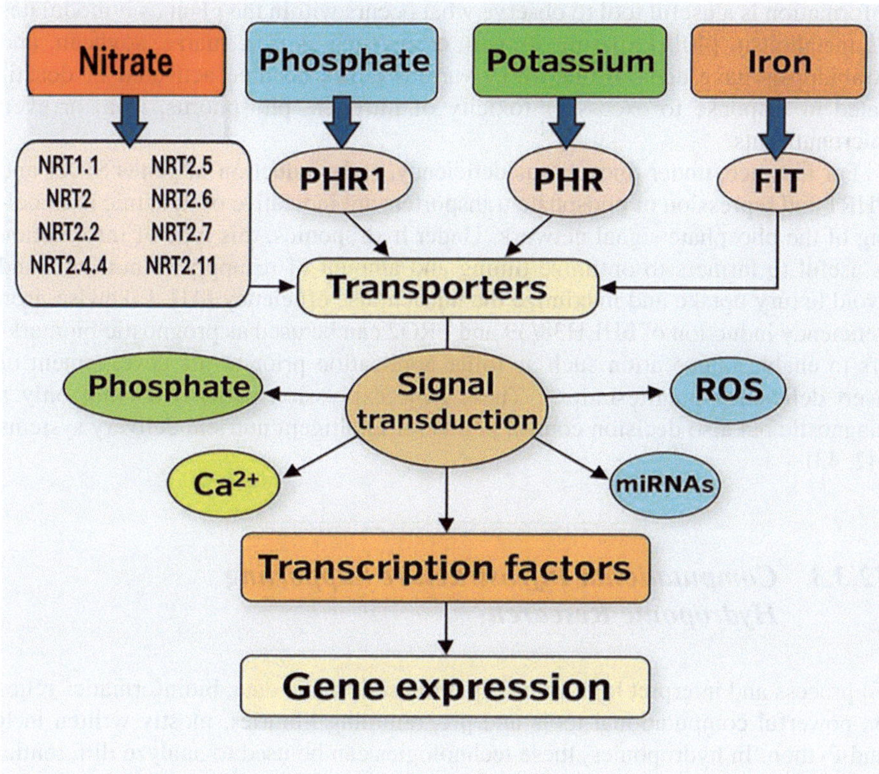

Fig. 12.2 Molecular integration of nutrient signals in plants. Key nutrients (nitrate, phosphate, potassium, and iron) are sensed via specific signaling proteins (e.g., PHR1, FIT), activating transporters and initiating intracellular signal transduction. Secondary messengers such as Ca^{2+}, reactive oxygen species (ROS), microRNAs (miRNAs), and hormones interact to regulate transcription factors, ultimately modulating gene expression in response to nutrient status

In hydroponics, where nutrient inputs can be isolated and precisely controlled, researchers can induce and monitor specific nutrient conditions to dissect plant responses. For instance, increased expression of NRT2.1 under nitrogen-rich conditions or suppression of PHT1 family genes under phosphorus sufficiency can validate nutrient availability or deficiency before any phenotypic symptoms appear [39]. These molecular signatures serve as early indicators of nutrient sufficiency or stress and can be incorporated into real-time monitoring systems.

12.3.2 Utilizing Transcriptomics for Nutrient Optimization

Transcriptomics, particularly through RNA-Seq, enables a genome-wide view of how plants adjust their gene expression in response to fluctuating nutrient conditions. In hydroponics, under the regulated environment, transcriptomic information is a useful tool to observe what occurs within the plant as it modulates its metabolism [40]. Experiments with crops such as rice, maize, soybean, and Arabidopsis have demonstrated how various genes become activated or deactivated in response to excess or toxicity of nitrogen, phosphorus, iron, or even micronutrients.

For instance, under phosphorus deficiency, early induction of genes SPX1 and PHR1 and repression of phosphate transporters are indicative of dynamic remodeling of the phosphate signal network. Under hydroponics, this type of information is useful to farmers to optimize timing and amount of resupply of nutrients and avoid luxury uptake and maximize the nutrient use efficiency [41]. Likewise, iron deficiency induction of bHLH38/39 and FRO2 can be used as prognostic biomarkers to enable intervention such as foliar application prior to the development of overt deficiency manifestations. These gene expression patterns are not only a diagnostic but also decision control points for intelligent nutrient delivery systems [42, 43].

12.3.3 Computational Infrastructure Supporting Hydroponic Research

To process and interpret high-throughput transcriptomic data, bioinformatics relies on powerful computational tools and programming libraries, mostly written in R and Python. In hydroponics, these technologies can be used to analyze differential gene expression as well as visualize nutrient-responsive circuits [44].

12.3.3.1 Computational Tools for Gene Expression Analysis

Bioinformatic computational packages like DESeq2, edgeR, and limma are often utilized to identify statistically significant gene expression changes between different nutritional conditions. ClusterProfiler and GSEA facilitate the identification of enriched pathways, which connect gene activity to metabolic or signaling pathways that are meaningful in nutritional homoeostasis [45]. Python adds to this analysis workflow with data handling libraries such as Pandas, NumPy, and Biopython for data analysis, sequence analysis, and communication with third-party databases such as KEGG and NCBI. With machine learning libraries such as scikit-learn, TensorFlow, or XGBoost added to these processes, predictive modeling is facilitated, the capacity to identify potential stress responses or nutritional requirements from previous and current expression profiles [46, 47]. Most significantly, these models can be updated in real time as new data are acquired, rendering them highly responsive to dynamic hydroponic systems.

12.3.3.2 Visualizing Molecular Responses for Decision-Making

Visualization is crucial to tunecomplex bioinformatics data into actionable decisions in hydroponic systems. Heatmaps, volcano plots, PCA, and pathway enrichment maps are common tools utilized to plot gene expression patterns and determine substantial regulatory changes to nutrient treatment [48]. In hydroponics, such plots enable gene expression variations to be attributed to specific fertilizer inputs and environmental perturbations. For instance, a time-course plot of induction of iron acquisition gene expression after transcriptional decrease in Fe content can be utilized to assess sensor and molecular accuracy [49]. Interaction network plots between transcription factors, transporters, and metabolic enzymes give a system-level view of nutrient regulation and allow researchers to detect regulatory bottlenecks. These visualizations are not only useful to research but can be developed into grower-focused tools to present difficult data in an understandable and readable form, finally aiding in nutrient management decisions.

12.3.3.3 From Data to Cultivation Strategy: Toward Adaptive Hydroponics

Integration of bioinformatics in hydroponic farming is revolutionizing nutrient management from a predetermined schedule-based practice to an adaptive plant-sensitive system. Hydroponic farming has thus relied on fixed formula recipes of nutrients supplied uniformly over time and development phases. Plants' nutritional requirements, however, are not constant, they vary with developmental phases, environmental conditions, and internal metabolic health [50]. Bioinformatics offers the tool to monitor such changes in real time and provide molecular access to the plant's physiological demands. By linking gene expression profiles with conditions of individual nutrients, growers and scientists can establish data-driven practices in cultivation. For instance, early induction of stress response gene transcription, e.g., those in phosphate starvation or iron deficiency, can detect nutrient deficiencies well before apparent signs [51]. Such molecular markers serve as an early warning system to enable anticipatory corrective actions like the alteration of nutrient concentration or the addition of specific ions to be undertaken.

Furthermore, machine learning models trained on historical transcriptomic and environmental data can predict plant response to future changes in the growth environment. Such prediction models enable prediction of nutrient imbalances, optimal dosing regimens, and minimize overuse of resources. Long short-term memory (LSTM) networks and decision tree-based frameworks are powerful in learning temporal patterns of gene expression and relating them to temporal nutrient regimes [52]. This convergence of transcriptomics, bioinformatics, and real-time monitoring gives rise to the development of adaptive hydroponic systems where selection is not anymore made based on merely pre-programmed timers or sensor readings but on real biological feedback from the plant itself. Growth is thus more efficient, resource-conservative, and customized according to the specific demands of the crop at each phase of development. Subsequently, bioinformatics transforms hydroponics into a smart and responsive type of agriculture in which plants take an active role in controlling their own nutrition [53, 54].

12.3.4 Case Studies

Recent pilot trials and experimental setups demonstrate the tangible impact of combining bioinformatics and nutrient management in hydroponics. The precision lettuce hydroponics trial is a case in point, where sensor readings of nitrate concentration were combined with monitoring of gene expression of key nitrogen-response genes, NRT2.1 and GS1. Although nitrate sensors had shown that there was enough availability, transcript concentrations of the two genes were below optimal levels indicating inefficiencies in the uptake of nitrate. This imbalance led to the system adapting root-zone oxygenation and infusion of specific micronutrients, followed by recovery from typical nitrogen assimilation and growth. This strategy translated raw sensor data into useful information with gene-level confirmation, demonstrating the value added by transcriptomics for latent nutrient stress diagnosis [55].

In another study, phosphorus optimization experiment, researchers employed RNA-Seq to follow expression dynamics under low-phosphorus stress. Induction of SPX1 and PHR1 regulatory genes preceded repression of phosphate transporter genes. By interpreting this dynamic transcriptomic pattern, phosphorus resupply was strategically delayed until transporter activity resumed, thereby avoiding luxury uptake and enhancing phosphorus use efficiency [56]. Further, a predictive model enabled early detection of iron deficiency through expression of bHLH38/39, FRO2, and ROS detoxification genes. The system provided dashboard alert prior to the time phenotypic symptoms occurred and enabled timely foliar iron application and prevented growth retardation. These cases indicate the practical promise of gene expression biomarkers as early warning systems—moving hydroponic nutrient management from a reactive mode to a proactive, biologically informed strategy [55].

12.3.5 Integrative Dashboards and Real-Time Interfaces

To realize the full potential of transcriptome-directed hydroponics, intricate multi-source information needs to be reduced to clear real-time decisions. This has promoted integrative dashboards that merge the outputs of bioinformatics and IoT sensor streams via user-friendly interfaces. R Shiny, Plotly Dash, and Streamlit platforms enable dynamic correlation visualization of gene expression, nutrient uptake, and sensor readings, and ThingsBoard or Node-RED backend systems handle real-time data flow, storage, and retrieval [48, 57–59]. Automated alerts of gene expression outside optimal profiles are delivered by the dashboards so growers can take timely corrective action. Overlays of visual data showing gene–nutrient–sensor linkages enable tracing of underlying biological reasons for sensor abnormalities or dipping growth indicators [60]. Automated reporting of data also enables longitudinal modeling of plant response to inform predictive algorithms that learn from past trends and use them to predict future nutrient requirements [61].

Machine learning plays a central role in this paradigm shift from descriptive to predictive analytics. Random Forests and Gradient Boosting machine learning algorithms are used in classifying gene expression responses under varying nutrient conditions. Recurrent Neural Networks such as Long Short-Term Memory (LSTM) are employed to model temporal trends in gene expression to predict future plant needs in subsequent growth stages [62, 63]. Bayesian networks, however, offer probabilistic frameworks to describe the interactions among genes, nutrients, and environmental stress factors. These are all contributing to the foundation of adaptive hydroponic systems where the nutrient recipes are no longer fixed recipes but dynamically optimized procedures as informed by prediction models based on plant biology [64].

12.3.6 Future Directions and Integration with Other Omics

The application of RNA-Seq in hydroponics is widening with its integration into other omics platforms like metabolomics, proteomics, and epigenomics. The multilevel approach provides systems-level information on plant growth and stress response with respect to nutrient dynamics. Transcriptomic information in smart hydroponic systems may be aligned with metabolite profiles and protein expression to confirm nutrient-dependent alterations, enabling better accuracy in nutrient intervention [65, 66]. For instance, reduced expression of nitrate-reactive genes, supported by metabolomic alteration, can initiate specific nutrient supplementation by way of automated regulation. Emerging research points to epigenetic regulation and, reveals that chromatin conformation and memory of gene expression are influenced by nutrition status introducing an extra dimension to plant adaptability under controlled environment. In the near future, multi-omics and sensor-trained machine learning algorithms are being created to predict nutrient requirements and dynamically optimize delivery [67]. This integrated approach is transforming hydroponics from precision agriculture to fully adaptive, self-adjusting systems.

12.4 Real-Time Integration: Bioinformatics Meets IoT

The blend of IoT and bioinformatics is a revolution in precision agriculture and hydroponic system control. The delivery of nutrients to hydroponics has been controlled conventionally based on physical properties including ion concentrations, electrical conductivity (EC), and pH measured by sensors [68]. But bringing gene expression information into this process adds a higher-order biological component, enabling growers to react not only to environmental cues, but also to the plant's internal physiological condition. In this closed-loop smart system, gene expression patterns by transcriptomic profiling serve as biomarkers for nutrient sufficiency or stress (Fig. 12.3). For instance, if there are sufficient sensors measuring nitrate, yet expression of glutamine synthetase (GS1) is below normal, this could be a sign of poor nitrate assimilation rather than availability. It could be caused by inadequate oxygen levels, enzyme insufficiency, or lack of cofactors [45]. The system can then

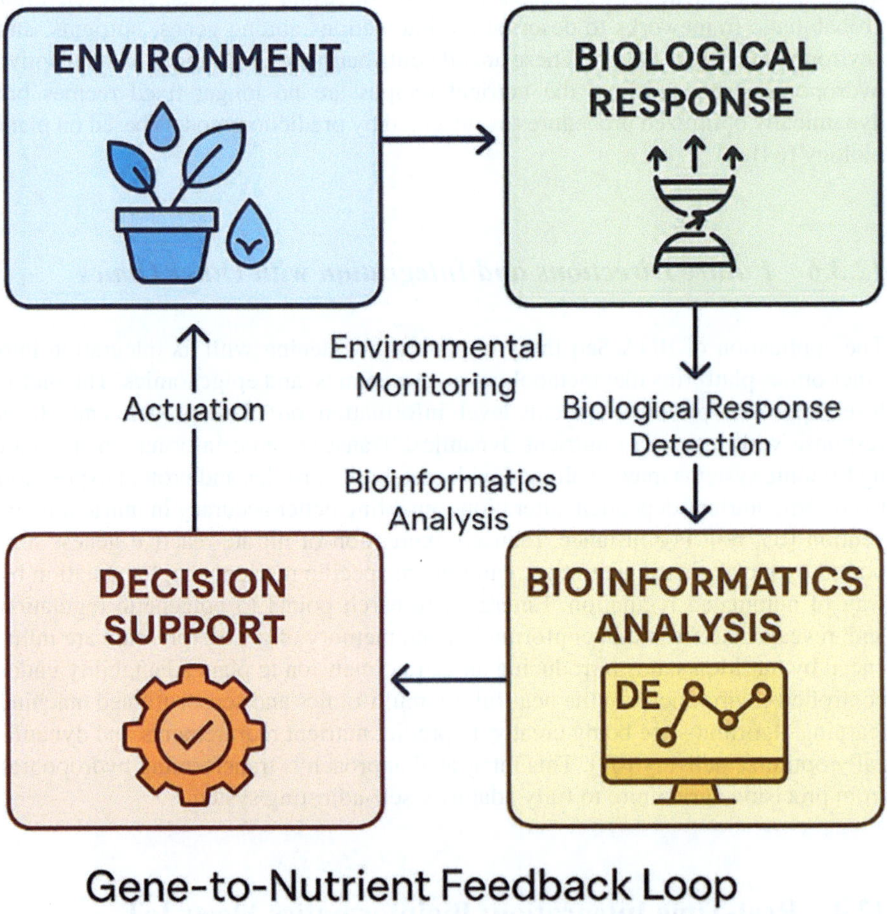

Fig. 12.3 Gene-to-nutrient feedback loop in a bioinformatics-integrated hydroponic system. The cycle begins with environmental monitoring via sensors (e.g., nutrient levels, pH), followed by the detection of biological responses (e.g., gene expression changes) in the plant. Bioinformatics analysis interprets molecular data to inform nutrient management decisions, which are implemented through automated actuators closing the feedback loop and enabling adaptive, plant-informed nutrient delivery

trigger compensatory actions, for instance, increased root zone aeration, pH modification, or micronutrient application such as molybdenum or iron that are part of nitrogen metabolism.

This dynamic decision-making calls for real-time computational infrastructure. Tools like R Shiny, Dash (Python), or Streamlit provide the capability to create interactive dashboards plotting sensor data as well as bioinformatics output (e.g., heatmaps, expression curves, pathway diagrams). These dashboards present not just real-time graphical feedback but also historic logging to study trends, early warning signals, and machine learning predictions [69]. For example, the system may be trained to alert significant deviations, e.g., suppressed expression of NRT2.1 or PHT1 genes upon detecting normal EC readings indicative of bottlenecks in nutrient uptake. Alerts may be sent through SMS, email, or app alerts to facilitate instant human action if required. Additionally, communication protocols such as MQTT allow IoT devices such as microcontrollers or smart nutrient pumps to act autonomously [54, 70]. With integration of expression data in real time, these systems have the ability to perform operations like manipulating levels of nutrients, powering actuators, or adjusting environmental parameters (e.g., light cycles, CO_2 level) based on the molecular feedback loop.

This degree of integration fosters adaptive, self-tuning hydroponic systems, where plant biology directly informs the system control. It moves agriculture from a reactive to a proactive, plant-centric paradigm. Over time, continuous integration of transcriptomic and sensor data supports the development of predictive models that can anticipate nutrient imbalances before symptoms manifest, enabling preventive rather than corrective interventions [51, 71, 72]. In summary, the real-time integration of bioinformatics with IoT transforms hydroponics into a living system where plants "communicate" their needs through gene expression, and intelligent systems respond with precise, timely adjustments—maximizing yield, resource efficiency, and plant health.

12.5 Designing an Integrated Nutrient Monitoring Framework

An integrated nutrient monitoring framework forms the technological backbone of real-time, adaptive, and biologically informed hydroponic cultivation systems (Fig. 12.4). It unifies data streams from physical sensors and molecular assays, enabling the dynamic interpretation and adjustment of nutrient regimes in response to both environmental inputs and plant physiological states [73]. The design of such a framework demands a multidisciplinary approach, drawing upon expertise in plant physiology, bioinformatics, embedded systems, cloud computing, and data analytics.

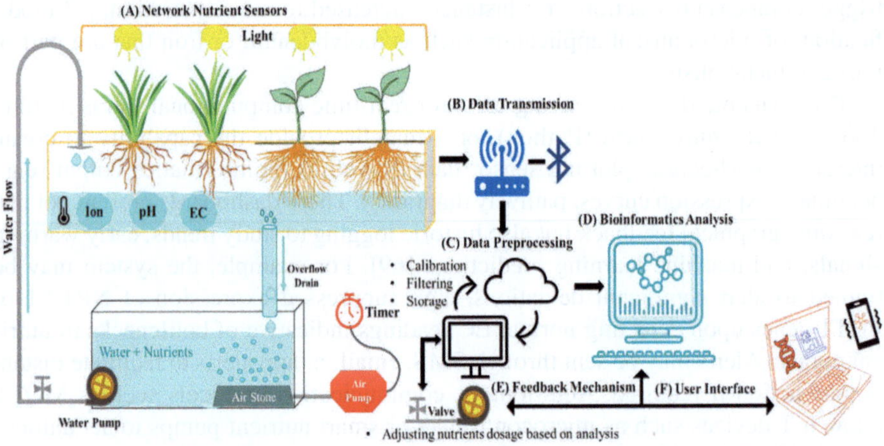

Fig. 12.4 Schematic representation of the IoT–bioinformatics integration for hydroponics. (**a**) Networked nutrient sensors (pH, EC, ion-specific) continuously monitor the hydroponic solution. (**b**) Sensor data are transmitted wirelessly or via wired networks to a central server or cloud platform. (**c**) A data preprocessing pipeline (including calibration, filtering, time-alignment and storage) integrates all sensor streams into a unified database. (**d**) A bioinformatics analysis module uses plant gene expression data to identify nutrient-responsive genes and pathways. (**e**) Closed-loop actuators (pumps/valves) adjust nutrient dosing in real time based on sensor feedback and analysis. (**f**) A smart dashboard (web/mobile UI) visualizes system parameters and allows user control of environmental setpoints

12.5.1 System Architecture for Real-Time Nutrient Monitoring in Hydroponics

The integrated framework for nutrient monitoring in hydroponics is designed around four closely linked layers, each playing a critical role in enabling data-driven crop management. It begins with the Sensor Layer, where instruments such as pH meters, electrical conductivity (EC) probes, and nutrient-specific sensors continuously capture environmental and nutrient-related data. This real-time information is then transmitted via the Communication Layer, which typically relies on microcontrollers and cloud-based platforms to ensure seamless data flow. The Bioinformatics Layer adds a deeper biological perspective by analyzing plant molecular responses through tools like RNA sequencing and gene expression profiling. These insights help interpret how plants are physiologically responding to their nutrient environment. Finally, the decision-making layer integrates all this data into user-friendly dashboards that not only allow for visualization and monitoring but can also trigger automated nutrient adjustments based on plant needs. Below, we discuss each layer in detail, highlighting its role in enabling responsive, data-driven nutrient management within hydroponic systems.

12.5.1.1 Sensor Layer: Environmental Monitoring and Data Acquisition

The foundational layer of the framework comprises a network of sensors that continuously monitor key environmental variables affecting nutrient dynamics (Fig. 12.5). These include pH, electrical conductivity (EC), ion concentrations (via ion-selective electrodes), temperature, humidity, and dissolved oxygen. Sensor selection must be guided by criteria such as accuracy, sensitivity, range, and resistance to corrosive, nutrient-rich hydroponic solutions. Sensors interface with microcontrollers (e.g., ESP32, Arduino Uno, Particle Photon) or single-board computers (e.g., Raspberry Pi), which serve as data acquisition units [74]. These units facilitate real-time environmental monitoring, enabling precise regulation of the cultivation environment.

12.5.1.2 Communication Layer: Data Transmission and Connectivity

Once acquired, sensor data must be transmitted efficiently to processing units and storage platforms. The communication layer employs protocols such as Wi-Fi or Ethernet for high-bandwidth local networks, MQTT for low-latency and resource-efficient data exchange, and LoRaWAN or GSM for deployments in infrastructure-poor areas. Edge processing on microcontrollers allows preliminary filtering and noise reduction, while batch uploads and real-time streaming support both long-term trend analysis and immediate feedback [75]. Storage solutions include redundant local storage (e.g., SD cards, SSDs) and scalable cloud databases such as InfluxDB, Firebase, or AWS DynamoDB.

12.5.1.3 Bioinformatics Layer: Molecular Diagnostics and Gene Expression Analysis

Complementing the environmental sensor data, the bioinformatics layer captures the plant's internal physiological responses through transcriptomic profiling. Plant samples, typically roots, leaves, or whole tissues, are collected at specific time points that align with environmental fluctuations or alerts triggered by the sensor network [14, 76]. Such samples are flash-frozen for the maintenance of RNA integrity and subsequently are subjected to RNA extraction using techniques like TRIzol reagent or column extraction kits. RNA extracted is then prepared for high-throughput sequencing on platforms like Illumina, Oxford Nanopore, or Ion Torrent based on the desired read depth, read length, and availability of facilities. The bioinformatics pipeline starts with the preprocessing steps such as quality control with tools like FastQC and trimming of adapters with tools such as Trimmomatic [4, 77]. After cleanup of reads, alignment to the reference transcriptome or genome is performed by aligners such as HISAT2, STAR, or pseudo-aligners such as Salmon, depending on the individual experimental requirements. Quantitation of transcript

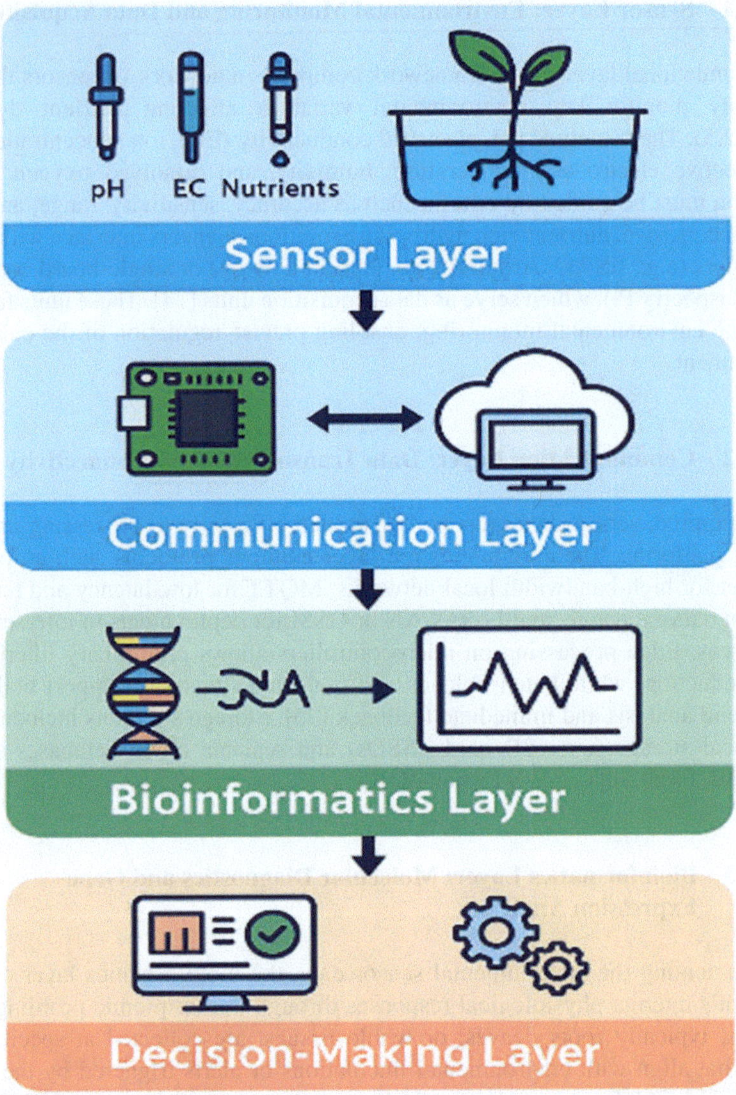

Fig. 12.5 System architecture diagram of an integrated nutrient monitoring framework in hydroponics. The framework comprises four interconnected layers: the Sensor Layer (pH, EC, nutrient-specific sensors), Communication Layer (microcontroller and cloud interface for data transmission), Bioinformatics Layer (RNA-seq analysis and interpretation of gene expression), and Decision-Making Layer (dashboard visualization and automated actuation). The flow from sensor to decision-making supports real-time, biologically informed nutrient management

abundance is then done by using tools such as HTSeq or featureCounts, which provide count data amenable for subsequent statistical analysis.

Differential gene expression analysis is carried out using statistical platforms like DESeq2, edgeR, or limma-voom, which determine genes being significantly up-regulated or downregulated under shifting nutrient or environmental stresses. GO or KEGG enrichment [47] analysis using packages like clusterProfiler is used for functional annotation of differentially expressed genes. Additionally, protein-protein interactions and gene co-expression networks are explored using databases and resources like STRING [78] and EggNOG Mapper [79] and to identify the regulatory patterns and putative biomarkers. The output of this layer is biologically actionable knowledge, e.g., the identification of stress-related transcription factors or nutrient-responsive genes. These results are then provided to the decision-making level such that the system can correlate external environmental measures with internal molecular signals to adaptively, plant-based manage nutrients.

12.5.1.4 Decision Layer: Visualization, Feedback, and Automated Actuation

The last layer converts complicated datasets into easily understandable dashboards and automated responses. Visualization libraries like Shiny, Dash, and Node-RED facilitate live exploration of both environmental as well as transcriptomic data [59]. Users can explore temporal trends, filter by gene or nutrient category, and set alerts for deviations. Automated actuators can be programmed to change levels of nutrients, pH, or aeration according to sensor and gene expression data in real-time, creating a closed-loop feedback control system for optimal plant growth.

12.5.2 Data Fusion Strategies

In hydroponic systems integrating IoT and bioinformatics, data fusion refers to the strategic merging of sensor-based environmental data with molecular datasets such as transcriptomics. IoT integration provides an integrated analysis to facilitate decision-making for informed nutrient management [23, 80]. Sensor readings as continuous and quantitative enable real-time measurement of nitrate concentration, pH, and electrical conductivity. In contrast, transcriptomic information gives snapshots of gene expression, collected at defined intervals. A common approach to link them is to employ the sensor-induced transcriptome response model, where specific changes in environmental readings prompt the collection of plant tissue for molecular profiling. This guarantees that the data of gene expression are understood in the environmental context that causes the response.

Alternatively, transcriptomic feedback may give inputs to environmental adjustments. For example, if nutrient assimilation genes are downregulated, despite normal sensor readings, it may indicate a problem with nutrient uptake, prompting changes in parameters such as pH or aeration [81]. More advanced approaches involve the integration of transcriptomic data with metabolomics or proteomics to validate molecular responses functionally. Standardized data structures—aligning timestamps, treatment IDs, and sample metadata—enable integrative analysis through techniques like regression or clustering [82]. Effective data integration drives hydroponics from passive observation to active, biologically interactive growth. It offers dynamic control as a reaction to environmental cues and current molecular feedback from the plant.

12.5.3 Validation Protocols

Validation is crucial in integrating sensor data with molecular diagnostics for system accuracy and biological significance. In hydroponic systems, this includes verifying environmental sensors and transcriptomic responses and ensuring their alignment. pH, EC, and ion sensor readings are all calibrated with standard solutions on a regular basis and kept in good working order to avoid drift or fouling. Reliable sensor readings are vital for reliable environmental monitoring [83]. Transcriptomic data undergo quality control from the very beginning with RNA integrity assays. Important gene expression findings such as those involving nitrate or phosphate uptake are confirmed using qPCR for the assessment of sequencing efficiency and biological response. Synchronization of sensor data and sampling is crucial. Correct timestamping and metadata facilitate correlation analysis between gene expression and environmental conditions. High correlation verifies system integration, and divergence indicates the sites where optimization is required [84]. Lastly, cross-validation across cultivars, conditions, and replicates verifies the system robustness and scalability, supporting broader agricultural applications.

12.5.4 Simulated Dataset Case for Demonstration

Simulated datasets offer an adequate and secure means of testing its operation before implementation on real IoT-bioinformatics hydroponic systems. Simulation allows researchers to validate data fusion, system reaction, and decision-making rules without the necessity of using real plants or sequencing [31]. A typical scenario could simulate nitrate starvation in lettuce. Artificial environmental data are generated to mimic declining nitrate concentrations over time to correlate with sensor readings. At the same time, artificial transcriptome profiles replicate gene expression responses to nitrate stress, e.g., NRT2.1 and GS1 induction. Datasets can be generated with statistical software such as R or Python, with realistic noise and appropriately structured metadata [85]. Synthetic data is then imported into the interface of the system (developed with Dash, Shiny, or Streamlit) and shown in real-time and tested for making decisions.

The capability of the platform to synchronize sensor and gene expression data, alert, and mimic the response like the amount of nutrients is being studied actively [86]. Simulations are also used in the creation of machine learning models for prediction of nutritional status or time-series and training users and demonstrating them to stakeholders. These simulations accelerate development, reduce risks, and condition the system to be well-suited for deployment in the real world by testing technical integration and biological logic within a controlled virtual environment [87].

12.5.5 Case Study: Pilot Implementation

An IoT-driven control system utilizes a Mamdani fuzzy inference system to control pH and total dissolved solids (TDS) levels in an NFT hydroponic system in real-time. Real-time sensor feedback and fuzzy logic were utilized by the system to control the parameters of the nutrient solution with high precision. It successfully adjusted both pH and TDS 1–3 control cycles following deviations with outstanding success, demonstrating the effectiveness of fuzzy control in nutrient management. The integration of cloud platforms and microcontrollers brings remote notification and monitoring, presenting an example of how the hydroponic stability may be improved by intelligent automation without adding further human intervention [29].

Stevens et al. created a new system to track nutrients in addition to AS7265x spectroscopic sensors and Internet-of-Things technology based on a mini micro-hydroponic system. The study focused on nitrogen dynamics by analyzing spectral reflectance data in real time using a Raspberry Pi running Node-RED and visualizing it on Grafana dashboards. The system effectively monitored nitrogen concentration variations and provided an alternative to conventional ion-selective electrodes. This study highlighted the possibility of non-invasive, light-based sensors for nutrient surveillance and proposed novel approaches to low-cost, high-throughput nutrient diagnostics [44].

Guzman and associates created a smart greenhouse system that combines IoT-enabled environmental detection with aeroponic nutrient delivery. Their method continuously monitored temperature, humidity, pH, EC, and light while managing fertilizer spray intervals with microcontrollers. A mobile application enabled users to remotely monitor plant growth parameters and ambient variables. The technology displayed accurate environmental control and water-use efficiency, resulting in higher lettuce biomass production. Notably, their implementation facilitates future integration with AI and multi-omics data, setting the framework for physiologically responsive greenhouse automation [64].

12.6 Benefits and Limitations

Bringing together IoT and bioinformatics in hydroponic farming is opening up exciting possibilities for smarter, more sustainable agriculture. One of the major benefits is the ability to monitor nutrient conditions with remarkable accuracy and in real time. Sensors can now track important variables like pH, electrical conductivity, and nutrient ion levels giving growers a clear picture of what's happening in their systems at any given moment [88]. What makes this even more powerful is the added layer of bioinformatics. By analyzing how plants respond at the molecular level such as changes in gene expression in response to nutrient fluctuations—we can start treating the plant itself as a kind of living sensor [89, 90]. This shift allows farmers to move from simply reacting to sensor alerts, to making proactive decisions based on what the plant is actually experiencing.

These technologies don't just stop at monitoring they also offer powerful tools for automation and prediction. With cloud-based platforms, data collected from sensors can be synced, stored, and analyzed to detect early signs of nutrient stress or growth issues [91]. Over time, patterns emerge that can guide smarter nutrient delivery schedules, reduce waste, and help optimize water and energy use [92]. For growers operating in cities or regions with limited arable land, these systems provide a highly efficient way to produce food without depending on traditional soil-based farming [93].

However, like any technology, there are some real challenges to consider. Sensor accuracy can drift over time, especially in variable environments, and keeping them calibrated particularly in large or remote setups can be difficult [94]. Another issue is data compatibility. Many farms use different types of devices and software that don't always "speak the same language," making integration a frustrating task [95]. Even when the data is all in one place, making sense of it can be tough. Interpreting plant gene expression data or running predictive models often requires a background in data science or bioinformatics skills that many farmers or even agricultural advisors may not yet have [96]. There's also the question of cost. Setting up these integrated systems, especially those involving omics-based analysis, can be expensive and might not be feasible for small-scale farmers without financial support or technical guidance [97].

Addressing these limitations will be key if we want this technology to benefit a wider range of growers. Investments in training, user-friendly tools, and low-cost infrastructure will be essential. At the same time, collaborations between tech developers, biologists, and farmers can help create systems that are not only cutting-edge but also practical and accessible. When done right, integrating IoT and bioinformatics holds great promise for building a more resilient and responsive food production system.

12.7 Future Perspectives

Looking ahead, the integration of IoT and bioinformatics in hydroponics is poised to become even more sophisticated, enabling a shift from smart farming to truly intelligent agriculture. As machine learning and artificial intelligence continue to evolve, there is growing potential to develop predictive models that can anticipate nutrient deficiencies before they manifest visually or physiologically [92]. These models could draw from real-time sensor data, historical environmental trends, and even genomic traits of specific plant varieties to make highly personalized nutrient recommendations [98].

We may also see greater use of CRISPR-based technologies and genomic selection to develop crop varieties that are inherently more nutrient-efficient and better adapted to soilless systems [99]. These improvements could reduce reliance on external inputs and make hydroponics more viable in regions with extreme climate conditions or limited access to agricultural resources. Meanwhile, advances in cloud computing and blockchain may improve data transparency and traceability, helping producers and consumers better understand how crops are grown, managed, and distributed [91].

The future of hydroponics lies in systems that are not just automated, but also biologically aware capable of learning from plant feedback and adapting over time. However, realizing this vision will require continuous collaboration across disciplines. Scientists, technologists, farmers, and policy makers must work together to ensure that emerging tools are not only cutting-edge, but also practical and equitable for growers at all scales.

12.8 Conclusion

The convergence of IoT and bioinformatics is transforming hydroponics from a controlled cultivation method into a dynamic, responsive, and data-rich farming system. With the ability to continuously monitor environmental and biological signals, growers can now make decisions that are both timely and tailored to the unique needs of their crops [100]. This integration supports a more sustainable approach to food production one that maximizes resource efficiency, improves yield, and reduces waste [97].

What sets this new model apart is its ability to connect plant-level biological insights with system-level management strategies. Rather than relying solely on external sensors, farmers can now incorporate transcriptomic and other molecular data to understand how plants are responding internally [89]. This enables a proactive style of farming, where interventions are guided not only by environmental readings but also by the plant's own molecular cues [101].

While the road ahead includes challenges such as data complexity, integration issues, and the need for technical training the opportunities are immense. By combining engineering, biology, and computational science, we are building a new blueprint for agriculture that is both intelligent and adaptable. As global food systems face mounting pressure, innovations like these will play a crucial role in creating resilient, high-performing farming practices for the future.

References

1. Zhang Y, Liu X, La S et al (2025) Transcriptome analysis reveals nutrient deprivation reduces nitrate content in lettuce (Lactuca sativa var. ramosa Hort.) and enhances nitrogen metabolism. Front Plant Sci 16:1585955
2. Dhal SB, Mahanta S, Moore JM, Kalafatis S (2024) Machine learning-based analysis of nutrient and water uptake in hydroponically grown soybeans. Sci Rep 14:24337
3. Velazquez-Gonzalez RS, Garcia-Garcia AL, Ventura-Zapata E et al (2022) A review on hydroponics and the technologies associated for medium-and small-scale operations. Agriculture 12:646
4. Rao PV (2024) The Routledge companion to the history of education in India, 1780–1947. Taylor & Francis
5. Lim D, Keerthi K, Perumbilavil S et al (2024) A real-time on-site precision nutrient monitoring system for hydroponic cultivation utilizing LIBS. Chem Biol Technol Agric 11:111
6. Keerthi K, Antony MM, Perumbilavil S, Matham MV (2024) LIBS: a rapid nutrition monitoring tool for hydroponic crops and nutrient supply. In: Remote sensing for agriculture, ecosystems, and hydrology XXVI. SPIE, pp 245–248
7. Rawal VK, Lal H, Shanker K et al (2025) Improving fertilizer use efficient method and strategies: a sustainable approach for future. International Journal of Environment and Climate Change 15:39–50
8. Kumar P, Subhash B, Gopika B, Jaisuriyan K (2024) Hydroponic system: hope and hype. In: Hydroponics: the future of sustainable farming. Springer, pp 43–69
9. Ibrahim LA, Shaghaleh H, El-Kassar GM et al (2023) Aquaponics: a sustainable path to food sovereignty and enhanced water use efficiency. Water 15:4310
10. Kanaujia VK, Sah CP (2023) Smart agriculture system. In: International conference on computer vision, high-performance computing, smart devices, and networks. Springer, pp 285–301
11. Stein EW (2024) The digital transformation of agriculture: industry 4.0 advances in hydroponic vertical farms. Digit Transform Soc 7:65
12. Thilakarathne NN, MSA B, Abas PE, Yassin H (2025) Internet of things enabled smart agriculture: current status, latest advancements, challenges and countermeasures. Heliyon 11(3)
13. Senapaty MK, Ray A, Padhy N (2023) IoT-enabled soil nutrient analysis and crop recommendation model for precision agriculture. Computers 12:61
14. Mansoor S, Iqbal S, Popescu SM et al (2025) Integration of smart sensors and IOT in precision agriculture: trends, challenges and future prospectives. Front Plant Sci 16:1587869
15. Lobanov V (2023) Nutrient transfer in Aquaponic systems–optimizing microbial processes for greater circularity and economic viability
16. Putra SD, Heriansyah H, Cahyadi EF et al (2024) Development of smart hydroponics system using AI-based sensing. J Inf Secur 16:474–485
17. Hasith Deshan HL, Mahima KTY, Divyanjalee M, et al (2024) Hyposense: an integrated sensor device for hydroponics farm monitoring

18. Miras M, García MS, Martínez V, Ortuño JÁ (2021) Inexpensive ion-selective electrodes for the simultaneous monitoring of potassium and nitrate concentrations in nutrient solutions. Anal Methods 13:3511–3520
19. Pal A, Dubey SK, Goel S, Kalita PK (2024) Portable sensors in precision agriculture: assessing advances and challenges in soil nutrient determination. TrAC Trends Anal Chem 117981
20. Kumar A, Trivedi A, Nandeha N, et al (2024) A comprehensive analysis of technology in aeroponics: presenting the adoption and integration of technology in sustainable agriculture practices. Int J Environ Clim Chang 14:872–882
21. Mulia SB (2024) Control and monitoring system of hydroponic nutrients in different hydroponic racks based on IoT in Cibodas Village Lembang District. Int J Smart Mater Mechatron 11:304–313
22. Yetgin A, Srivastava RK, Mandal N (2025) Insights into plant hormone signaling networks for environmental responses. Mitig Adapt Strateg Against Clim Chang Nat Syst:505–523
23. Niţulescu I-V, Korodi A (2020) Supervisory control and data acquisition approach in node-RED: application and discussions. IoT 1:5
24. Tesfahunegn M, Makonnen W (2019) Introduction to emerging technologies course module
25. Leijonberg C (2021) The viability of using trusted execution environments to protect data in node-RED: a study on using AMD-SEV and Intel SGX to protect sensitive data when node-RED is deployed on the cloud
26. Selarka T, Gajjar H, Talavia M (2023) Enhancing home automation: a smart system with M5 Stack and multiple control interfaces. In: 2023 IEEE 11th Region 10 Humanitarian Technology Conference (R10-HTC). IEEE, pp 335–340
27. Lamsal RR, Karthikeyan P, Otero P, Ariza A (2023) Design and implementation of internet of things (IoT) platform targeted for smallholder farmers: from Nepal perspective. Agriculture 13:1900
28. Chan JJ (2024) Development of an IOT-integrated app for monitoring hydroponic farming systems
29. Agustian I, Prayoga BI, Santosa H et al (2022) Nft hydroponic control using mamdani fuzzy inference system. arXiv Prepr arXiv220800364
30. Stevens JD, Murray D, Diepeveen D, Toohey D (2023) Development and testing of an IoT spectroscopic nutrient monitoring system for use in micro indoor smart hydroponics. Horticulturae 9:185
31. Atapattu AJ, Perera LK, Nuwarapaksha TD et al (2024) Challenges in achieving artificial intelligence in agriculture. In: Artificial intelligence techniques in smart agriculture. Springer, pp 7–34
32. Idrees Z, Zou Z, Zheng L (2018) Edge computing based IoT architecture for low cost air pollution monitoring systems: a comprehensive system analysis, design considerations & development. Sensors 18:3021
33. Sicari S, Rizzardi A, Coen-Porisini A (2019) How to evaluate an Internet of Things system: models, case studies, and real developments. Softw Pract Experience 49:1663–1685
34. Dhaka VS, Kundu N, Rani G et al (2023) Role of internet of things and deep learning techniques in plant disease detection and classification: a focused review. Sensors 23:7877
35. Raj M, Gupta S, Chamola V et al (2021) A survey on the role of Internet of Things for adopting and promoting Agriculture 4.0. J Netw Comput Appl 187:103107
36. Perera C (2020) Internet of Things: systems design lab book
37. Méndez-Guzmán HA, Padilla-Medina JA, Martínez-Nolasco C et al (2022) IoT-based monitoring system applied to aeroponics greenhouse. Sensors 22:5646
38. Nasar S, Muhammad M, Siddiqi EH et al (2024) OMICS-based knowledge for achieving food and nutritional security. Omi Tech Glob Food Secur:67–90
39. Fan X, Zhou X, Chen H et al (2021) Cross-talks between macro-and micronutrient uptake and signaling in plants. Front Plant Sci 12:663477
40. Figueroa-Macías JP, García YC, Núñez M et al (2021) Plant growth-defense trade-offs: molecular processes leading to physiological changes. Int J Mol Sci 22:693

41. Shoaib M, Shah B, Ei-Sappagh S et al (2023) An advanced deep learning models-based plant disease detection: a review of recent research. Front Plant Sci 14:1158933
42. Ijaz S, Haq IU, Babar M, Nasir B (2022) Disease resistance genes' identification, cloning, and characterization in plants. In: Cereal diseases: nanobiotechnological approaches for diagnosis and management. Springer, pp 249–269
43. Khan N, Bano A, Ali S, Babar MA (2020) Crosstalk amongst phytohormones from planta and PGPR under biotic and abiotic stresses. Plant Growth Regul 90:189–203
44. Wang W, Hu B, Li A, Chu C (2020) NRT1. 1s in plants: functions beyond nitrate transport. J Exp Bot 71:4373–4379
45. Sharma M, Sudheer S, Usmani Z et al (2020) Deciphering the omics of plant-microbe interaction: perspectives and new insights. Curr Genomics 21:343–362. https://doi.org/10.2174/1389202921999200515140420
46. Badenhorst M, Barry CJ, Swanepoel CJ, et al (2019) Workflow for data analysis in experimental and computational systems biology: using python as 'glue.' PRO 7:460
47. Yang T, Liu J, Li X et al (2022) Transcriptomic analysis of fusarium oxysporum stress-induced Pathosystem and screening of Fom-2 interaction factors in contrasted melon plants. Front Plant Sci 13. https://doi.org/10.3389/fpls.2022.961586
48. Lane M, Maiocco A, Bhatia SK, Climer S (2020) Eyeing the patterns: data visualization using doubly-seriated color heatmaps. In: Advances in computers. Elsevier, pp 121–156
49. Babicki S, Arndt D, Marcu A et al (2016) Heatmapper: web-enabled heat mapping for all. Nucleic Acids Res 44:W147–W153
50. Gao F, Dubos C (2021) Transcriptional integration of plant responses to iron availability. J Exp Bot 72:2056–2070
51. Chao J, Jin J, Wang D et al (2014) Cytological and transcriptional dynamics analysis of host plant revealed stage-specific biological processes related to compatible rice-Ustilaginoidea virens interaction. PLoS One 9. https://doi.org/10.1371/journal.pone.0091391
52. Ngongoma MSP, Kabeya M (2023) Applied sciences a review of plant disease detection systems for farming applications
53. Prabha K (2020) Disease sniffing robots to apps fixing plant diseases: applications of artificial intelligence in plant pathology—a mini review. Indian Phytopathol. https://doi.org/10.1007/s42360-020-00290-3
54. Shaikh TA, Rasool T, Lone FR (2022) Towards leveraging the role of machine learning and artificial intelligence in precision agriculture and smart farming. Comput Electron Agric 198:107119
55. Nitu OA, Ivan EŞ, Tronac AS, Arshad A (2024) Optimizing lettuce growth in nutrient film technique hydroponics: evaluating the impact of elevated oxygen concentrations in the root zone under LED illumination. Agronomy 14:1896
56. Osorio MB, Ng S, Berkowitz O et al (2019) SPX4 acts on PHR1-dependent and-independent regulation of shoot phosphorus status in Arabidopsis. Plant Physiol 181:332–352
57. McDermaid A, Monier B, Zhao J et al (2019) Interpretation of differential gene expression results of RNA-seq data: review and integration. Brief Bioinform 20:2044–2054
58. Nusrat S, Harbig T, Gehlenborg N (2019) Tasks, techniques, and tools for genomic data visualization. In: Computer graphics forum. Wiley Online Library, pp 781–805
59. Mullan KA, Bramberger LM, Munday PR et al (2021) ggVolcanoR: a shiny app for customizable visualization of differential expression datasets. Comput Struct Biotechnol J 19:5735–5740
60. Obradovic Z (2020) US-Serbia and West Balkan data science workshop. In: NSF workshop reports
61. Wratten L, Wilm A, Göke J (2021) Reproducible, scalable, and shareable analysis pipelines with bioinformatics workflow managers. Nat Methods 18:1161–1168
62. Pazhamala LT, Kudapa H, Weckwerth W et al (2021) Systems biology for crop improvement. Plant Genome 14:e20098

63. Steinegger A, Wolfbeis OS, Borisov SM (2020) Optical sensing and imaging of pH values: spectroscopies, materials, and applications. Chem Rev 120:12357–12489
64. Zhan J, Mcdonald BA (2013) Field-based experimental evolution of three cereal pathogens using a mark-release-recapture strategy. Plant Pathol 62:106–114. https://doi.org/10.1111/ppa.12130
65. Satya S, Sharma S, Choudhary G, Kaushik G (2024) Advances in environmental microbiology: a multi-omic perspective. In: Microbial omics in environment and health. Springer, pp 175–204
66. John Martin JJ, Song Y, Hou M et al (2024) Multi-omics approaches in oil palm research: a comprehensive review of metabolomics, proteomics, and transcriptomics based on low-temperature stress. Int J Mol Sci 25:7695
67. Cembrowska-Lech D, Krzemińska A, Miller T et al (2023) An integrated multi-omics and artificial intelligence framework for advance plant phenotyping in horticulture. Biology (Basel) 12:1298
68. Djaffardjy M, Marchment G, Sebe C et al (2023) Developing and reusing bioinformatics data analysis pipelines using scientific workflow systems. Comput Struct Biotechnol J 21:2075–2085
69. Mulla DJ (2013) Twenty five years of remote sensing in precision agriculture: key advances and remaining knowledge gaps. Biosyst Eng 114:358–371
70. Meshram V, Patil K, Meshram V et al (2021) Machine learning in agriculture domain: a state-of-art survey. Artif Intell Life Sci 1:100010
71. Andersen EJ, Ali S, Byamukama E et al (2018) Disease resistance mechanisms in plants. Genes (Basel) 9. https://doi.org/10.3390/genes9070339
72. Sahu SK, Waseem M, Aslam MM (2023) Bioinformatics, big data and agriculture: a challenge for the future. Front Plant Sci 14:1271305
73. Rosca C-M, Stancu A, Popescu M (2025) The impact of cloud versus local infrastructure on automatic IoT-driven hydroponic systems. Appl Sci 15
74. Kansara M (2021) Cloud migration strategies and challenges in highly regulated and data-intensive industries: a technical perspective. Int J Appl Mach Learn Comput Intell 11:78–121
75. Reza MN, Lee K-H, Karim MR, et al (2025) Trends of soil and solution nutrient sensing for open field and hydroponic cultivation in facilitated smart agriculture. Sensors 25:453
76. Jolly CM, Nyandat B, Yang Z et al (2023) Dynamics of aquaculture governance. J World Aquac Soc 54:427–481
77. Liakos KG, Busato P, Moshou D et al (2018) Machine learning in agriculture: a review. Sensors 18:2674
78. Szklarczyk D, Gable AL, Lyon D et al (2019) STRING v11: protein–protein association networks with increased coverage, supporting functional discovery in genome-wide experimental datasets. Nucleic Acids Res 47:D607–D613
79. Cantalapiedra CP, Hernández-Plaza A, Letunic I et al (2021) eggNOG-mapper v2: functional annotation, orthology assignments, and domain prediction at the metagenomic scale. Mol Biol Evol 38:5825–5829
80. Rao PBS, Venkatesh RT, Murthy SG et al (2024) Automated IoT solutions for efficient hydroponic farming: nutrients, PH and lighting management. J Eur des Systèmes Autom:57, 1273–1283
81. Mokhtar A, El-Ssawy W, He H et al (2022) Using machine learning models to predict hydroponically grown lettuce yield. Front Plant Sci 13:706042
82. Zhang M, Zhou J, Sudduth KA, Kitchen NR (2020) Estimation of maize yield and effects of variable-rate nitrogen application using UAV-based RGB imagery. Biosyst Eng 189:24–35
83. Constas MA, d'Errico M, Hoddinott JF, Pietrelli R (2021) Resilient food systems–a proposed analytical strategy for empirical applications: background paper for the state of food and agriculture 2021. In: FAO agricultural development economics working paper 21–10. Food & Agriculture Org

84. Aroge T, Zhu Y, Jin D-N et al (2024) Omics and CRISPR-Cas9 molecular perception: a progressive review approach for powdery mildew disease management. Physiol Mol Plant Pathol 102217. https://doi.org/10.1016/j.pmpp.2024.102217
85. Tripoli M, Schmidhuber J (2018) Emerging opportunities for the application of blockchain in the agri-food industry
86. Sharmin S, Hossan MT, Uddin MS (2025) A review of machine learning approaches for predicting lettuce yield in hydroponic systems. Smart Agric Technol 100925
87. Mohmed G, Hasanaliyeva G, O'Mahony R, Lu C (2025) Optimising nutrient formulations through artificial intelligence model to reduce excessive fertigation in lettuce grown in hydroponic systems. IEEE Access
88. Brewster C, Roussaki I, Kalatzis N et al (2017) IoT in agriculture: designing a Europe-wide large-scale pilot. IEEE Commun Mag 55:26–33
89. Rhee SY, Mutwil M (2014) Towards revealing the functions of all genes in plants. Trends Plant Sci 19:212–221
90. Campos EG de (2020) Novel drugs of abuse and bioanalytical strategies in Forensic Toxicology: metabolomics, analysis in blood, urine or DBS and human performance and post mortem cases
91. Tzounis A, Katsoulas N, Bartzanas T, Kittas C (2017) Internet of Things in agriculture, recent advances and future challenges. Biosyst Eng 164:31–48
92. Kamilaris A, Kartakoullis A, Prenafeta-Boldú FX (2017) A review on the practice of big data analysis in agriculture. Comput Electron Agric 143:23–37
93. Kalantari F, Tahir OM, Joni RA, Fatemi E (2018) Opportunities and challenges in sustainability of vertical farming: a review. J Landsc Ecol 11:35–60
94. Sishodia RP, Ray RL, Singh SK (2020) Applications of remote sensing in precision agriculture: a review. Remote Sens 12:3136
95. Wolfert S, Ge L, Verdouw C, Bogaardt M-J (2017) Big data in smart farming–a review. Agric Syst 153:69–80
96. Bhuiyan MMR, Rahaman MM, Aziz MM et al (2023) Predictive analytics in plant biotechnology: using data science to drive crop resilience and productivity. J Environ Agric Stud 4:77–83
97. Santolini E, Pulvirenti B, Benni S et al (2018) Numerical study of wind-driven natural ventilation in a greenhouse with screens. Comput Electron Agric 149:41–53
98. Pantazi XE, Moshou D, Alexandridis T et al (2016) Wheat yield prediction using machine learning and advanced sensing techniques. Comput Electron Agric 121:57–65
99. Chen K, Wang Y, Zhang R et al (2019) CRISPR/Cas genome editing and precision plant breeding in agriculture. Annu Rev Plant Biol 70:667–697
100. Miorandi D, Sicari S, De Pellegrini F, Chlamtac I (2012) Internet of things: vision, applications and research challenges. Ad Hoc Netw 10:1497–1516
101. Michelmore RW, Christopoulou M, Caldwell KS (2013) Impacts of resistance gene genetics, function, and evolution on a durable future. Annu Rev Phytopathol 51:291–319. https://doi.org/10.1146/annurev-phyto-082712-102334

Chapter 13
Machine Learning-Based Management of Climatic Factors in Hydroponic Greenhouses

Susai Arul Selvan and Tarun Pal

Abstract Advancements in artificial intelligence (AI) and Internet of Things (IoT) technologies have revolutionized climate management in hydroponic agriculture by enabling real-time monitoring, predictive analytics, and adaptive environmental control. This chapter provides a comprehensive overview of how integrated sensor networks, machine learning (ML) algorithms, and cloud-edge computing architectures collectively transform greenhouse microclimate regulation. A detailed exploration of sensor technologies including thermocouples, quantum light sensors, and electrochemical probes demonstrates their role in capturing dynamic environmental parameters such as temperature, humidity, light intensity, and nutrient composition. The application of supervised learning, neural networks, and reinforcement learning techniques is critically examined in the context of predictive modeling and autonomous decision-making for optimized plant growth. Emphasis is placed on model validation metrics such as RMSE, R^2, and MAPE, alongside real-world challenges like data variability, sensor drift, model interpretability, and sustainability concerns. Figures and flowcharts enhance understanding of ML pipelines, validation strategies, and system integration. The chapter concludes by addressing ethical implications, equity in access, and the importance of explainable AI (XAI) frameworks to facilitate trust and adoption in agricultural communities. Overall, this chapter serves as a foundational reference for researchers and practitioners aiming to implement intelligent, scalable, and sustainable climate control in modern hydroponic systems.

Keywords Hydroponics · Machine learning · IoT sensors · Climate control · Smart agriculture

S. A. Selvan
School of Biotechnology, Faculty of Applied Sciences & Biotechnology, Shoolini University, Bajhol, Himachal Pradesh, India

T. Pal (✉)
School of Bioengineering and Food Technology, Faculty of Applied Sciences & Biotechnology, Shoolini University, Bajhol, Himachal Pradesh, India

P. Kumar, A. R. Warghat (eds.), *Transforming Medicinal Plant Agriculture*, Smart Agriculture 16, https://doi.org/10.1007/978-981-95-3963-5_13

13.1 Introduction

Hydroponic greenhouses represent a transformative approach to modern agriculture, enabling year-round, high-density crop production in controlled environments. By eliminating the reliance on soil and utilizing nutrient-rich aqueous solutions, these systems offer enhanced water and nutrient use efficiency, reduced land footprint, and the potential for significant yield improvements compared to traditional open-field farming [1]. The global hydroponics market has witnessed exponential growth, driven by rising consumer demand for fresh produce, urban agriculture initiatives, and the imperative for climate-resilient food production. Recent estimates suggest that the global hydroponics market will reach over USD 22 billion by 2028, underscoring its economic significance and potential for global food security [2].

Central to the success of hydroponic greenhouse systems is the precise management of climatic factors temperature, relative humidity, light intensity, and carbon dioxide (CO_2) concentration. These parameters collectively determine photosynthetic efficiency, plant metabolism, nutrient uptake, and crop yield [3]. Achieving optimal microclimatic conditions is not only critical for maximizing productivity but also for minimizing resource inputs and environmental impact. However, maintaining these conditions in dynamic greenhouse environments is a complex challenge due to interactions among multiple climatic variables and external factors such as fluctuating weather patterns.

Traditional climate control strategies in greenhouses, typically based on manual adjustments or rule-based logic controllers, often fall short in addressing the nonlinear, dynamic relationships among climatic variables and plant physiological responses [4]. These conventional approaches are constrained by their inability to adapt rapidly to changing conditions, leading to suboptimal climate management, increased disease susceptibility, and diminished resource efficiency.

In response, machine learning (ML) has emerged as a powerful and adaptive tool for managing climatic factors in hydroponic greenhouses. ML algorithms are capable of learning intricate, nonlinear patterns within large datasets, enabling predictive modeling, real-time anomaly detection, and dynamic optimization of climate control systems [5]. By integrating real-time data from advanced sensors with historical datasets, ML-based frameworks can autonomously adjust environmental parameters to maintain optimal conditions for plant growth, ensuring improved yield predictability and resource efficiency [6].

The global research landscape is increasingly focused on leveraging ML to enhance greenhouse climate control. Recent studies have demonstrated that ML-based climate management can achieve substantial energy savings, reduce greenhouse gas emissions, and improve overall sustainability [7]. Furthermore, the convergence of ML with emerging technologies such as digital twins, edge computing, and renewable energy systems is poised to revolutionize greenhouse operations, bridging the gap between precision agriculture and climate-resilient food production systems [8].

This chapter aims to provide a comprehensive overview of ML-based climatic management in hydroponic greenhouses, highlighting both foundational principles and recent innovations. We begin by discussing the critical climatic factors and their interactions, followed by an exploration of data acquisition and integration frameworks that underpin ML applications. Subsequent sections examine advanced ML algorithms, their practical applications in climate control, and robust validation methods to ensure model reliability. We also delve into emerging trends including digital twins and generative AI and outline the persistent challenges and knowledge gaps that shape future research directions. By synthesizing cutting-edge insights and best practices, this chapter underscores the potential of ML to transform hydroponic greenhouse climate management, supporting sustainable, efficient, and climate-resilient food production.

Furthermore, the global hydroponics market is experiencing robust growth, with market estimates projecting it to surpass USD 22 billion by 2028 [2]. This rapid expansion is driven by the need for climate-resilient and resource-efficient food production systems, particularly in regions facing water scarcity and erratic weather patterns. Compared to open-field farming, hydroponic greenhouses can reduce water usage by up to 90% and offer significant yield improvements, especially when integrated with precision climate control [9]. In this context, greenhouse climate management is not only a tool for ensuring high productivity but also a critical strategy for climate change adaptation. Extreme weather events, including heatwaves and droughts, have underscored the vulnerability of traditional agriculture. Hydroponic greenhouses, coupled with advanced ML algorithms, present a viable solution for climate-smart agriculture, mitigating risks and stabilizing food production [10]. This convergence of economic opportunity, climate resilience, and technological innovation sets the stage for exploring the intricacies of ML-driven climate management in hydroponic systems.

13.2 Climatic Factors in Hydroponic Systems

Climatic factors are fundamental determinants of plant growth, physiological processes, and overall productivity in hydroponic greenhouse systems. In these controlled environments, the ability to finely regulate temperature, relative humidity (RH), light intensity, and carbon dioxide (CO_2) concentration is pivotal for optimizing photosynthesis, nutrient uptake, and stress resilience [11]. Unlike traditional open-field cultivation, hydroponic greenhouses offer unparalleled precision in environmental control, yet achieving and maintaining these optimal conditions is a complex and dynamic challenge.

Temperature is a primary driver of metabolic and enzymatic activity in plants. Optimal temperature ranges vary by crop species, typically spanning 18–25 °C for most leafy vegetables and 20–28 °C for fruiting crops such as tomatoes and cucumbers [12]. Deviations beyond these ranges can adversely affect photosynthetic rates, nutrient assimilation, and crop development. Additionally, temperature influences root-zone processes in hydroponic systems, including nutrient solubility, microbial community dynamics, and oxygen availability within the nutrient solution [13]. Fluctuations in temperature can also exacerbate plant stress responses, affecting both growth and disease susceptibility.

Relative humidity (RH) plays a critical role in transpiration, stomatal conductance, and nutrient transport. An RH level that is too low can lead to excessive water loss and nutrient imbalances, while excessively high RH (>85%) can favor fungal pathogen proliferation and reduce gas exchange efficiency [14]. Maintaining RH within an optimal range typically 60%–80% is essential for balancing water-use efficiency and plant health.

Light intensity and spectral quality are equally crucial in hydroponic greenhouses, where both natural sunlight and artificial lighting systems (such as LEDs) are used to meet photosynthetic requirements. Photosynthetically active radiation (PAR), within the 400–700 nm range, is directly linked to photosynthetic rates and biomass accumulation [15]. Light intensity, photoperiod, and spectral distribution must be carefully managed to regulate photomorphogenesis and flowering responses. Dynamic light management strategies that adjust spectral composition can further enhance resource use efficiency and crop quality [16].

Carbon dioxide (CO_2) concentration is a key modulator of photosynthetic efficiency and plant productivity. Enriching greenhouse air with CO_2 typically targeting concentrations of 800–1200 ppm can significantly boost photosynthetic carbon assimilation and yield potential [17]. However, elevated CO_2 must be balanced with optimal temperature and light conditions to avoid negative impacts on plant respiration and metabolic efficiency.

Importantly, these climatic factors do not operate in isolation. They interact in nonlinear, synergistic ways that can either amplify or mitigate each other's effects. For example, temperature and RH jointly define the vapor pressure deficit (VPD) a critical parameter influencing transpiration rates and water-use efficiency [18]. Similarly, light intensity modulates the efficacy of CO_2 enrichment, while temperature impacts both photochemical and biochemical reactions within photosynthetic pathways [19]. These interactions underscore the need for integrated climate management strategies that consider the complexity and interdependencies inherent in hydroponic greenhouse environments.

Quantitative studies have shown that the optimal temperature range for leafy greens, such as lettuce and spinach, lies between 18 and 22 °C, while fruiting crops like tomatoes and peppers perform best between 20 and 28 °C [20]. Relative humidity (RH) control is equally vital; maintaining RH between 60% and 80% has been demonstrated to reduce fungal disease incidence by up to 30% compared to higher RH environments [21]. Vapor pressure deficit (VPD) has emerged as a crucial metric linking temperature and RH, directly influencing plant transpiration and water-use efficiency. Research has shown that optimizing VPD in hydroponic systems can improve nutrient uptake by 15%–20% and boost yield consistency [22].

Similarly, light spectrum manipulation has become an area of active research. LED-based lighting systems that adjust red-blue light ratios have been reported to enhance chlorophyll synthesis and photosynthetic efficiency by up to 25%, depending on crop type [23]. This dynamic interaction between light quality, temperature, and CO_2 enrichment requires holistic management to avoid trade-offs and unintended stress responses.

These quantitative insights emphasize the complexity of climate management in hydroponic greenhouses and highlight the need for integrated, data-driven strategies capable of dynamically adapting to both external weather fluctuations and internal greenhouse microclimate variations. Beyond climatic parameters, non-climatic factors such as nutrient solution composition, pH, and electrical conductivity (EC) are indirectly affected by environmental conditions. For instance, temperature fluctuations can influence nutrient solubility and microbial stability in the root zone, while RH variations can modulate the risk of foliar disease outbreaks [24]. Recognizing these linkages is crucial for holistic climate management that supports plant health and resource efficiency.

Traditional climate control systems often rely on linear, rule-based logic that struggles to capture these multivariate, dynamic interactions [25]. In contrast, machine learning (ML) approaches offer an adaptive and predictive framework capable of modeling these complex relationships, enabling real-time optimization of climatic conditions in hydroponic greenhouses. In the following sections, we will explore the data acquisition frameworks and advanced ML algorithms that underpin these innovations, highlighting their potential to transform climate management and drive the next generation of precision-controlled, sustainable greenhouse systems (Fig. 13.1).

13.3 Data Acquisition and Integration

Effective climate management in hydroponic greenhouses hinges on robust and continuous data acquisition systems that capture the nuanced dynamics of the growing environment. These data streams provide the foundation for predictive modeling, real-time anomaly detection, and adaptive control using machine learning (ML) algorithms. The advent of advanced sensor technologies, Internet of Things (IoT) frameworks, and edge computing has revolutionized the capacity for real-time environmental monitoring and dynamic climate control in hydroponic systems [26].

13.3.1 Sensor Technologies and IoT Frameworks

Modern hydroponic greenhouses employ a sophisticated suite of sensors to monitor key climatic parameters such as temperature, humidity, light intensity, and CO_2 concentration. Typical sensors include thermocouples and Resistance Temperature Detectors (RTDs) for temperature monitoring, providing high accuracy measurements crucial for regulating air and nutrient solution temperatures [27]. Capacitive and resistive humidity sensors are used for real-time RH monitoring, enabling

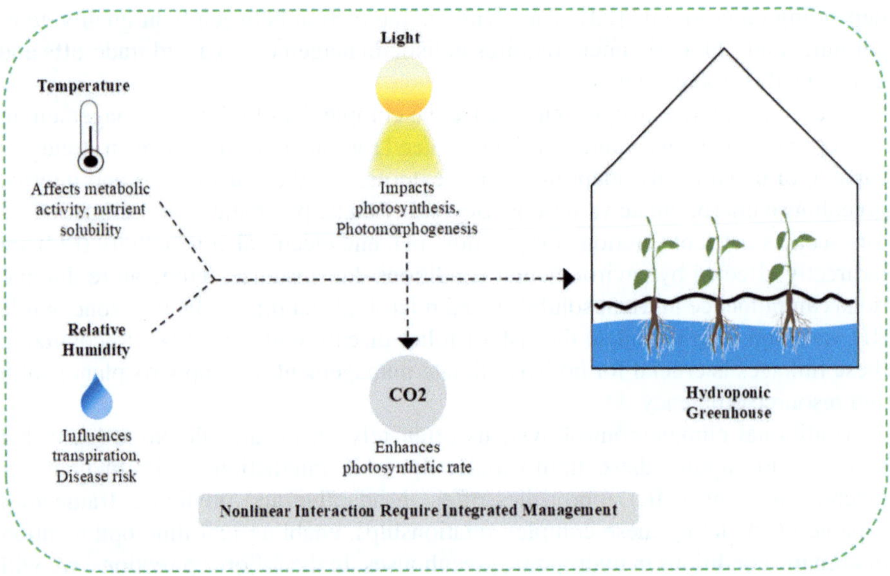

Fig. 13.1 Conceptual schematic illustrating the synergistic interactions of key climatic factors in hydroponic greenhouse systems. Temperature, relative humidity, light intensity, and carbon dioxide (CO_2) concentration dynamically influence plant metabolic processes, nutrient solubility, and photosynthetic efficiency. These factors do not operate in isolation; rather, they interact in nonlinear ways such as the temperature-humidity linkage defining vapor pressure deficit (VPD) and the light-CO_2 synergy enhancing photosynthetic carbon assimilation requiring integrated climate management to optimize plant health, yield, and resource efficiency

precise control of transpiration rates and disease prevention [28]. Infrared gas analyzers are implemented to track CO_2 concentrations, facilitating dynamic CO_2 enrichment protocols (Add Ref). Quantum and spectral sensors measure photosynthetically active radiation (PAR), ensuring optimal light conditions for photosynthesis [15]. Additionally, multi-sensor platforms integrate pH, electrical conductivity (EC), and dissolved oxygen (DO) sensors to provide comprehensive monitoring of nutrient solution properties [29].

Integration of these sensors within IoT frameworks facilitates seamless data acquisition and transmission. Wireless sensor networks (WSNs) using protocols like Zigbee, LoRa, and Wi-Fi ensure real-time data streaming even in large-scale greenhouse operations [30]. These frameworks support scalability and interoperability, connecting environmental sensors with cloud-based data storage and analytics platforms.

In real-world deployments, sensor calibration is a critical yet often overlooked aspect. For instance, temperature sensors like the DHT22 require periodic recalibration to maintain their ±0.5 °C accuracy, especially in high-humidity greenhouse environments prone to condensation [31]. pH sensors and EC meters, integral to nutrient solution monitoring, must also be recalibrated weekly to avoid drift that can compromise nutrient balance [32]. Edge computing has emerged as a promising solution to address the challenges of data latency and reliability. In commercial hydroponic systems, edge devices such as industrial-grade microcontrollers and low-power computing units are deployed alongside sensors to process data locally before transmission to cloud platforms [33]. For example, studies have demonstrated that integrating edge computing with cloud-based ML models can reduce climate control response times by up to 30%, ensuring rapid adjustments to environmental changes [34].

Data fusion is another area of growing interest. Beyond standard climatic sensors, modern greenhouses often incorporate external weather data, historical crop performance metrics, and even image-based canopy monitoring to create comprehensive datasets for ML model training [35]. Such integrated datasets can capture not only the immediate microclimate variations but also longer-term trends in crop development and stress responses. Ensuring data integrity in these complex environments requires robust data cleaning and standardization protocols. Techniques like Kalman filtering and wavelet-based denoising have been employed to filter out sensor noise and correct for transient data dropouts, particularly in wireless sensor networks that may be subject to electromagnetic interference [36].

These developments in real-world data acquisition and integration highlight the foundational role of high-quality, multi-modal data in enabling reliable and responsive ML-driven climate management systems (Table 13.1).

Table 13.1 Sensors and IoT frameworks in hydroponic greenhouses

Sensor type	Measurement principle	Typical accuracy	Application	Reference
Temperature & Humidity (DHT22, DHT11)	Capacitive sensing	±0.5 °C (temp), ±2–5% RH	Monitoring ambient conditions	[37]
Water temperature (DS18B20)	Digital thermometry	±0.5 °C	Ensuring optimal root zone temperature	[38]
pH sensor	Electrochemical	±0.1 pH	Monitoring nutrient solution acidity	[39]
EC sensor	Conductivity measurement	±2%	Assessing nutrient concentration	[40]
Light sensor (LDR, TSL2561)	Photodiode-based	±5%	Measuring light intensity for photosynthesis	[41]
CO_2 sensor (MH-Z19)	Non-dispersive infrared (NDIR)	±50 ppm	Monitoring CO_2 levels for plant growth	[42]
Leaf sensor (e.g., AgriHouse SG-1000)	Measures leaf thickness	Variable	Detecting plant water stress	[43]
Soil moisture sensor	Dielectric constant measurement	±3%	Monitoring moisture levels in substrates	[44]

13.3.2 Data Collection, Preprocessing, and Management

Data collection in hydroponic greenhouses typically involves high temporal resolution often minute-level data capturing rapid microclimatic fluctuations and transient environmental changes [45]. However, raw sensor data are prone to noise, drift, and occasional data gaps due to hardware malfunctions or network outages.

To ensure data integrity, preprocessing steps are essential:

1. Noise filtering: Techniques such as moving average filters and wavelet-based denoising reduce sensor noise and improve signal fidelity [46].
2. Outlier detection and removal: Algorithms identify anomalous readings (e.g., sudden spikes) to maintain data reliability [47].
3. Interpolation and smoothing: Methods like linear interpolation fill data gaps and produce smoother datasets for downstream ML applications [48].

Data management relies on robust databases often cloud-based platforms that offer secure, scalable storage and real-time data accessibility [49]. These platforms facilitate data standardization and integration across diverse sensor types and greenhouse zones, supporting advanced analytics and decision-making processes.

13.3.3 Challenges in Data Quality and Granularity

1. Despite technological advances, challenges remain in ensuring high-quality data acquisition and processing:
2. Sensor drift and calibration: Over time, sensor readings can deviate due to environmental stresses or aging, necessitating regular recalibration to maintain accuracy [50].
3. Environmental interference: Factors such as dust accumulation, water vapor, or electromagnetic interference can degrade sensor performance [51].

Data granularity and computational trade-offs: While high-frequency data capture reveals subtle climate variations, it can also create large datasets that strain computational resources [52]. Identifying the optimal temporal resolution is crucial for balancing precision and operational efficiency.

For example, minute-level data might reveal transient microclimate fluctuations that hourly data could miss. However, excessively high data rates may require edge computing solutions or data reduction techniques to ensure real-time ML-based control without latency [53].

13.3.4 Importance of Real-Time Monitoring and Edge Computing

Real-time monitoring is indispensable for dynamic climate control in hydroponic greenhouses, where environmental changes and plant physiological responses occur rapidly [54]. Edge computing processing data locally near the source offers a solution to latency and bandwidth limitations associated with cloud-only architectures. By deploying ML models directly on edge devices (e.g., microcontrollers, gateways), greenhouses can achieve low-latency climate adjustments and enhanced resilience to network disruptions [55].

Edge computing also enables distributed intelligence, where localized data processing occurs in parallel with centralized cloud-based analytics. This hybrid architecture supports real-time climate optimization while ensuring comprehensive data analytics for long-term system improvements [56]. In the subsequent sections, we will explore how these rich, high-frequency data streams are harnessed by advanced ML algorithms to enable predictive, adaptive, and sustainable climate management in hydroponic greenhouses.

13.4 Machine Learning Techniques in Climatic Management

Machine learning (ML) has rapidly emerged as a transformative approach for managing the complex and dynamic climatic conditions in hydroponic greenhouses. Unlike conventional rule-based climate control systems which rely on predefined setpoints and linear responses ML algorithms possess the ability to learn intricate, nonlinear relationships from large datasets. This capability enables them to predict climatic variations, identify anomalies, and autonomously optimize environmental parameters in real time [57].

Traditional climate control methods, such as proportional-integral-derivative (PID) controllers and heuristic rule-based logic, are often limited by their static nature and inability to adapt to rapidly changing greenhouse conditions [58]. These approaches typically operate on simple threshold-based rules, struggling to incorporate the synergistic interactions between temperature, humidity, light, and CO_2 concentration that define greenhouse microclimates [59]. In contrast, ML models can continuously learn from new data, capturing complex dependencies and evolving climatic patterns to enable precision-controlled management.

ML techniques can be broadly categorized into supervised learning, artificial neural networks (ANNs), and reinforcement learning (RL), each offering unique capabilities for greenhouse climate control.

13.4.1 Supervised Learning Algorithms

Supervised learning algorithms use historical, labeled data to build predictive models of climatic parameters. In greenhouse applications, algorithms such as Random Forests, support vector machines (SVM), and gradient boosting machines (GBMs) have been extensively studied for their ability to forecast temperature, humidity, and light intensity [60].

Random Forests are ensemble models that construct multiple decision trees and aggregate their predictions, effectively reducing overfitting and improving generalization. Studies in hydroponic lettuce greenhouses have demonstrated the effectiveness of Random Forest models in predicting temperature and humidity fluctuations, thereby enabling proactive adjustments to heating and ventilation systems [61]. Similarly, support vector machines (SVMs) excel at modeling complex, nonlinear relationships by mapping input data to higher-dimensional feature spaces. Applications in hydroponic tomato greenhouses have shown that SVM models can predict microclimatic variations with high accuracy, outperforming conventional regression methods [62]. Supervised learning models are particularly valuable for tasks where ground-truth data are available, offering robust predictive performance in climate control applications.

13.4.2 Artificial Neural Networks (ANNs)

Artificial neural networks (ANNs) are powerful tools for capturing nonlinear and multivariate interactions in greenhouse environments. Among these, convolutional neural networks (CNNs) and recurrent neural networks (RNNs) particularly long short-term memory (LSTM) networks have gained significant attention. CNNs, originally developed for image analysis, have been adapted to interpret spatial and spectral data in greenhouse systems. For instance, CNNs have been used to analyze thermal imagery from greenhouse sensors, enabling early detection of microclimatic hotspots and targeted climate control interventions [35]. In parallel, LSTM models are designed to handle sequential, time-series data, making them ideal for predicting dynamic climatic trends. Studies in hydroponic leafy green systems have demonstrated that LSTM models outperform traditional linear models in predicting diurnal temperature and humidity variations, leading to enhanced yield predictability and energy efficiency [63]. These ANN-based models offer superior performance in capturing the dynamic and temporal complexities of greenhouse microclimates, enabling real time, data-driven control strategies.

13.4.3 Reinforcement Learning (RL) Algorithms

Reinforcement learning (RL) represents a paradigm shift in climate control strategies, enabling autonomous, self-improving management of greenhouse environments. In RL frameworks, an agent learns to optimize its control actions by interacting with the greenhouse environment and receiving feedback in the form of rewards or penalties [64].

1. Model-Free RL: Algorithms such as Q-learning and Deep Q-Networks (DQNs) have shown promise in dynamically adjusting heating and ventilation systems to maintain desired temperature and humidity setpoints while minimizing energy use [65].
2. Model-Based RL: More advanced frameworks incorporate predictive models of greenhouse microclimates, allowing the RL agent to simulate and evaluate multiple control strategies before deployment. This approach enhances safety and stability, particularly in complex hydroponic environments [66].

Recent studies have provided detailed performance comparisons for these ML techniques in hydroponic greenhouse contexts. For example, a 2023 study using Random Forest models to predict microclimatic variations in tomato greenhouses reported an RMSE of 0.9 °C for temperature forecasting, outperforming linear regression models by 20% [67]. Similarly, LSTM-based models have shown strong time-series prediction capabilities in hydroponic lettuce systems, achieving MAE values below 0.5 °C for 24-hour humidity predictions [38]. CNNs have been successfully applied to interpret thermal imagery data in large-scale greenhouses, detecting microclimate hotspots and informing spatially differentiated climate control strategies [66]. These approaches are particularly valuable for preventing localized plant stress and optimizing resource distribution within the greenhouse. Reinforcement learning (RL) applications in hydroponic greenhouses have demonstrated the ability to develop multi-objective control policies that maximize yield while minimizing energy consumption. For instance, a 2024 pilot study using a DQN-based climate control framework in a strawberry hydroponic system achieved a 15% yield increase and a 20% reduction in energy use compared to PID-based systems [68]. In practice, these ML models are often tailored to the unique operational parameters of each greenhouse, including crop type, lighting configuration, and regional climatic conditions. As a result, hyperparameter tuning such as adjusting the number of decision trees in Random Forests or the number of LSTM layers in deep learning architectures is critical to achieving robust and generalizable performance [69] (Table 13.2). These examples underscore the flexibility of ML algorithms to adapt to specific greenhouse microclimates and highlight the need for tailored implementation strategies that consider the operational nuances of each system (Fig. 13.2)

In hydroponic greenhouses, RL models have been experimentally validated for optimizing climate control policies, achieving multi-objective goals such as yield

Table 13.2 Strengths and limitations of ML models in greenhouse climate control

ML approach	Strengths	Limitations	Reference
Supervised learning (e.g., random Forest, SVM)	High accuracy with labeled data- effective for specific prediction tasks	Requires extensive labeled datasets- limited adaptability to new conditions	[70]
Artificial neural networks (ANNs)	Capable of modeling complex nonlinear relationships- suitable for time-series data	Prone to overfitting- requires significant computational resources	[71]
Reinforcement learning (RL)	Learns optimal control policies through interaction- adapts to dynamic environments	Requires extensive training- potentially unstable during learning phase	[72]

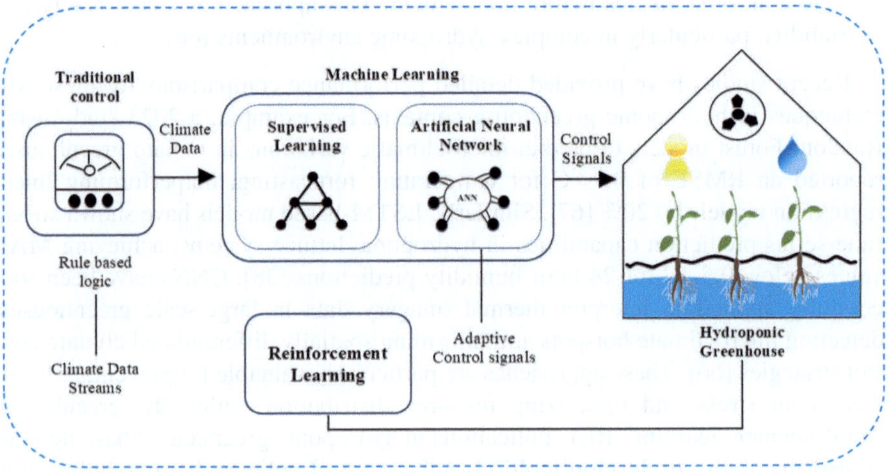

Fig. 13.2 Conceptual flowchart illustrating how traditional rule-based climate control systems rely on static thresholds, while ML-based approaches including supervised learning, artificial neural networks (ANNs), and reinforcement learning (RL) enable dynamic and adaptive management of greenhouse microclimates. RL uniquely interacts with the greenhouse environment in a feedback loop, continuously optimizing control policies based on real-time observations

maximization, resource conservation, and sustainability [73]. Their capacity to continuously adapt to changing greenhouse conditions makes them particularly well-suited for long-term, autonomous climate management.

A key strength of ML models across these categories lies in their ability to capture the synergistic interactions among climatic variables. For example, ML algorithms can model how temperature and humidity jointly influence vapor pressure deficit (VPD) and how light intensity interacts with CO_2 enrichment to impact photosynthesis rates [74]. Such insights go beyond the linear assumptions of traditional controllers, offering holistic, data-driven optimization of hydroponic greenhouse performance.

13.5 ML Applications for Climatic Factor Management

The practical integration of machine learning (ML) algorithms into hydroponic greenhouse management has led to significant advances in precision-controlled agriculture. These data-driven approaches have demonstrated the ability to enhance climatic factor stability, optimize resource use, and improve overall crop productivity. This section explores key real-world applications of ML in managing climatic factors within hydroponic greenhouses, showcasing their transformative potential for sustainable food production. One of the most prominent applications of ML is predictive climate control, where models anticipate short-term fluctuations in temperature, humidity, and CO_2 levels based on historical and real-time data. For example, Random Forest models have been deployed to forecast hourly temperature and humidity variations in hydroponic lettuce greenhouses, enabling proactive heating and ventilation adjustments to maintain ideal microclimates [75]. Similarly, LSTM networks have demonstrated superior performance in predicting daily temperature and humidity cycles, outperforming traditional autoregressive models and leading to reduced energy consumption [76].

Another crucial application is the dynamic optimization of resource use. CNNs have been applied to interpret thermal and optical imagery within greenhouses, identifying microclimatic variations that inform localized climate control measures [66]. These targeted interventions not only reduce energy demand for heating and cooling but also minimize water loss through more precise humidity management.

Disease and stress detection is another area where ML has shown substantial promise. Deep learning-based CNNs have been used to detect early signs of plant stress and disease outbreaks based on subtle climatic deviations and crop physiological signals [77]. By correlating these signals with environmental data, greenhouse managers can adjust temperature and humidity regimes to suppress disease development and mitigate yield losses. Reinforcement Learning (RL) models represent a particularly exciting frontier in hydroponic climate management. RL agents can learn optimal climate control policies by interacting with the environment, balancing competing objectives such as maximizing yield and minimizing energy costs [78]. In simulated and pilot-scale greenhouses, RL-based control strategies have demonstrated up to 25% reductions in energy consumption compared to static PID controllers, while maintaining or improving crop productivity [58]. Recent studies highlight the diverse ML algorithms and applications now emerging in hydroponic climate management. Table 13.3 summarizes selected examples of ML applications, illustrating the range of algorithmic approaches and their associated benefits.

Additional real-world case studies continue to highlight the practical impact of ML-driven climate management. In a vertical farming hydroponic greenhouse in Singapore, LSTM models integrated with LED lighting control reduced energy consumption by 25% during peak photoperiods, while maintaining uniform temperature and humidity profiles across growing zones [83]. In North America, CNNs have been deployed to analyze real-time infrared imagery from strawberry greenhouses, detecting canopy-level stress patterns and enabling localized climate control adjustments that reduced disease incidence by 18% [84].

Table 13.3 Summary of ML applications in climate-controlled hydroponic systems

ML algorithm	Target parameter(s)	Crop/system	Key findings	Reference
LSTM (Long short-term memory)	Temperature and humidity prediction	Indoor hydroponic greenhouse	Achieved lowest MAE and RMSE, outperforming DNN and 1D-CNN models.	[79]
CNN (convolutional neural network)	Growth index estimation (e.g., fresh weight, leaf area)	Greenhouse lettuce	Provided R^2 values from 0.88 to 0.95 for various growth indices.	[80]
Random Forest regression	Indoor air temperature forecasting	Smart greenhouse	Demonstrated mean absolute error of less than 1 °C for climate precision.	[67]
CLSTM-CNN-BP (hybrid model)	Microclimate forecasting (temp, RH, PAR)	Tomato greenhouse	Enabled proactive environmental control and optimized photosynthesis.	[81]
Reinforcement learning (RL)	Climate control optimization	Simulated greenhouse environment	RL-based model predictive control outperformed traditional strategies.	[82]

Another innovative application involves combining ML models with near-infrared (NIR) spectroscopy sensors to monitor plant nutrient status and water uptake in real time [85]. This integrated approach has shown promise in achieving fine-grained climate-nutrient feedback loops, further enhancing yield predictability. These applications collectively demonstrate that ML models are not just theoretical constructs they are powerful, field-tested tools that offer measurable benefits in operational hydroponic greenhouses. By continuously integrating new sensor data, these models can refine climate control recommendations and adapt to emerging environmental stressors or shifts in crop growth patterns.

These examples highlight the flexibility and adaptability of ML-based climate management systems, which can be tailored to diverse crop types, greenhouse designs, and operational priorities. The ability of ML to integrate multiple data streams including sensor data, crop growth data, and external weather forecasts positions it as a key enabler of climate-resilient and resource-efficient hydroponic agriculture.

13.6 Model Validation and Performance Metrics

The reliability and effectiveness of machine learning (ML) models in managing climatic factors within hydroponic greenhouses hinge on rigorous validation and performance evaluation. Without comprehensive validation, even sophisticated

ML algorithms risk poor generalizability, overfitting, and suboptimal decision-making when deployed in real-world greenhouse environments [86]. This section explores the critical metrics and methods used to assess ML model performance, along with the challenges associated with ensuring robust, reliable climate control outcomes.

13.6.1 Common Metrics for Model Evaluation

Multiple performance metrics are utilized to evaluate ML models in greenhouse climate control. The coefficient of determination (R^2) measures the proportion of variance in the observed data explained by the model; R^2 values close to 1.0 indicate strong predictive power and good model fit, whereas low R^2 values suggest poor alignment with real-world patterns [87]. Root mean squared error (RMSE) quantifies the magnitude of prediction errors by taking the square root of the mean squared differences between observed and predicted values, with lower RMSE values indicating better model performance and accuracy [88]. Mean absolute error (MAE) assesses the average magnitude of prediction errors without regard to direction, providing a straightforward interpretation of model error in real-world applications [89]. Mean absolute percentage error (MAPE) expresses prediction errors as a percentage of actual values, enabling easy comparison across diverse data scales and greenhouse systems [90]. Confusion matrices, while more applicable to classification tasks such as stress or disease detection, detail the number of correct and incorrect predictions across different classes and support the calculation of accuracy, precision, recall, and F1-score [91]. These quantitative metrics are foundational for selecting, refining, and benchmarking ML models prior to deployment in greenhouse climate control systems.

In real-world deployments, robust validation frameworks are essential to ensure ML models remain accurate and reliable under dynamic greenhouse conditions. Common strategies include k-fold cross-validation, where data are split into k subsets to ensure performance consistency across multiple partitions [92]. For time-series climate data, sliding window cross-validation methods are often employed to simulate real-world forecasting challenges, preserving the temporal structure of data while evaluating model robustness [93].

Beyond quantitative metrics like RMSE and R^2, confusion matrix-based metrics (precision, recall, and F1-score) are particularly relevant for stress or disease classification tasks. These metrics provide nuanced insights into how well ML models detect early stress symptoms critical for preventing yield losses in high-value crops [94]. Finally, real-time retraining and continuous performance monitoring using feedback from operational climate control outcomes are essential for adapting ML models to evolving greenhouse environments and maintaining predictive accuracy over time [95].

13.6.2 Challenges in Model Validation

Despite the availability of well-established metrics, validating ML models in dynamic hydroponic greenhouse environments poses several challenges:

1. Overfitting and Underfitting: Overfitting occurs when a model becomes too tailored to the training dataset, capturing noise and specificities rather than general patterns. Underfitting arises when the model lacks complexity to adequately represent underlying relationships. Both phenomena reduce the reliability of predictions in operational environments [96].
2. Real-World Data Variability: Hydroponic greenhouses experience dynamic environmental changes, sensor drift, and sudden system perturbations (e.g., equipment malfunctions, external weather events). Static training datasets often fail to capture this variability, limiting model robustness [97].
3. Generalizability Across Contexts: ML models developed for one greenhouse system may not generalize effectively to others due to differences in crop type, greenhouse design, or geographic location [98]. Addressing this requires domain adaptation and transfer learning approaches.
4. Model Interpretability: Many high-performing ML models (e.g., deep neural networks) operate as "black boxes," making it challenging for growers to understand how predictions are generated. Enhancing interpretability is crucial for building trust and ensuring adoption in practical settings [99].

13.6.3 Continuous Validation and Model Updating

Given these challenges, continuous validation and periodic model updating are essential for maintaining high performance and reliability in ML-driven climate management systems. Online learning and real-time retraining allow modern ML pipelines to integrate new data as they become available, enabling continuous model refinement and adaptation to changing greenhouse conditions [100]. This approach ensures that models remain current and responsive to new climatic patterns or operational adjustments. Edge-cloud hybrid architectures, by deploying lightweight validation and retraining routines at the edge while maintaining deeper analytical capabilities in the cloud, help greenhouse managers balance low-latency decision-making with robust long-term model evolution [101]. Additionally, the integration of domain knowledge such as crop physiological thresholds and historical climate performance enhances model interpretability and improves the practical relevance of ML-driven recommendations [102].

Continuous validation is not merely a technical requirement; it is a foundational practice to ensure that ML algorithms deliver actionable, reliable, and scientifically sound climate management strategies in dynamic hydroponic greenhouse environments. In the next section, we explore emerging trends and future directions that will shape the next generation of ML-based climate control systems.

13.7 Emerging Trends and Future Directions

The integration of machine learning (ML) into hydroponic greenhouse climate management is undergoing a rapid evolution, driven by technological innovations, sustainability imperatives, and the increasing complexity of global food systems. This section explores the most promising emerging trends and identifies key future directions that will shape the next generation of ML-based climate control systems in hydroponic greenhouses.

13.7.1 Digital Twins and Virtual Greenhouse Simulations

One of the most transformative trends in greenhouse climate management is the convergence of ML with digital twins virtual, real-time replicas of physical greenhouse environments that continuously mirror their operational states [103]. Digital twins enable real-time simulation of climate scenarios, resource use, and crop responses, offering growers a dynamic platform for testing control strategies without disrupting actual production. By synchronizing sensor data with digital models, ML algorithms can optimize climate control policies, simulate "what-if" scenarios, and identify the most efficient pathways to yield maximization and resource conservation [8].

13.7.2 Generative AI and Predictive Scenario Modeling

The rise of generative AI models, including generative adversarial networks (GANs) and large transformer-based frameworks, is expanding the horizons of predictive climate control. Generative models can simulate new climatic scenarios, anticipate rare events (such as extreme weather fluctuations), and generate synthetic data to enhance ML model robustness in data-scarce environments [104]. For hydroponic greenhouses, this means the ability to forecast complex microclimate dynamics and preemptively adjust environmental parameters for resilience and sustainability.

Blockchain technologies are also gaining traction in greenhouse climate control as secure, transparent data management frameworks. For instance, blockchain-based ledgers can track calibration histories, sensor maintenance logs, and control system interventions, ensuring data integrity and enabling transparent climate data sharing [105]. Another emerging trend is the convergence of ML with computer vision-based plant health monitoring. By integrating CNN-based image analysis of leaf surfaces with climate data, future greenhouse systems could develop even more precise stress mitigation strategies [35].

Multi-objective optimization frameworks that simultaneously balance yield, energy efficiency, and resource use are already being tested in pilot-scale greenhouses. These systems leverage reinforcement learning models to autonomously identify trade-offs and adapt control strategies to seasonal variations and external weather forecasts [106]. However, realizing these trends requires addressing lingering knowledge gaps in standardizing data collection protocols across greenhouse operations and establishing universal benchmarks for ML performance evaluation in climate control contexts.

13.7.3 Integration with Renewable Energy Systems

As sustainability becomes a central pillar of agricultural innovation, integrating ML-driven climate control systems with renewable energy sources such as solar, wind, and geothermal energy has emerged as a key research priority [107]. ML models can dynamically balance energy demand and renewable generation, adjusting climate control strategies to optimize resource use while minimizing greenhouse gas emissions [108]. This alignment of climate management with renewable energy sources positions hydroponic systems as a cornerstone of low-carbon, climate-resilient agriculture.

13.7.4 Multi-objective Optimization and Sustainable Metrics

Future ML applications will increasingly prioritize multi-objective optimization, simultaneously balancing yield, resource efficiency, energy consumption, and sustainability metrics [109]. Reinforcement learning (RL) and advanced optimization algorithms are uniquely suited for these complex, multi-criteria decisions, moving beyond single-parameter climate control to holistic environmental management. For example, RL frameworks can learn to optimize plant growth while minimizing water usage and carbon footprint, offering an integrated pathway to sustainable food production [110].

13.7.5 Explainable and Trustworthy AI

As ML becomes integral to operational decision-making in greenhouses, ensuring explainability and transparency of algorithms is paramount [111]. Black-box ML models present challenges for grower adoption and trust, especially when managing high-value crops and critical climate interventions. Emerging research in explainable AI (XAI) aims to provide growers with interpretable insights such as how specific environmental changes influence crop performance thereby enhancing user trust and fostering broader adoption of ML-driven solutions [112].

13.7.6 Blockchain and Data Governance

Data security and transparency are increasingly important in ML-driven greenhouse operations. Blockchain technologies offer robust, decentralized solutions for tracking climate data, sensor calibration records, and operational decisions, ensuring data integrity and facilitating fair access to digital climate control platforms [113]. Integration of blockchain with ML pipelines can also enhance data sharing across multiple greenhouse sites while preserving confidentiality and ownership.

13.7.7 Challenges and Research Gaps

Despite these advances, several challenges and knowledge gaps persist:

1. Standardization of Data Protocols: Lack of standardized sensor data protocols hampers model portability and integration across greenhouse systems [114].
2. Cross-System Generalizability: ML models often remain specific to particular greenhouse environments, limiting transferability to new systems or climatic zones [115].
3. Ethical Implications: Balancing automation with human expertise and addressing concerns around job displacement and digital inequity in greenhouse management require thoughtful governance frameworks [116].

In light of these trends and challenges, future research should focus on integrating domain expertise with advanced ML frameworks, promoting equitable access to climate-resilient technologies, and aligning ML-driven climate management with broader sustainability and food security goals.

13.8 Challenges and Knowledge Gaps

While machine learning (ML)-driven climatic management in hydroponic greenhouses offers promising solutions for precision agriculture and sustainability, significant challenges and knowledge gaps remain that must be addressed for widespread adoption and operational success.

13.8.1 Data Quality and Availability

High-quality data is foundational to the success of machine learning (ML) models in greenhouse climate control. However, acquiring such data is fraught with several challenges. One of the primary issues is sensor drift and calibration. Over time, sensors used to monitor key environmental parameters such as temperature, humidity, light intensity, and CO_2 concentration can experience drift or a gradual loss of accuracy, necessitating regular calibration and maintenance to ensure reliable measurements [117]. Additionally, greenhouse environments are often subject to transient disturbances such as condensation forming on sensor surfaces or interference from other equipment which introduce environmental noise and create gaps in the data streams [118]. Another significant limitation is the lack of standardized protocols for data collection across different greenhouse operations. This variability in methodology reduces the generalizability of ML models and complicates the development of universally applicable frameworks for climate management [119]. Addressing these challenges requires establishing standardized data protocols, robust data cleaning pipelines, and continuous sensor validation strategies.

13.8.2 Model Generalizability and Adaptability

Another critical limitation in the application of machine learning (ML) for greenhouse climate control lies in the generalizability of models across diverse operational contexts. A key challenge is system-specific bias, wherein many ML models are developed and trained using datasets tailored to specific greenhouse layouts, crop types, or climatic zones. As a result, these models often exhibit reduced performance when applied to different or unfamiliar environments, limiting their broader applicability and scalability [120]. Additionally, greenhouse environments are inherently dynamic, characterized by rapid and complex fluctuations in microclimatic conditions influenced by external weather variations and internal plant growth dynamics. These changes necessitate the use of adaptive ML models capable of continuous learning and real-time adjustment to maintain accuracy and reliability under evolving conditions [121]. Emerging solutions such as transfer learning and domain adaptation offer pathways to enhance model robustness across varied greenhouse environments, but require further research and refinement [122].

13.8.3 Integration with Existing Greenhouse Control Systems

Translating machine learning (ML)-based insights into actionable control decisions within real-world greenhouse environments presents several operational challenges. One of the foremost issues is compatibility with legacy systems [123]. Many greenhouse facilities still depend on conventional control infrastructures, such as proportional-integral-derivative (PID)-based heating and ventilation systems, which often lack the flexibility to directly accommodate or interpret ML-generated climate recommendations. Furthermore, successful deployment of ML frameworks requires effective edge-cloud integration. While edge computing enables low-latency, real-time decision-making at the greenhouse site, it must be effectively synchronized with cloud-based platforms that handle long-term data storage and advanced analytical tasks [124]. Achieving seamless communication, reliability, and scalability across this hybrid architecture remains a critical engineering challenge for the widespread adoption of ML-driven climate management systems.

13.8.4 Interpretability and Trust

Many high-performing machine learning (ML) algorithms particularly deep neural networks function as black-box models, offering limited interpretability regarding how predictions and decisions are made [125]. This inherent opacity presents a significant barrier to trust and adoption among greenhouse growers. One of the primary concerns is limited explainability: without transparent insights into the decision-making logic of ML models, growers may be reluctant to implement algorithmic recommendations for critical climate control operations [126]. Beyond technical concerns, ethical considerations further complicate adoption. Issues such as data ownership, digital equity, and the preservation of human expertise in increasingly automated systems have emerged as important dimensions of responsible AI deployment in agriculture [127]. To address these concerns, it is essential to integrate explainable AI (XAI) frameworks into greenhouse ML pipelines. Doing so not only enhances transparency and user trust but also ensures alignment with regulatory expectations and broader ethical standards governing AI-driven agricultural innovations.

13.8.5 Sustainability and Societal Impacts

Finally, although ML-driven climate control offers substantial potential for reducing resource consumption and greenhouse gas emissions, broader sustainability considerations must be thoughtfully addressed to ensure responsible implementation. One important concern is the energy consumption associated with advanced ML algorithms. Especially in cases where models require frequent retraining or real-time processing, the computational demand can become significant, thereby increasing the overall energy footprint of greenhouse operations [128]. Another crucial consideration is equity in access. The benefits of ML-enabled climate control systems should not be confined to technologically advanced or resource-rich regions. To prevent the exacerbation of digital inequalities, it is imperative that adoption strategies are accompanied by inclusive policies, capacity-building initiatives, and support mechanisms that ensure smallholder and marginalized agricultural communities can equally benefit from these technological advancements [129].

Socioeconomic factors also influence the adoption of ML-driven climate control in hydroponic systems. High upfront costs of sensor networks and digital infrastructure can be prohibitive for small- and medium-scale growers [130]. Additionally, digital literacy and access to technical expertise remain barriers, particularly in emerging economies [131]. Environmental trade-offs are another important consideration. While ML-based climate control can reduce water and fertilizer use, the computational demands of training and retraining deep learning models can add to a greenhouse's carbon footprint [86]. Addressing these challenges requires integrated policies and capacity-building initiatives that democratize access to ML technologies and ensure sustainable deployment in diverse agricultural contexts [132].

13.8.6 Future Research Needs

To bridge these challenges and gaps, future research should prioritize:

1. Developing standardized, open-access climate datasets for hydroponic systems.
2. Designing lightweight, adaptive ML models suitable for real-time edge deployment.
3. Advancing XAI techniques tailored to greenhouse environments.
4. Exploring integrated energy models that couple renewable energy sources with climate control algorithms.
5. Establishing inclusive data governance frameworks to ensure transparent, ethical, and equitable ML adoption.

13.9 Summary

This chapter has provided a comprehensive exploration of machine learning (ML)-driven climatic management in hydroponic greenhouse systems, underscoring its transformative potential for precision agriculture and sustainable food production. We began by examining the essential climatic parameters temperature, humidity, light intensity, and carbon dioxide (CO_2) that define the microclimatic environment in hydroponic systems. Their intricate, nonlinear interactions underscore the complexity of achieving and maintaining optimal growing conditions, particularly in the face of external climate variability and internal crop dynamics.

Conventional climate control strategies, constrained by static, rule-based logic, are increasingly unable to meet the dynamic demands of modern greenhouse operations. In contrast, ML algorithms offer adaptive, data-driven solutions capable of predicting microclimate trends, identifying anomalies, and dynamically optimizing environmental parameters in real time. We explored the main classes of ML algorithms supervised learning models, artificial neural networks (ANNs), and reinforcement learning (RL) detailing their respective strengths, operational trade-offs, and real-world performance metrics in managing greenhouse microclimates.

Practical applications of these algorithms have been demonstrated across a range of crops and greenhouse types, yielding measurable improvements in resource efficiency, energy conservation, and crop yield stability. These advancements are supported by rigorous validation protocols, leveraging metrics such as R^2, RMSE, and MAE to ensure predictive reliability and real-world applicability. The inclusion of tables and figures throughout this chapter has further illustrated the state-of-the-art in ML deployment, providing readers with a visual synthesis of key concepts and application case studies.

Looking forward, we identified several emerging trends poised to reshape the field:

1. Integration of digital twins for real-time climate simulation and predictive scenario modeling.
2. Deployment of generative AI frameworks to anticipate extreme events and climate variability.
3. Coupling ML-based climate control with renewable energy systems to enhance sustainability.
4. Emphasis on multi-objective optimization, balancing yield, resource use, and environmental impact.
5. Advances in explainable and trustworthy AI to foster broader adoption and user confidence.

However, challenges persist in ensuring data quality, enhancing model generalizability, integrating ML with existing greenhouse infrastructure, and addressing ethical concerns around data governance and automation. Moreover, socioeconomic barriers such as digital infrastructure costs, technical expertise gaps, and equitable technology access remain critical considerations for translating ML innovations into practice.

Addressing these gaps will require not only focused research on lightweight, interpretable ML models and inclusive data practices but also supportive policy frameworks and capacity-building initiatives that democratize access to these transformative technologies. In sum, ML-driven climatic management in hydroponic greenhouses represents a paradigm shift in modern agriculture, enabling growers to achieve unprecedented levels of precision, resilience, and sustainability. As this field evolves, integrating domain expertise, technological innovation, and ethical stewardship will be crucial for realizing climate-resilient, equitable food systems capable of meeting global challenges.

References

1. Lakhiar IA et al (2025) Soilless agricultural systems: opportunities, challenges, and applications for enhancing horticultural resilience to climate change and urbanization. Horticulturae 11(6):568
2. Sharma T et al (2025) Hydroponics farming: a holistic perspective for crop production. Oilseed Crop 2025:309–332
3. Badji A et al (2022) Design, technology, and management of greenhouse: a review. J Clean Prod 373:133753

4. Nagarsheth S et al (2025) The advancements in agricultural greenhouse technologies: an energy management perspective. Sustainability 17(8):3407
5. Zhao X et al (2024) Applications of machine learning in real-time control systems: a review. Meas Sci Technol. (in press)
6. Eze VHU et al (2025) Integrating IoT sensors and machine learning for sustainable precision agroecology: enhancing crop resilience and resource efficiency through data-driven strategies, challenges, and future prospects. Discov Agric 3(1):1–34
7. Javed H et al (2025) Sustainable energy management in the AI era: a comprehensive analysis of ML and DL approaches. Computing 107(6):1–64
8. Awais M et al (2025) Advancing precision agriculture through digital twins and smart farming technologies: a review. AgriEngineering 7(5):137
9. Reza MN et al (2025) Trends of soil and solution nutrient sensing for open field and hydroponic cultivation in facilitated smart agriculture. Sensors 25(2):453
10. Abdelhafez AA et al (2024) Climate smart greenhouses: innovations and impacts. BoD–Books on Demand
11. Sun C (2024) Application of smart sensors to monitor the interactive effects of temperature and lighting on plant growth in a simulated controlled environment facility. [unpublished/journal not specified]
12. Bhattacharya A (2022) Physiological processes in plants under low temperature stress. Springer
13. Levine CP et al (2023) Controlling root zone temperature improves plant growth and pigments in hydroponic lettuce. Ann Bot 132(3):455–470
14. Ray RV (2024) Effects of pathogens and disease on plant physiology. In: Agrios' plant pathology. Elsevier, pp 63–92
15. Liu J et al (2025) Advances in research and application of techniques for measuring photosynthetically active radiation. Remote Sens 17(10):1765
16. Slattery RA, Ort DR (2021) Perspectives on improving light distribution and light use efficiency in crop canopies. Plant Physiol 185(1):34–48
17. Ainsworth EA, Long SP (2021) 30 years of free-air carbon dioxide enrichment (FACE): what have we learned about future crop productivity and its potential for adaptation? Glob Chang Biol 27(1):27–49
18. Rezvani Z, Rezvani F (2023) A review of recent advances in water vapor deficit sensor technology for improving plant water usage efficiency. Biomech Bioenergy Res 2(2):65–86
19. Hussain S et al (2021) Photosynthesis research under climate change. Photosynth Res 150:5–19
20. Chaudhary S et al (2022) Physiological and molecular approaches for developing thermotolerance in vegetable crops: a growth, yield and sustenance perspective. Front Plant Sci 13:878498
21. Guarnieri G et al (2023) Relative humidity and its impact on the immune system and infections. Int J Mol Sci 24(11):9456
22. Fathidarehnijeh E et al (2023) Current perspective on nutrient solution management strategies to improve the nutrient and water use efficiency in hydroponic systems. Can J Plant Sci 104(2):88–102
23. Rahman MM et al (2021) LED illumination for high-quality high-yield crop growth in protected cropping environments. Plants 10(11):2470
24. Kumar D, Mukhopadhyay R (2025) Climate change and plant pathogens: understanding dynamics, risks and mitigation strategies. Plant Pathol 74(1):59–68
25. Sathya D et al (2024) Fuzzy logic and its applications in mechatronic control systems. Comput Intell Tech Mechatron:211–241
26. Thilakarathne NN et al (2025) Internet of things enabled smart agriculture: current status, latest advancements, challenges and countermeasures. Heliyon 11(3):e42136
27. Stephenson RJ et al (2023) Temperature measurement. In: Mechanical variables measurement—solid, fluid, and thermal. CRC Press, pp 11–13

28. Lee G et al (2021) Emerging wearable sensors for plant health monitoring. Adv Funct Mater 31(52):2106475
29. Lin JY et al (2021) An integrated wireless multi-sensor system for monitoring the water quality of aquaculture. Sensors 21(24):8179
30. Mowla MN et al (2023) Internet of things and wireless sensor networks for smart agriculture applications: a survey. IEEE Access 11:145813–145852
31. Miry AH, Aramice GA (2024) Agriculture based on wireless sensor network: literature survey. Wasit J Eng Sci 12(4):171–190
32. Ruiz-Gonzalez A et al (2022) In vivo sensing of pH in tomato plants using a low-cost and open-source device for precision agriculture. Biosensors 12(7):447
33. Lea P (2020) IoT and edge computing for architects: implementing edge and IoT systems from sensors to clouds with communication systems, analytics, and security. Packt Publishing Ltd
34. Trigka M, Dritsas E (2025) Edge and cloud computing in smart cities. Futur Internet 17(3):118
35. Akbar JUM et al (2024) A comprehensive review on deep learning assisted computer vision techniques for smart greenhouse agriculture. IEEE Access 12:4485–4522
36. Kavalcıoğlu C (2020) De-noising method for continuous glucose monitoring (CGM) signal with improved Morlet wavelet filter. Near East University
37. Anuar NA et al (2024) Temperature and humidity monitoring system. Prog Eng Appl Technol 5(1):259–266
38. Chua SJ (2024) IoT-enabled hydroponic farming: a solution for high-temperature regions. UTAR
39. Ameer S et al (2024) Real-time detection and measurements of nitrogen, phosphorous & potassium from soil samples: a comprehensive review. J Soils Sediments 24(7):2565–2583
40. Karimzadeh S et al (2025) Machine learning-based fault detection and diagnosis of electrical conductivity and pH sensors in hydroponic systems. Comput Electron Agric 237:110544
41. Wong CYS et al (2024) Hyperspectral and photodiode retrievals of nighttime LED-induced chlorophyll fluorescence (LEDIF) for tracking photosynthetic phenology in a vineyard. J Geophys Res Biogeosci 129(1):e2023JG007742
42. Taştan M (2022) A low-cost air quality monitoring system based on internet of things for smart homes. J Ambient Intell Smart Environ 14(5):351–374
43. Lazzoni V et al (2024) Development of a microwave sensor for the non-invasive detection of plant responses to water stress: a practical application on maize (Zea mays L.). Biosyst Eng 246:191–203
44. Zaccarin AM et al (2023) Fabrication and characterization of soil moisture sensors on a biodegradable, cellulose-based substrate. IEEE Sensors J 24(6):7235–7243
45. Gopi G et al (2024) Investigating the effect of lower body local radiant warming on occupant thermal comfort in battery electric vehicles during cold conditions. Sustain Cities Soc 111:105535
46. Alaeiyan H et al (2023) Hybrid noise removal to improve the accuracy of inertial sensors using lifting wavelet transform optimized by genetic algorithm. Alex Eng J 80:326–341
47. Alghushairy O et al (2020) A review of local outlier factor algorithms for outlier detection in big data streams. Big Data Cogn Comput 5(1):1
48. Niedzielski T, Halicki M (2023) Improving linear interpolation of missing hydrological data by applying integrated autoregressive models. Water Resour Manag 37(14):5707–5724
49. Karunamurthy A et al (2023) Cloud database: empowering scalable and flexible data management. Quing Int J Innov Res Sci Eng
50. Nalakurthi NVSR et al (2024) Challenges and opportunities in calibrating low-cost environmental sensors. Sensors 24(11):3650
51. Butt MA et al (2022) Environmental monitoring: a comprehensive review on optical waveguide and fiber-based sensors. Biosensors 12(11):1038
52. Montillet JP et al (2024) How big data can help to monitor the environment and to mitigate risks due to climate change: a review. IEEE Geosci Remote Sens Mag (In press)

53. Murshed MGS et al (2021) Machine learning at the network edge: a survey. ACM Comput Surv 54(8):1–37
54. Ragaveena S et al (2021) Smart controlled environment agriculture methods: a holistic review. Rev Environ Sci Biotechnol 20(4):887–913
55. Kolhar A (2025) Energy-efficiency strategies for wireless sensor networks in IoT. In: Energy-efficient deep learning approaches in IoT, fog, and green Blockchain revolution. IGI Global Scientific Publishing, pp 167–192
56. George J (2022) Optimizing hybrid and multi-cloud architectures for real-time data streaming and analytics: strategies for scalability and integration. World J Adv Eng Technol Sci 7(1):10–30574
57. Han H et al (2024) Challenges in remote sensing based climate and crop monitoring: navigating the complexities using AI. J Cloud Comput 13(1):1–14
58. Adesanya MA et al (2024) Deep reinforcement learning for PID parameter tuning in greenhouse HVAC system energy optimization: a TRNSYS-python cosimulation approach. Expert Syst Appl 252:124126
59. Barve P et al (2025) Sustainable automated vertical farming. [unpublished/journal not specified]
60. Cai W et al (2022) A method for modelling greenhouse temperature using gradient boost decision tree. Inf Process Agric 9(3):343–354
61. Beckers MA (2024) Machine learning applied to indoor hydroponic lettuce optimization. Instituto Politecnico de Braganca (Portugal)
62. Cletus F, John AE (2024) Comparative analysis of machine learning models for greenhouse microclimate prediction. Brill Res Artif Intell, pp 162–175
63. He Z et al (2022) Gated recurrent unit models outperform other machine learning models in prediction of minimum temperature in greenhouse based on local weather data. Comput Electron Agric 202:107416
64. Wang L et al (2020) Deep reinforcement learning for greenhouse climate control. In: In: 2020 IEEE international conference on knowledge graph (ICKG). IEEE, pp 474–480
65. Chatterjee A, Khovalyg D (2023) Dynamic indoor thermal environment using reinforcement learning-based controls: opportunities and challenges. Build Environ 244:110766
66. Ojo MO, Zahid A (2022) Deep learning in controlled environment agriculture: a review of recent advancements, challenges and prospects. Sensors 22(20):7965
67. Hosseini Monjezi P et al (2023) Prediction of greenhouse indoor air temperature using artificial intelligence (AI) combined with sensitivity analysis. Horticulturae 9(8):853
68. Singh RR, Hati AJ (2024) Soilless smart agriculture systems for future climate. In: digital agriculture: a solution for sustainable food and nutritional security. Springer, pp 61–111
69. Zainab A et al (2020) Performance evaluation of tree-based models for big data load forecasting using randomized hyperparameter tuning. In: 2020 IEEE international conference on big data (big data). IEEE, pp 5332–5339
70. Jiang T et al (2020) Supervised machine learning: a brief primer. Behav Ther 51(5):675–687
71. Muñoz-Zavala AE et al (2024) A literature review on some trends in artificial neural networks for modeling and simulation with time series. Algorithms 17(2):76
72. Li SE (2023) Reinforcement learning for sequential decision and optimal control. [Publisher not specified]
73. Ghiasi M et al (2023) A systematic review of optimal and practical methods in design, construction, control, energy management and operation of smart greenhouses. IEEE Access 12:2830–2853
74. Ghiat I et al (2023) Hyperspectral-physiological based predictive model for transpiration in greenhouses under CO2 enrichment. Comput Electron Agric 213:108255
75. Sun W, Chang FJ (2023) Empowering greenhouse cultivation: dynamic factors and machine learning unite for advanced microclimate prediction. Water 15(20):3548
76. Wang JQ et al (2020) LSTM based long-term energy consumption prediction with periodicity. Energy 197:117197

77. Ngugi HN et al (2024) Revolutionizing crop disease detection with computational deep learning: a comprehensive review. Environ Monit Assess 196(3):302
78. Stavrev S, Ginchev D (2024) Reinforcement learning techniques in optimizing energy systems. Electronics 13(8):1459
79. Eraliev O, Lee CH (2023) Performance analysis of time series deep learning models for climate prediction in indoor hydroponic greenhouses at different time intervals. Plants 12(12):2316
80. Gang MS et al (2022) Estimation of greenhouse lettuce growth indices based on a two-stage CNN using RGB-D images. Sensors (Basel) 22(15):5499
81. Lee MH et al (2024) An artificial intelligence-powered environmental control system for resilient and efficient greenhouse farming. Sustainability 16(24):10958
82. Platero-Horcajadas M et al (2024) Enhancing greenhouse efficiency: integrating IoT and reinforcement learning for optimized climate control. Sensors 24(24):8109
83. Paradiso R, Proietti S (2022) Light-quality manipulation to control plant growth and photomorphogenesis in greenhouse horticulture: the state of the art and the opportunities of modern LED systems. J Plant Growth Regul 41(2):742–780
84. Zheng C et al (2021) Remote sensing and machine learning in crop phenotyping and management, with an emphasis on applications in strawberry farming. Remote Sens 13(3):531
85. Zahir SADM et al (2024) Review–plant nutritional status analysis employing the visible and near-infrared spectroscopy spectral sensor. Spectrochim Acta A Mol Biomol Spectrosc 304:123273
86. Farooq O, Khan A (2025) The impact of machine learning on climate change modeling and environmental sustainability. Artif Intell Mach Learn Rev 6(1):8–16
87. Sharma PN et al (2021) Prediction-oriented model selection in partial least squares path modeling. Decis Sci 52(3):567–607
88. Hodson TO (2022) Root mean square error (RMSE) or mean absolute error (MAE): when to use them or not. Geosci Model Dev Discuss 2022:1–10
89. Rasheed AA (2023) Improving prediction efficiency by revolutionary machine learning models. Mater Today Proc 81:577–583
90. Uyeh DD et al (2022) Grid search for lowest root mean squared error in predicting optimal sensor location in protected cultivation systems. Front Plant Sci 13:920284
91. Priya A et al (2020) Predicting anxiety, depression and stress in modern life using machine learning algorithms. Procedia Comput Sci 167:1258–1267
92. Inan O, Uzer MS (2021) A method of classification performance improvement via a strategy of clustering-based data elimination integrated with k-fold cross-validation. Arab J Sci Eng 46(2):1199–1212
93. Bergmeir C, Benítez JM (2012) On the use of cross-validation for time series predictor evaluation. Inf Sci (NY) 191:192–213
94. Udu AG (2025) Machine learning in aero-engine fault and structural damage predictions University of Leicester
95. Morales-García J et al (2023) Data-driven evaluation of machine learning models for climate control in operational smart greenhouses. J Ambient Intell Smart Environ 15(1):3–17
96. Aliferis C, Simon G (2024) Overfitting, underfitting and general model overconfidence and under-performance pitfalls and best practices in machine learning and AI. Artif Intell Mach Learn Health Care Med Sci best Pract pitfalls:477–524
97. Shaif A (2021) Predictive maintenance in smart agriculture using machine learning: a novel algorithm for drift fault detection in hydroponic sensors. [Publisher not specified]
98. Hu G, You F (2022) Renewable energy-powered semi-closed greenhouse for sustainable crop production using model predictive control and machine learning for energy management. Renew Sust Energ Rev 168:112790
99. Hong SR et al (2020) Human factors in model interpretability: industry practices, challenges, and needs. Proc ACM Hum Comput Interact 4(CSCW1):1–26

100. Antony J et al (2024) Adapting to changes: a novel framework for continual machine learning in industrial applications. J Grid Comput 22(4):7
101. Xu X et al (2024) Intelligent architecture and platforms for private edge cloud systems: a review. Futur Gener Comput Syst (in press)
102. Pukrongta N et al (2024) Enhancing crop yield predictions with PEnsemble 4: IoT and ML-driven for precision agriculture. Appl Sci 14(8):3313
103. Agostinelli S (2024) Optimization and management of microgrids in the built environment based on intelligent digital twins. [unpublished/no journal specified]
104. Ma Z et al (2024) Generative deep learning for data generation in natural hazard analysis: motivations, advances, challenges, and opportunities. Artif Intell Rev 57(6):160
105. Hossain MI et al (2024) Enhancing data integrity and traceability in industry cyber physical systems (ICPS) through Blockchain technology: a comprehensive approach. arXiv Prepr arXiv240504837
106. Gheibi O et al (2021) Applying machine learning in self-adaptive systems: a systematic literature review. ACM Trans Auton Adapt Syst 15(3):1–37
107. Ukoba K et al (2024) Harnessing machine learning for sustainable futures: advancements in renewable energy and climate change mitigation. Bull Natl Res Cent 48(1):99
108. Ukoba K et al (2024) Optimizing renewable energy systems through artificial intelligence: review and future prospects. Energy Environ 35(7):3833–3879
109. Qiu L et al (2025) Machine learning-driven multi-objective optimization for sustainable, cost-effective, and low-emission gold mining. J Clean Prod 2025:145621
110. Cohen AR et al (2021) Dynamically controlled environment agriculture: integrating machine learning and mechanistic and physiological models for sustainable food cultivation. ACS ES&T Eng 2(1):3–19
111. Tahir HA et al (2025) A federated explainable AI framework for smart agriculture: enhancing transparency, efficiency, and sustainability. IEEE Access. (in press)
112. Roy P et al (2025) Machine learning for precision agriculture and crop yield optimization: techniques and applications. In: The role of artificial intelligence in advancing applied life sciences. IGI Global Scientific Publishing, pp 189–234
113. Gade PK (2023) AI-driven blockchain solutions for environmental data integrity and monitoring. NEXG AI Rev Am 4(1):1–16
114. Farooq MS et al (2022) Internet of things in greenhouse agriculture: a survey on enabling technologies, applications, and protocols. IEEE Access 10:53374–53397
115. Graham EB et al (2021) Toward a generalizable framework of disturbance ecology through crowdsourced science. Front Ecol Evol 9:588940
116. George AS, George ASH (2024) Towards a super smart society 5.0: opportunities and challenges of integrating emerging technologies for social innovation. Partners Univ Int Res J 3(2):1–29
117. González Rivero RA et al (2023) A low-cost calibration method for temperature, relative humidity, and carbon dioxide sensors used in air quality monitoring systems. Atmosphere (Basel) 14(2):191
118. Bhujel A et al (2020) Sensor systems for greenhouse microclimate monitoring and control: a review. J Biosyst Eng 45:341–361
119. Srivastava A, Maity R (2023) Assessing the potential of AI–ML in urban climate change adaptation and sustainable development. Sustainability 15(23):16461
120. Bommasani R et al (2021) On the opportunities and risks of foundation models. arXiv Prepr arXiv210807258
121. Bello HO et al (2024) Adaptive machine learning models: concepts for real-time financial fraud prevention in dynamic environments. World J Adv Eng Technol Sci 12(2):21–34
122. Pylianidis C et al (2024) Domain adaptation with transfer learning for pasture digital twins. Environ Data Sci 3:e8
123. Mohamed M (2023) Agricultural sustainability in the age of deep learning: current trends, challenges, and future trajectories. Sustain Mach Intell J 4:1–2

124. Sathya D et al (2024) The revolution of edge computing in smart farming. In: Intelligent robots and drones for precision agriculture. Springer, pp 351–389
125. Esna-Ashari M (2025) Beyond the black box: a review of quantitative metrics for neural network interpretability and their practical implications. Int J Sustain Appl Sci Eng 2(1):1–24
126. Shams MY et al (2024) Enhancing crop recommendation systems with explainable artificial intelligence: a study on agricultural decision-making. Neural Comput & Applic 36(11):5695–5714
127. Tzachor A et al (2022) Responsible artificial intelligence in agriculture requires systemic understanding of risks and externalities. Nat Mach Intell 4(2):104–109
128. Iddio E et al (2020) Energy efficient operation and modeling for greenhouses: a literature review. Renew Sust Energ Rev 117:109480
129. Hernandez K et al (2024) Towards digital inclusion in rural transformation. Food & Agriculture Org
130. Makini FM et al (2020) The status of ICT infrastructure, innovative environment and ICT4AG services in agriculture. Food Nutr Kenya 5(11):75
131. Neumeyer X et al (2020) Overcoming barriers to technology adoption when fostering entrepreneurship among the poor: the role of technology and digital literacy. IEEE Trans Eng Manag 68(6):1605–1618
132. Mekonnen SA et al (2024) Building resilience in sub-Saharan Africa's food systems: diversification, traceability, capacity building and technology for overcoming challenges. Food Energy Secur 13(4):e563

Chapter 14
IoT-Based Nutrient Monitorization in Hydroponics Using Bioinformatics Tools

Manvi Sharma, Shrutika Surada, Vinayak Sirkeck, Tarun Pal, and Chandresh Kumari

Abstract Hydroponics is a successful effort in contemporary agriculture for more efficient crop production in a much smaller area. The incorporation of the Internet of Things (IoT) enhanced the accessibility to address various issues, and with much more precision, such a revolutionized pattern of parameter monitoring leads to the optimization of nutrients and environmental conditions in real time, thereby increasing the yield with sustainable resource management. This chapter has included the complementarity among IoT, bioinformatics, and hydroponics systems in agriculture, mainly focusing on the real-time data collection and analysis for quicker action and advanced preventive measures, and efficient management. Sensor calibration, data integration, and analysis in association with precise computation techniques help in the process of analysis and interpretation of results. All these techniques and software put together into action have harnessed the power of the computational world with agriculture for nutrient profiling, in an attempt to lead the path of resilient and resource-efficient agriculture and encounter the challenges of global food security.

Keywords Wireless communication · QTL mapping · Regression tree · AI automation · Nutrient Film Technique

M. Sharma · S. Surada · V. Sirkeck · C. Kumari (✉)
School of Biotechnology, Faculty of Applied Sciences and Biotechnology, Shoolini University, Solan, H.P., India
e-mail: chandreshkumari@shooliniuniversity.com

T. Pal
School of Bioengineering and Food Technology, Faculty of Applied Sciences and Biotechnology, Shoolini University, Solan, H.P., India

P. Kumar, A. R. Warghat (eds.), *Transforming Medicinal Plant Agriculture*, Smart Agriculture 16, https://doi.org/10.1007/978-981-95-3963-5_14

14.1 Introduction

Soilless agriculture enables farming in a controlled environment with efficient use of resources as well as flexibility in the space, considering the uneven ratio of population to agricultural land. Hydroponics is an advanced form of one of such soilless agricultural practices that rely on water solution enriched with essential nutrients to nourish plants, increasing the need to monitor nutrients for managing the system and efficient plant growth. A well-calculated concentration and regulation of the nutrients promote maximum growth while reducing nutrient wastage [1]. In conventional methods, nutrient monitoring and management are time-consuming and are often very static, while the adaptation of Internet of Things (IoT) devices incorporated with bioinformatics tools and software helps in the automation of the framework with real-time adaptability, monitorization, and optimization of nutrients in hydroponics [2].

Wireless sensors and other IoT devices monitor various parameters like pH, electrical conductivity (EC), total dissolved solids (TDS), temperature, dissolved oxygen, and so on, in real time and transmit precisely to the system with the help of a cloud platform. A vast stream of data is formed, which can be interpreted using various bioinformatics tools for proper analysis [3]. The unstructured data is very complex and wide, which requires highly advanced computational methods for precise interpretation. Bioinformatics, traditionally rooted in molecular biology and genomics, has evolved into a multidisciplinary field capable of handling vast datasets from diverse biological and environmental sources. Bioinformatics tools are used in hydroponics for the integration of data and organizing it for extracting useful information helpful in altering the response of the plant for better growth. Statistical modeling, machine learning, network analysis, and many more such techniques enable bioinformatics to interpret the unrefined, raw data by analyzing the hidden patterns and converting them into information applicable to crop improvement and nutrient enhancement, for example, altering the nutrient combination and delivery timings according to specific plant species and nutrient requirements [4]. Besides contributing to the development of an efficient model to predict the response of varying nutrients, real-time data can also help in the detection of early signs of stress or nutrient deficiency, leading to a course of action at the initial stage. System biology also plays a crucial role in the process by constructing a network for interaction between genes and nutrients, including transcriptomics, proteomics, and genomics datasets, and elucidating the environmental influence on the expression of genes or proteins and therefore the metabolic pathways that utilize those nutrients [5]. By using genomics to help with precision farming, it is possible to create or modify plants that are more durable, nutrient-efficient, and suited to hydroponic settings. Various bioinformatics tools assist in data management and planning with molecular profiling and sensors to detect environmental changes, as well as nutrient variation. Tools like MATLAB, R, Python (with bioinformatics libraries), Cytoscape, and cloud-based platforms analyze and interpret the wide range of datasets [6], and databases like KEGG, PlantCyc are used for mapping the metabolic pathways in association with the nutrients [7], while the repositories like NCBI GEO are the archives that are referred to for a comparative analysis [8].

Both long-term agricultural planning and short-term nutrient management decisions are influenced by this systems-level understanding, however, there are significant obstacles in the way of this integrated approach and its expanding potential. The efficacy of IoT-bioinformatics integration may be constrained by the variety of data sources, sensor calibration problems, and the requirement for domain-specific algorithms designed for hydroponics. This union of IoT and bioinformatics for the advancement of agriculture is a very powerful addition to the system of development in the scientific community and biotechnology. There is a significant enhancement in the productivity of crops, with a considerable enhancement in the nutrients due to the integration of real-time monitoring and prediction of data [9]. This chapter consists of an efficient addition of smart farming techniques with the help of bioinformatics tools, working significantly for the emergence of groundbreaking devices to advance agricultural practices with higher efficiency, eradicating crop loss and world hunger in the near future.

14.2 Fundamentals of Hydroponics

Hydroponics is a system of cultivating plants without soil, using high nutrient water solutions or chemically inactive media such as coco coir, perlite, sand, or gravel. Unlike conventional agriculture, where soil acts as the predominant nutrient source, hydroponic systems provide nutrients abruptly to plant roots through water, intervening rapid growth and expanded productivity. Common hydroponic techniques include the Nutrient Film Technique (NFT), deep water culture, and drip systems, are all intended at improving nutrients adsorption and water optimizing efficiency [10]. In the recent years, hydroponics has sustained significant attention resulting in rising agricultural challenges, such as degrading soil quality, reduced arable lands, water shortage, and escalating global food needs. As the global population is expected to attain nearly 9.8 billion by 2050, adopting innovative farming strategies has become increasingly important. Hydroponics farming provides a streamlined alternative by ensuring continuous crop production in controlled environments, independent of climate and soil availability. One of the main edges of hydroponics is distinctive water efficiency, using up to 90% less than traditional farming practices. Additionally, since the soil is not present, it eliminates soil-borne diseases and pests, diminishes the dependence on chemical interventions, and permit accurate management over nutrient supply. This system is adaptable enough to be installed in varied environments, including urban rooftops to uncultivated lands, making them intensely robust and enduring [11]. Hydroponics therefore represents as a transformative and innovative system to agriculture. Its ability to yield high-quality crops with lowest resource expenditure demonstrate it as a core technology for securing future food integrity and improving sustainable agricultural approaches [12].

14.2.1 Types of Hydroponic Systems

Hydroponic systems determine the classification based on the method employed to supply nutrients to plant roots. These systems vary in complexity, scalability, and suitability for different crop types. Each system offers unique advantages and limitations, making them adaptable to a wide range of environmental and operational settings. The primary hydroponic systems include:

14.2.1.1 Wick System (Passive System)

It is the simplest and most accessible form of hydroponics system. Through capillary action, it uses a fiber wick made of absorbent material, nylon which passively pulls the nutritional solution from the reservoir to the root zone. This method is inexpensive and simple to maintain because it does not require complex equipment's. However, only small, low demanding plants with less growth cycles may use it due to limited food supply [13].

14.2.1.2 Drip System

Widely used in both soil-based farming and hydroponic cultivation, the drip system uses emitters to supply nutrient solutions straight to each plants base. By ensuring that roots stay moist without using excessive amount of water, the exact measured application lowers runoff and leaching. It is highly adaptable and can support various types of crops, particularly those that are grown in slightly porous media [12].

14.2.1.3 Ebb and Flow (Flood and Drain) System

This adaptive system comprises periodically flooding the grow bed with nutrient solution and then draining it back into a reservoir. Controlled by a timer, this cyclical process ensures optimal moisture content and nutrient absorption. Although it requires a reliable pump and sequence, it is favored for its high productivity and is ideal for a wide range of plants [14].

14.2.1.4 Deep Water Culture (DWC)

This dynamic system, periodically floods the grow bed with nutritional solution before draining it back into the reservoir. This cyclical process, which is timed, guarantees the best possible adsorption of nutrient and moisture. An air pump inputs dissolved oxygen to the root zone, promoting rapid nutrient absorption. DWC systems are uncomplicated and effective for short-cycle crops and are less suited for long-term cultivation due to challenges in maintaining pH and nutrient stability [10].

14.2.1.5 Nutrient Film Technique (NFT)

NFT comprises a continuous flow of nutrient solution throughout a sloped passage in which plants are placed. Only the root ends touch the circulating solution, while the top parts of the plants remain exposed to air. This technique is extremely efficient and well adapted for small anchored crops such as leafy greens and herbs [10].

14.2.1.6 Aeroponics

This is the most technologically advanced among hydroponic systems, aeroponics suspends plant roots in air within a closed chamber. Nutrients are delivered as a fine mist at regular intervals using elevated pressure sprayers. While soilless cultivation ensures maximum oxygenation and efficient resource use, it demands precise control and frequent maintenance, making it more suitable for research and high yield crops than for extensive commercial farming [12, 15].

14.2.2 Essential Nutrients and Their Function in Plant Physiology

Essential nutrients are necessary for plant growth and are broadly classified into macronutrients and micronutrients based on the quantities required. Macronutrients such as nitrogen, phosphorus, potassium, calcium, magnesium, and sulfur are abundant [15]. Micronutrients such as iron, manganese, zinc, boron, and copper, albeit only in trace amounts. In hydroponics, where soil is not present, a well-balanced nutrient solution is only the source of minerals. Maintaining ionic balance and nutrient specificity tailored to each variety of crop is crucial for optimal growth. Regular monitoring of pH and total dissolved solids assures nutrient availability, therefore precise nutrient management is a cornerstone of successful hydroponic production [10] (Table 14.1).

14.2.3 Challenges in Maintaining Nutrient Balance

Maintaining nutrient equilibrium in throughout the hydroponic systems is crucial yet demanding due to the absence of soil as a natural buffer. Unlike soil-based cultivation, hydroponic systems respond rapidly to any nutrient instability, by ensuring continuous monitoring of nutrient solution composition. Multiple factors lead to imbalance, including excessive nutrient use, water pollution, salinization, and imprecise pH levels. Readily soluble salts, often leads to inadequate leaching or evaporation, chemically driven drought, induce wilting and leaf burn despite

Table 14.1 Functions of macronutrients and micronutrients

Macronutrients	Function	Micronutrients	Function
Potassium	Supports growth, reproduction, osmotic balance, and chlorophyll production.	**Iron**	Supports respiration, photosynthesis, and chlorophyll synthesis.
Nitrogen	Key for amino acids and disease resistance.	**Manganese**	Aids photosynthesis, enzyme activity, disease resistance, and pollen development.
Phosphorus	Involved in energy transfer, photosynthesis, and genetic processes.	**Zinc**	Prevents stunted growth, supports enzymes, and promotes hormone production.
Magnesium	Activates enzymes vital for growth.	**Boron**	Regulates cell development, protein synthesis and stomatal function
Calcium	Maintains cell structure and supports cell division.	**Molybdenum**	Regulates transpiration, water balance, and aids in nitrogen fixation in legumes and related plants.
Sulfur	Helps in the formation of proteins and amino acids	**Copper**	Activates enzymes and supports plant metabolic growth.
Carbon	Essential for healthy and productive plant growth.		

adequate irrigation. Essential nutrient deficiency such as nitrogen, calcium, magnesium, and iron arises symptoms like chlorosis, stunted growth, necrotic tissue, and toxicity issues, such as excess boron, highlight the narrow margin between sufficiency and toxicity in hydroponic systems. These imbalances can spread quickly in interconnected systems, making them vulnerable to systemic nutrient. Additionally, hydroponics demands precise formulation of nutrient solutions, considering ionic interactions and crop-specific requirements. Power outages, incorrect formulations, or contaminated water can abruptly impact plant health and yield. Thus, hydroponic nutrient management is a technically intensive process requiring regular EC and pH monitoring, tailored nutrient mixes, and prompt corrective actions to sustain optimal plant growth and system efficiency [10].

14.3 Role of IoT in Smart Hydroponics

The incorporation of IoT in hydroponics provides innovative approaches for contemporary agriculture by facilitating real time, remote monitoring, and AI automation of fundamental growth parameters such as pH, electrical conductivity, temperature, humidity, and light. Unlike traditional non-digital lab-based nutrient assessment, IoT-based systems improve reliability and efficacy through continuous real-time data gathering and low-latency control [16]. Systems like iPONICS exhibits cost-effective wireless sensor networks and user-friendly interfaces can make smart hydroponics attainable even to small-scale farmers. This specific environmental control not only enhances crop productivity and resource proficiency but also supports scalable and sustainable food production in controlled environments [17].

14.3.1 Key IoT Components for Smart Hydroponics

Smart hydroponic setups rely on IoT-based sensor networks to optimize plant growth, reduce human effort, and mitigate environmental exposure. Sensors serve as the base of this system, ensuring real-time monitoring of physical and chemical parameters essential for healthy crop development. These devices constantly collect and transmit environmental data for timely analysis and decision-making, improving resource optimization and crop yield [18] (Table 14.2).

14.3.2 Sensors in Smart Hydroponics

Sensors are key tools that detect specific physiochemical conditions in the hydroponic environment and convert this inflow into detectable signals. These measurements allow growers to automate systems and make informed decisions [18]. Key sensor types include:

14.3.2.1 Temperature Sensors

It plays a vital role in managing ideal thermic conditions, as uneven temperature dispersion can harm plant growth. Proper placement typically 15–30 cm above the planter is crucial for accurate readings. In environments with electromagnetic hinderance, resistance temperature detectors (RTDs) are preferred over thermocouples due to their enhanced steadiness [14].

14.3.2.2 Humidity Sensors

Measure the moisture content in the air, altered by plant transpiration and soil evaporation. They work by sensing changes in resistance or electric potential storage and are vital for preventing fungal diseases. Integrated temperature and humidity sensors enhance efficiency [18].

Table 14.2 Process pathway of IoT-based sensor and cloud-integrated data collection and monitoring system for the Hydroponics system

Process	Description	Components involved	Data/signal type	Next step/result
Data acquisition	Sensors continuously measure critical physical and chemical parameters of the hydroponic system in real-time.	**Sensors** (DHT22, TDS sensor, ultrasound sensors, EC meter, pH meter, DHT11, DS1820)	**Real-time measurements** (temperature, humidity, EC/TDS, pH levels, water level, nutrient supply level)	Raw data is sent to the **microcontroller**.
Local processing and transmission	The microcontroller receives raw sensor data, may perform initial processing or execute part of the automation algorithm, and transmits the data to the cloud platform.	**Microcontroller** (ESP32, NodeMCU)	Raw or processed sensor data. Data formatted into **MQTT messages** for transmission.	Data is published to the **cloud platform** (MQTT broker/Adafruit IO).
Cloud handling and visualization	The cloud platform receives the data, stores it in a database, and hosts a dashboard for users to monitor system status and view historical data.	**Cloud platform** (Adafruit IO, MQTT broker), **dashboard** (Adafruit IO, node-RED)	**Sensor data** (for storage and display). **Dashboard visualizations** (gauges, graphs of temperature, humidity, EC, pH, etc.).	Data is available for user viewing and system decision-making. Historical data is stored.
Decision-making logic	Necessary actions are determined based on current sensor data compared to optimal ranges/thresholds. This logic can reside in the cloud or on the microcontroller: **A. Cloud logic:** Pre-programmed logical conditions on the cloud platform, such as threshold checking. **B. Local algorithm:** An automation algorithm running on the microcontroller performs calculations based on sensor data, optimal ranges, tank volume, etc.	**Cloud platform** (Adafruit IO logic) **microcontroller** (automation algorithm)	Sensor values. Optimal ranges, thresholds, tank size, current volume. **Action required** (e.g., add nutrient, add water, pump ON time).	An action command or signal is generated. This leads to actuator control determination.

Table 14.2 (continued)

Process	Description	Components involved	Data/signal type	Next step/result
Actuator control	The microcontroller sends control signals to driver circuits or relay modules to activate or deactivate specific actuators based on the decision made in the previous step (either receiving a command from the cloud or acting on its local algorithm's calculation).	**Microcontroller** (ESP32, NodeMCU), **driver circuits/ relay modules**, motor drives (for pump speed)	**Control signals** (logic signals sent to drivers, commands to turn pumps ON/ OFF, duration for pump activation).	Actuators are triggered.
Parameter adjustment	Actuators perform physical actions to modify the environmental conditions or nutrient solution parameters within the hydroponic system, aiming to bring them back into the optimal range.	**Actuators** (water pumps, nutrient pumps, pH-up/ pH-down pumps, lighting system)	**Physical actions** (e.g., circulating water, dispensing nutrient solution, adding pH adjustment solutions, turning lights on/off).	Hydroponic conditions are adjusted. The cycle repeats with data acquisition as parameters change.
User notifications	The system generates and sends alerts to the user to notify them of critical conditions (e.g., parameters outside optimal range) or operational anomalies.	**Cloud platform** (Adafruit IO), **dashboard system** (node-RED), **notification services** (email, LINE notify)	**Alert messages**.	User is informed of system status and potential issues.

14.3.2.3 CO_2 Sensors

It records carbon dioxide concentration, crucial for photosynthesis. Levels oscillate between day and night; sensors help maintain CO_2 stability through aeration and artificial supplementation if necessary [19].

14.3.2.4 Light Sensors

Assess light intensity and quality, both are important for photosynthesis. They help to regulate auxiliary lighting (e.g., LEDs) or shading systems to ensure optimal environmental influence [13].

14.3.2.5 Soil Moisture Sensors

These measure soil moisture level or tension. This manual irrigation systems by delivering real-time data, ensuring plants receive adequate water without waterlogging [20].

14.3.3 Actuators

Actuators are important components in IoT-based hydroponics systems, responsible for operating system elements such as valves, pumps, fans, lights, and greenhouse panes. Each actuator is linked to a specific environmental factor like temperature, humidity, water level, or CO_2, responding to control prompts by converting electrical, hydraulic, or pneumatic energy into kinetic motion. This function ensures real-time environmental adjustments and ideal plant growth. A control command and power source are required to operate them. Actuators execute user-defined or automated instructions, playing a vital role in maintaining the dynamic conditions of hydroponic systems through precise, responsive actions based on sensor data [18].

14.3.4 Microcontrollers

Microcontrollers form the central processing unit of automated IoT-based hydroponics systems. They coordinate various sensors and actuators to enable accurate monitoring and control of environmental parameters. These compact, low-power devices collect real-time sensor data, process it, and transmit it to other systems via wired or wireless networks. They also receive and decode control signals to activate corresponding actuators. Their programmability, energy efficiency, and cost-effectiveness make them ideal for smart farming applications. Microcontrollers enable seamless device communication, enhancing automation and decision-making in nutrient monitoring, and are especially suited for battery-powered systems, contributing to sustainable, scalable hydroponic operations [18].

14.3.5 Data Acquisition and Real-Time Monitoring

Data acquisition and real-time monitoring are vital units in the development of IoT-based hydroponic systems. Leveraging microcontrollers such as ESP8266 and ESP32, these systems combine multiple sensors including those measuring pH, total dissolved solids (TDS), water and air temperature, humidity, and water level to capture critical environmental and nutrient parameters [21]. Data is gathered employing both analog and digital sensors, analyzed through onboard firmware, and calibrated using correlation equations to ensure high accuracy, precision, and consistency. The transformed data is then transmitted at frequent intervals, to cloud platforms such as Google Firebase, where it is stored and aligned as structured datasets. This real-time data transmission ensures smooth interfacing with mobile and web-based applications, providing users intuitive dashboards to illustrate the operational condition of hydroponic installations [22].

The incorporation of real-time monitoring allows for immediate feedback and automatic control of hydroponic conditions. Modifiable threshold values prompt automated responses such as activating pumps for nutrient delivery, maintain pH levels, or managing water circulation to sustain ideal growing conditions. This not only reduces manual intervention but also enhances the consistency and reliability of crop cultivation [21]. Data visualization tools such as Android applications and cloud-based control boards empower users with remote access to sensor data and system alerts, allowing timely decisions and enhanced crop management. Moreover, the use of substitute energy sources, such as solar panels, advances the durability and adaptability of these smart systems. Data acquisition and real-time monitoring provide significant optimizing plant growth, resource efficiency, and operational in advanced hydroponic agriculture [22].

14.3.6 Cloud Computing and Wireless Communication Protocols

Wireless communication protocols and cloud computing ensure smooth data exchange, real-time monitoring, and intelligent decision-making. Cloud computing supports expandable data storage, processing, and remote approachability, allowing farmers and researchers to control hydroponic setups from anywhere. These technologies collectively improve efficiency, reduce manual labor, and enhance crop yields by ensuring consistent nutrient standard and environmental control [18]. Wireless communication technologies like MQTT, CoAP, LoRa, and Wi-Fi are essential for transmitting sensor data to controllers or cloud servers, where it can be retained, assessed, and used to enhance plant growth factors.

14.3.6.1 MQTT (Message Queuing Telemetry Transport)

MQTT is a communication convention that upholds a publish-subscribe model, making it optimal for hydroponic systems with limited internet bandwidth and intermittent connectivity. In this setup, sensors send data to a broker, which then provides it to subscribers. MQTT is an outstanding for uninterrupted monitoring climate control, water management, and real-time cautions. It provides timely amendments and data delivery, making it both consistent and resource efficacious [23].

14.3.6.2 Constrained Application Protocol (CoAP)

CoAP is a protocol specifically developed for devices with limited processing power and storage, small sensor nodes used in hydroponics. It uses User Datagram Protocol (UDP) and Transmission Control Protocol (TCP), which decreases data transmission and speeds up communication. CoAP follows a RESTful structure, comparable to how navigation tool communicates with websites, facilitating sensors to readily demand or send data. This makes it ideal for power saving hydroponic systems that need to function efficiently over extended periods [24].

14.3.6.3 Long Range Radio (LoRa)

LoRa is a wireless communication technology known for its competence to disseminate data over several kilometers while using very little power. This makes it an excellent choice for hydroponic farms in distant areas or large areas where internet access is confined or inaccessible. LoRa can transmit minimum amounts of data (like pH levels or humidity) consistently without exhausting battery sensors. Due to its extended coverage and energy efficiency, it is frequently used for monitoring nutrients, environmental conditions, or system collapses in domain installed hydroponic subunits. It maintains connectivity even in areas lacking conventional network infrastructure [25].

14.3.6.4 Wi-Fi (Wireless Fidelity)

Wi-Fi is an often-used wireless communication standard that extensively provide high speed internet connection and is widely available in indoor environments. In hydroponic systems, Wi-Fi permits real-time monitoring by conveying sensor data uninterruptedly to cloud sites for further study and management. Although it depletes additional power compared to standards like LoRa or MQTT, it is appropriate for concise hydroponic subunits with steady power sources. Wi-Fi facilitates large scale data processing such as visual based nutrient assessment, machine learning diagnostics, and user interactive interface dashboards [26, 27].

14.4 System Architecture

14.4.1 Design of an IoT-Based Hydroponic System

The most popular type of vertical farming system mentioned is hydroponic vertical farming. Key components of the system include a growing space, a reservoir for the nutrient solution, and mechanisms to maintain optimal levels of oxygenation, salinity, pH, and EC5. Other factors such as ambient temperature, nutritional solution temperature, photoperiod, and air humidity. The system design incorporates the concept of the Internet of Things (IoT) to monitor and control environmental parameters in real time. Sensors and actuators are installed within the system, enabling real-time measurements and control of parameters. The system can be divided into a control unit and sensor nodes (Fig. 14.1). A physical setup involves tanks for water, nutrient solution, pH-up and pH-down solutions, pumps, lights, and sensors. Experiments also utilized the Dynamic Root Floating Technique (DRFT) [28].

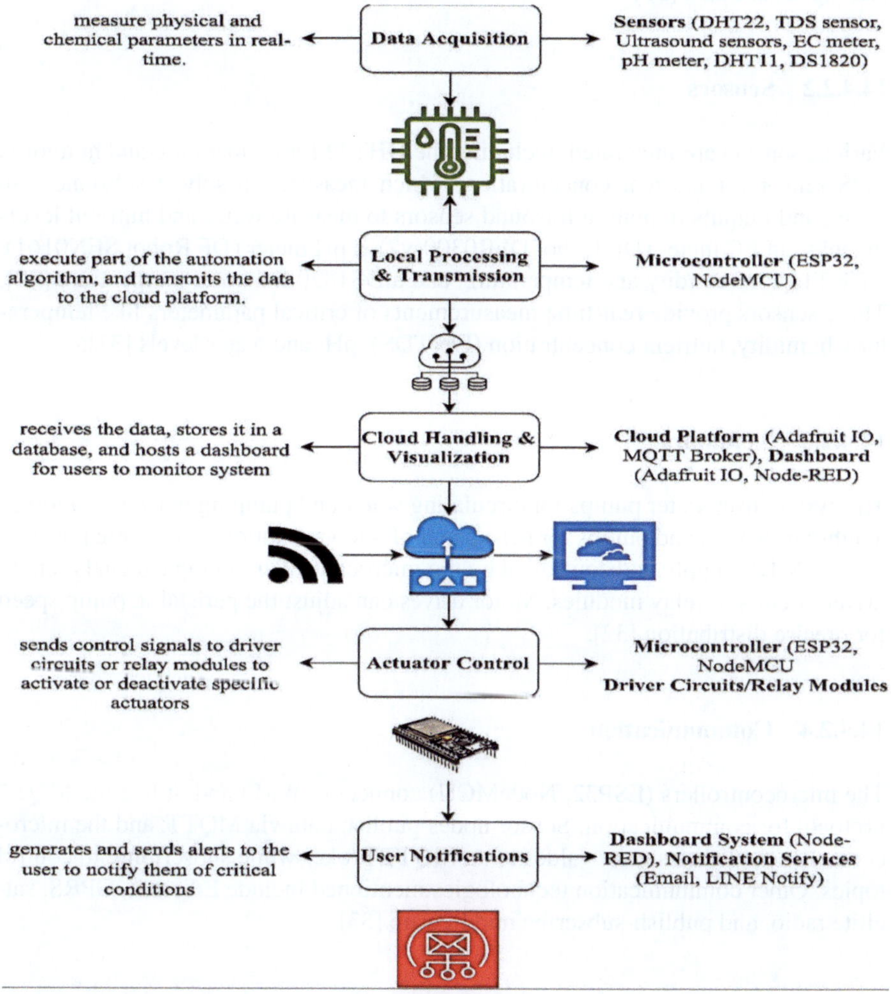

Fig. 14.1 IoT-based sensor data collection and imputation architecture for an integrated cloud-based hydroponic monitoring system consisting of a microcontroller: ESP32 and NodeMCU, Cloud platforms: Adafruit IO and Node-RED, and a Dashboard system: LINE Notify

14.4.2 *Hardware and Software*

14.4.2.1 Microcontrollers

The ESP32 is used as the primary controller in one system, chosen for its built-in Wi-Fi and MQTT support, and is responsible for interfacing with sensors and the cloud dashboard. Another system uses a NodeMCU microcontroller to execute the automation algorithm and receive data from sensors. The ESP32 is also used as the core processing unit in the control unit of a small-scale system, receiving and processing sensor data [29].

14.4.2.2 Sensors

Various sensors are integrated, including the DHT22 for temperature and humidity, TDS sensor for nutrient concentration, which measures dissolved substances in water, and outputs in ppm, ultrasound sensors to measure water and nutrient levels in tanks, an EC meter (DF Robot DFR0300-v2), a pH meter (DF Robot SEN0161), a DHT11 for humidity and temperature, and a DS1820 for water temperature [30]. These sensors provide real-time measurements of critical parameters like temperature, humidity, nutrient concentration (EC/TDS), pH, and water levels [31].

14.4.2.3 Actuators

The system uses water pumps for circulating water and pumping nutrient solutions, a lighting system, and pumps for pH-up and pH-down solutions. These are powered by a 12 V DC supply and controlled by the microcontroller via logic signals sent to driver circuits or relay modules. Motor drives can adjust the peristaltic pump speed for precise distribution [32].

14.4.2.4 Communication

The microcontrollers (ESP32, NodeMCU) connect to Wi-Fi and utilize the MQTT protocol for communication. Sensor nodes publish data via MQTT, and the microcontroller publishes sensor values to an MQTT broker while subscribing to control topics. Other communication technologies mentioned include Ethernet, GPRS, satellite radio, and publish-subscribe middleware [33].

14.4.2.5 Cloud Platform

Adafruit IO is used as a cloud platform, hosting the dashboard and storing data [34].

14.4.2.6 Software/Algorithms

An automation algorithm is implemented in the microcontroller (NodeMCU) to maintain and control EC and pH levels [35] which decides water level, nutrient density, or pH is necessary based on sensor readings and programmed optimal ranges. It calculates the required volume of solution/water and the duration for which the corresponding pump needs to be turned on. The algorithm uses correlation between EC and TDS for different water types and uses tank size and current volume in calculations. Logical condition checking is offered by Adafruit IO and used to automate actuator control based on threshold values. Visual programming through Node-RED is used for developing the dashboard in one system [35].

14.4.3 Data Flow Pipeline: Sensor→ Cloud→ Actionable Insights

14.4.3.1 Sensor Data Acquisition

Sensors continuously measure environmental and nutrient parameters in the hydroponic system. Microcontrollers (ESP32/NodeMCU) are interfaced with these sensors to acquire real-time values [36].

14.4.3.2 Transmission to Cloud

The microcontroller (ESP32) publishes the acquired sensor values to an MQTT broker, such as Adafruit IO. In systems with separate sensor nodes, these nodes publish data via MQTT to a control unit which then transmits it to the cloud. Sensor data is transmitted to the cloud for storage and monitoring [36].

14.4.3.3 Cloud Storage and Monitoring

The cloud platform like Adafruit IO stores the sensor data in a database. Real-time measurements are made available for viewing on a dashboard. Historical data, such as the past 24 h of readings, can also be viewed through graphs on the dashboard [29].

14.4.3.4 Data Processing and Actionable Insights

The sources mention that the Adafruit IO broker performs logical operations but explicitly state that no analytical operations are performed there. However, the automation algorithm implemented on the microcontroller (NodeMCU) performs calculations based on sensor data to determine necessary actions for maintaining parameters within optimal ranges. The control unit processes sensor data to manage EC and pH levels [37].

Based on the logical conditions set in the cloud (Adafruit IO) or the calculations performed by the microcontroller algorithm, actionable insights translate into automated control of actuators. For instance, if nutrient level falls below a threshold, the nutrient pump is automatically turned on. Similarly, if pH or EC/TDS values deviate from the ideal range, the system calculates the amount of solution/water needed and turns on the corresponding pump for the calculated duration [37, 38].

14.4.3.5 Notifications

Alert emails can be sent from Adafruit IO for critical condition. Notifications can also be sent via LINE Notify from the dashboard system to alert users to operational anomalies [39].

14.4.4 User Interface and Visualization Dashboards

An interactive dashboard is hosted on the cloud platform, such as Adafruit IO. Another source mentions a dashboard built using Node-RED. It allows users to view real-time measurements of various environmental parameters and actuator statuses. Instantaneous values of sensors and actuators are displayed on the dashboard. The status of water pumps, lights, and nutrient pumps. Visual representation of elements through gauges and graphical representations including temperature, humidity, EC, pH levels, and water flow rates. Historical data for parameters can be viewed in graphs over periods like 24 h. Some dashboards include intuitive controls, such as push buttons, allowing users to directly control components like pumps or enable/disable automated operations. Notifications are part of the user interface, alerting users to critical conditions via email from Adafruit IO or to operational anomalies through LINE Notify. The primary interface largely used is a web-based dashboard, other source mentions use of mobile application for remote monitoring and control. User inputs, such as ideal pH/TDS ranges or tank size, can be provided via an application [40].

14.4.5 Deep Learning Imputation Technique

Challenges arise in sensor faults and missing data in IoT-based hydroponic systems are compromise on accuracy of nutrient monitoring and the reliability of crop yield predictions so use of novel Deep Learning Precision Imputation Model (DLPIM) which is built upon a Generative Adversarial Network (GAN) architecture. Instead of a general "bioinformatics tool," a four-headed self-attention mechanism is used to capture complex temporal dependencies and preserve critical agricultural patterns such as photoperiod cycles, growth phase transitions, and nutrient interdependencies. The model was developed and evaluated using a dataset comprising 1092 data points related to tomato growth and nutrient analysis collected over 167 days [41]. DLPIM achieves a balanced computational efficiency with faster execution times (4.0–5.0 s) compared to other high-performing deep learning methods (8.3–12.1 s), making it more suitable for practical applications [17, 41] (Table 14.3).

14.5 Data Collection and Integration

Efficient methods of data collection in industrial or technological areas, as well as scientific settings in modern biotechnology, are essential for correct analysis and reliable results, which in turn help in taking appropriate measures to improve production. Integration of the obtained data is achieved using various smart tools, interpreting the heterogeneous data, leading to informed decision-making applicable in health monitoring, smart agriculture, laboratories, etc. [42]. The creation of an Internet of Things (IoT) network enabled remote parameter monitoring and control, which increased system stability. The sensor associated with the NFT framework helps in collecting useful information about the system applicable in analyzing and controlling the water level, EC, pH, etc., in a hydroponics system [43]. The precision in gathering and integrating data is one of the most important factors for a reliable decision applied in laboratories or factories, as well as in remote locations, smart farming, and in the healthcare sector. Calibration of data, structuring the data format, and dealing with the integration of enormous data of different types pose challenges that can be overcome using bioinformatics tools [44].

14.5.1 Sensor Calibration and Validation

For optimization of plant growth and health, sensor calibration is crucial for adjustments in the output of the sensor with respect to the standard values or the reference value for accuracy and consistency in readings over time. These sensors include pH value, EC, nutrient concentrations, or TDS in the hydroponics system [45]. There are different ways to calibrate a sensor. Single-point calibration is suitable when a rough baseline adjustment is adequate. Still, multi-point calibration provides a more accurate correction across a broader range of expected input values, ensuring the response remains linear, consistent, and repeatable [46].

Calibration is followed by validation, which ensures that the sensor readings are reliable and within the range of acceptable values for a particular standard. The data from the sensors is compared with the established standards for verification. In the case of hydroponics, nutrient sensors are validated by comparing the values with those of the laboratory analysis readings [47]. Testing all parameters like accuracy,

Table 14.3 Expanded applications of bioinformatics in nutrient monitoring and optimization

Sr. No.	Application	Bioinformatics tool / method	Crop/plant	Impact	References
1	**Nutrient uptake prediction**	Random Forest, regression trees	Wheat fields in India using yield + soil test data	25% N-fertilizer reduction, yield maintained	[78]
2	**Gene discovery for nutrient traits**	GWAS, QTL mapping	P-efficiency in rice (IRRI)	Identified OsPSTOL1 gene improving phosphorus uptake	[79]
3	**Transcriptomics in nutrient stress**	RNA-seq + DESeq2	Zn-deficiency in Arabidopsis	Identified ZIP family transporters; insights into transport pathways	[80]
4	**Image-based diagnosis**	CNN + Mobile app	Leaf nutrient status in sugarcane	Real-time classification of N, P, K deficiencies with 90% accuracy	[81]
5	**Pathway analysis of nutrient metabolism**	KEGG mapper + Cytoscape	Sulfur assimilation in soybean under stress	Identified key pathway bottlenecks; guided metabolic engineering	[82]
6	**Crop breeding programs**	Marker-assisted selection	Nitrogen-efficient bread wheat	Deployed SNP markers; enhanced NUE by 15%	[83]
7	**Data mining from historical trials**	PCA, clustering, decision trees	10-year wheat trials in sub-Saharan Africa	Revealed climate–soil–nutrient interaction patterns	[84]
8	**Multi-omics integration**	Genomics + metabolomics + ML	Iron uptake efficiency in biofortified beans	Developed predictive models for Fe-uptake and accumulation	[85]
9	**Field-level decision support**	DSS + machine learning	Nutrient expert for rice and maize (South Asia)	Increased profit margins by 10–15% through optimized fertilizer use	[86]

linearity, repeatability, and precision after calibration to confirm the reliability of the sensor. Validation is required under various environmental conditions, like temperature, humidity, etc., to ensure reliable and efficient performance in any condition. This validation process is critical not only during initial deployment but also as part of regular maintenance to detect sensor drift or wear over time [48].

Proper sensor calibration and validation are required for accurate nutrient management by maintaining the optimal nutrient level according to the requirements of the plant. This helps in preventing nutrient deficiency and combating the toxicity caused by excess nutrients or water levels [49]. It is very crucial that the data is reliable for correct interpretation and informed decision-making regarding the adjustments in the nutrient concentration and varying environmental conditions. This also prevents wastage of nutrients and resources by monitoring in real-time, in addition to saving the plant at the initial stage from nutrient deficiency or toxicity. Crop growth and yield are directly affected by this technique of nutrient monitoring [50].

14.5.2 Data Logging Frequency and Formats

The frequency of recording data influences the outcome through the data logging rate. Rapid changes in the data due to high-frequency logging are monitored closely for the reliability of readings, most preferably in the case of medical monitoring, where an enormous amount of high-frequency data needs to be stored and processed properly [51]. The low-frequency logging of the readings is applied for the observations that change slowly, like the parameters in agriculture or hydroponics, including moisture content, nutrient concentration, environmental conditions like temperature, etc. [52]. The data needs a particular format to be read and analyzed easily by the software. CSV (Comma-separated values) is considered compatible with most software, widely accepted for general-purpose logging, while other formats like JSON or XML are used for storing the complex, high-profile data that needs to analyze regarding various attributes [53]. In scientific research and high-volume industrial applications, binary data formats like HDF5 or NetCDF are often preferred for efficient storage of large data [54].

14.5.3 Real-Time Vs Batch Data Analysis

It is of the utmost significance to determine the time and method of analysis, real time or batch, that influence the time of action. In real-time analysis, the data is analyzed as it is collected at the same time, while batch analysis requires the storage of bulk data before analysis [55]. The immediate processing of data in real-time helps in quick response to the situation like in case of autonomous vehicles, critical healthcare systems etc. Processing of the data depends upon the computing, sensors and accuracy of the system, which is made possible with the help of optimized software systems associated with reliable sensors [56]. AI-based systems have been developed to regulate the controllers, logic algorithms, NFT hydroponics (Nutrient film technology), user-friendly interface, and IoT-based cloud environment, and nutrient sensor data monitoring with quick comparative analysis and decision-making in real-time analysis [57]. Batch analysis, in contrast, is slower and time-consuming and is generally used for big data where immediate action is not necessary. It involves a complex computation process in addition to advanced machine learning and statistics [58]. Batch analysis is excellent at finding intricate patterns and correlations in big datasets, while lacking the instantaneity of real-time systems. A lot of contemporary systems now employ a hybrid model that blends the two strategies—batch analysis for thorough assessment and real-time warnings for irregularities [59].

14.6 Application of Bioinformatics Tools

14.6.1 Overview of Bioinformatics Tools Relevant to Nutrient Modeling

Bioinformatics involves the processing and analysis of data assessed by computational modeling for easier, efficient, and precise interpretation of results/readings. The stimulation, analysis, and nutrient response estimation in the hydroponics system or the agriculture system are facilitated greatly by bioinformatics tools for nutrient modeling. The most important programs include:

- **Gene Expression Analysis Platforms:** There are genes in plants that express themselves according to the conditions, like the genes that show differential expression in conditions where nutrients are abundant, while a different expression is seen in a nutrient-deficient environment. To hunt for such differentially expressed genes, databases like NCBI GEO or Ensembl Plants have been a popular choice for a convenient and accurate search [60].
- **Tools for Pathway and Network Analysis:** The mapping of metabolic pathways aids in analyzing the slowest steps in the process, the rate-limiting steps, and thereby detecting the target gene involved in the process whose alteration could change the course of the mechanism for better expression. KEGG, BioCyc, MetaCyc and Plant Reactome are some of the platforms that are used for pathway mapping [61, 62]. It has been used for tracing the pathway of Nitrogen Absorption and for analyzing the transport of phosphate, including all the steps involved in the pathway [63].
- **Frameworks for modeling and simulation:** Plant models are created using the information gathered based on their adaptation to environmental stress conditions, their nutrient requirements, genetics, age of the plant, etc. These computational models with mathematical input are created by tools like MATLAB, Python, and R [64].
- **Visualization and Network Tools:** Genes associated with nutrient metabolism and their expression are recorded in the databases, where it is also possible to analyze the gene-nutrient interaction network by finding the regulatory points. Programs like Cytoscape simplify these visualizations, like gene regulatory network (GRN), protein–protein interaction (PPI), and multi-omics data [65].

- **Databases and Functional Annotation:** AgriGo, STRING, and PlantTFDB are used for the functional annotation that aids in connecting genes to the issues faced by plants due to nutrient deficiency or toxicity, like ion transportation, stress conditions, etc. [66].

14.6.2 Predictive Modeling of Nutrient Uptake Using Machine Learning

Machine learning (ML) has enabled the data to be analyzed without specific programming for the interpretation of the entered data. The models are used in machine learning that develop their datasets based on the previous readings and trials from the sensors, fields, and laboratory experiments, which include:

- **Classification Models**: The major features to categorize the requirements of plants based on soil characteristics, plant age, and climate, which influence the nutrient response, using algorithms like support vector machines (SVMs) [67].
- **Regression models:** There are some algorithms that are used to predict the fertilizer requirement of a plant based on its nutrient uptake. Such algorithms enable the user to conveniently decide the dosage of fertilizer required without any possible damage that is caused by chemical toxicity e.g., Random Forest Regression, Hydroponics system Gradient Boosting [68].
- **Neural Networks:** These networks are used to examine the disease or nutritional deficiency in a plant by analyzing the photograph of the affected leaf via the mobile application or neural networks like artificial and convolutional neural networks (ANNs and CNNs) [69].

14.6.3 Integration with Genomic Data to Understand Nutrient Efficiency Traits

By connecting plant genetics and nutrient behavior, genomic integration enables the creation of genetically enhanced cultivars and the discovery of nutrient-efficient cultivars using quantitative trait loci (QTL) mapping that helps in finding the genes linked to a particular trait and then selection of the superior genotypes using marker-assisted selection methods [70]. Genome-Wide Association Studies (GWASs) help in identifying the relationship of genes with a particular nutrient deficiency in plants using SNP data [71]. Transcriptomics helps in identifying the gene expression regarding certain nutritional changes, and the genetic variations that affect those gene expressions are identified by expression QTLs (eQTLs) [72].

14.6.4 Examples of Implemented Systems in Research

Use of bioinformatics for nutrient monitoring:

- For site-specific nutrition management in rice and maize cultivation, the International Plant Nutrition Institute (IPNI) has formed a tool to support the decision regarding nutrition expertise. It helps in deciding the best fertilizer with the accurate dosage using data that drives the algorithm [73].
- Smart Farming Platforms help farmers gain a handsome outcome from their crop by providing aids through application advice for the fertilizers and nutrients by collecting the data from Climate FieldView, Arable Labs, and John Deere, and combining the sensor data into a predictive model for accurate interpretation [74].
- Genomic-Assisted Breeding helps in creating a nutrient-efficient crop by analyzing the QTLs with the assistance of organizations like the International Rice Research Institute (IRRI) and CIMMYT (for maize) [75].
- **Economic Benefits**: The savings in fertilizer cost up to $50/ha or approximately INR 4500 because of the reduced fertilizer use as precise amounts and dosages are applied with accurate prediction using bioinformatics tools. This also improved the nutrient balance in the crop, quality improvement, and thereby an increase in the market value, which has a positive economic impact for both commercial and small-scale farmers, in field crops and in hydroponics [76].
- **Environmental Benefits**: Use of hydroponics with a balanced amount of fertilizers minimizes the emission of greenhouse gases that are associated with the manufacture of fertilizers. Minimal use of fertilizers like urea also decreases the discharge of heavy amounts of phosphorus and nitrogen into the water bodies [77].

14.7 Challenges and Limitations

Sustaining continuous sensor output, significantly when it comes to auditing nutrient absorption, is one of the most critical issues in the implementation of smart hydroponic systems. The limitation of Electrical Conductivity sensors is restricted for particular ionic nutrients; despite they are cost-effective and uncomplicated to incorporate. Moreover, these probes are likely to degrade with extended immersion, which restrict their use for continuous tracking in prolonged settings [87]. Ion specific Electrodes are more selective, but they are also substantially more expensive, vulnerable to environmental variations, and often need advanced setup and calibration [2]. Errors are integrated by sensor drift progressively, which makes constant calibration mandatory but complex. Advanced spectroscopic devices have illustrated potential in expanded hydroponic systems, but their expense and complication inhibit their use in compact or domestic systems. All of these limitations make it challenging to ensure consistent and stable nutrient management, primarily in dispersed or microscale environments where digital resources are potentially limited [88].

The challenges in operating and analyzing the continuous channels of digital input generated by these systems are equally essential. Whereas real-time monitoring is facilitated by Internet of Things (IoT) platforms, these systems intermittently cope with skewed and incomplete data due to inconsistent network access or malfunctioning sensors [89]. These variances can balance the standard of frameworks used by machine learning systems to estimate substantial metal tolerance or evaluate nutrient movements. Beyond that autonomously operated hydroponic systems are not able to handle the programmatic effort necessary to process such datasets, notably when incorporating genetic characteristics or bioinformatics work flows [90]. Models are required to be confined periodically in order to maintain their efficiency because of the variety in crop species and external factors. Accessibility is another limitation; many platforms have complex monitoring displays or user interfaces that make learning challenging for laypersons [91]. Another issue is integration capability, as multiple sensors and systems are not routinely standardized. These limitations demonstrate that future smart hydroponic systems can support a spectrum of dimensions and expertise levels requiring more resilient, versatile, and user-centered designs [13].

14.8 Summary

The decreased area of agricultural land is a concern for future generation to fulfill their nutritional requirements, which is put to ease with the hydroponics technique and the advanced methods and tools for accurate predictions and decision-making regarding the nutritional requirements of plants even in soilless agricultural practice. Accurate management of a hydroponics system is the result of highly reliable bioinformatics tools, and the Internet of Things associated with nutrient prediction. These technologies have facilitated an accurate, reliable, repeatable diagnosis of plants and their environment, nutrient uptake, and optimization of nutrient profile using computer modeling in real time. For the improvement in agricultural outcomes, these technologies and their predictions can be used to create a data-driven agri-system to face the world hunger and malnutrition issues, enabling food security in the future.

References

1. Khatri L, Kunwar A, Bist DR (2024) Hydroponics: advantages and challenges in soilless farming. Big Data Agric 6:81–88
2. Dhal SB, Mahanta S, Gumero J, O'Sullivan N, Soetan M, Louis J, Gadepally KC, Mahanta S, Lusher J, Kalafatis S (2023) An IoT-based data-driven real-time monitoring system for control of heavy metals to ensure optimal lettuce growth in hydroponic set-ups. Sensors 23(1):451

3. Pyingkodi M, Thenmozhi K, Karthikeyan M, Kalpana T, Palarimath S, Kumar GB (2022) IoT-based soil nutrients analysis and monitoring system for smart agriculture. In: In 2022 3rd International Conference on Electronics and Sustainable Communication Systems (ICESC), IEEE, pp 489–494
4. Jain SK, Jain PK (2024) Synergizing smart farming and human bioinformatics through IoT and sensor devices. In: Microbial data intelligence and computational techniques for sustainable computing singapore. Springer, Singapore, pp 139–149
5. Pazhamala LT, Kudapa H, Weckwerth W, Millar AH, Varshney RK (2021) Systems biology for crop improvement. Plant Genome 14(2):e20098
6. Zhang X, Ibrahim Z, Khaskheli MB, Raza H, Zhou F, Shamsi IH (2024) Integrative approaches to abiotic stress management in crops: combining bioinformatics educational tools and artificial intelligence applications. Sustainability16(17):7651
7. Hawkins C, Ginzburg D, Zhao K, Dwyer W, Xue B, Xu A, Rice S, Cole B, Paley S, Karp P, Rhee SY (2021) Plant metabolic network 15: a resource of genome-wide metabolism databases for 126 plants and algae. J Integr Plant Biol 63(11):1888–1905
8. Clough E, Barrett T, Wilhite SE, Ledoux P, Evangelista C, Kim IF, Tomashevsky M, Marshall KA, Phillippy KH, Sherman PM, Lee H (2024) NCBI GEO: archive for gene expression and epigenomics data sets: 23-year update. Nucleic Acids Res 52(D1):D138–D144
9. Bhuiyan MM, Rahaman MM, Aziz MM, Islam MR, Das K (2023) Predictive analytics in plant biotechnology: using data science to drive crop resilience and productivity. J Environ Agric Stud 4(3):77–83
10. Sanchez E, Berghage R, Ford T, Gioia FD (2021) Hydroponics systems and principles of plant nutrition: essential nutrients, function, deficiency, and excess. PennState Extension Fact Sheet:1–5
11. Brar GS, Kumar A, Singh G, Ishika I, Shubham S, Kaushal S (2024) Optimisation of advanced nutrient solutions for escalating crop production under hydroponic systems: a comprehensive review. Int J Plant Soil Sci 36(6):165–177
12. Thapa U, Hansda NN, Kundu S, Giri A, Tamang D, Rahaman AO (2024) Advancements in hydroponic systems: a comprehensive review. Arch Curr Res Int 24(11):317–328
13. Rao PB, Venkatesh RT, Murthy SG, Negavadi SB, Srinivasulu T, Nataraj YB, Satish V, Balenahalli VK (2024) Automated IoT solutions for efficient hydroponic farming: nutrients, PH and lighting management. Journal Europèen des Systèmes Automatisés 57(5):1273
14. Akter T, Mahmud T, Chakma R, Datta N, Hossain MS, Andersson K (2024) Smart monitoring and control of hydroponic systems using iot solutions. In: In2024 second international conference on inventive computing and informatics (ICICI), IEEE, pp 761–767
15. Monisha K, Kalai Selvi H, Sivanandhini P, Sona Nachammai A, Anuradha CT, Rama Devi S, Kavitha Sri A, Neya NR, Vaitheeswari M, Hikku GS (20230 Hydroponics agriculture as a modern agriculture technique. J Achiev Mater Manufact Eng 116(1)
16. Shukla S, Hassan MF, Tran DC, Akbar R, Paputungan IV, Khan MK (2023) Improving latency in Internet-of-Things and cloud computing for real-time data transmission: a systematic literature review (SLR). Clust Comput:1–24
17. Tatas K, Al-Zoubi A, Christofides N, Zannettis C, Chrysostomou M, Panteli S, Antoniou A (2022) Reliable IoT-based monitoring and control of hydroponic systems. Technologies 10(1):26
18. Hosny KM, El-Hady WM, Samy FM (2025) Technologies, protocols, and applications of internet of things in greenhouse farming: a survey of recent advances. Inform Process Agric 12(1):91–111
19. Villagran E, Espitia JJ, Amado G, Rodriguez J, Gomez L, Velasquez JF, Gil R, Baeza E, Aguilar CE, Akrami M, Arias LA (2025) CO2 enrichment in protected agriculture: a systematic review of greenhouses, controlled environment systems, and vertical farms—part 2. Sustainability 17(7):2809
20. Reza MN, Lee KH, Karim MR, Haque MA, Bicamumakuba E, Dey PK, Jang YY, Chung SO (2025) Trends of soil and solution nutrient sensing for open field and hydroponic cultivation in facilitated smart agriculture. Sensors 25(2):453

21. Hariono T, Putra MC (2021) Data acquisition for monitoring IoT-based hydroponic automation system using ESP8266. NEWTON: Networking and Information Technology 1(1):1–7
22. Megantoro P, Prastio RP, Kusuma HF, Abror A, Vigneshwaran P, Priambodo DF, Alif DS (2022) Instrumentation system for data acquisition and monitoring of hydroponic farming using ESP32 via Google firebase. Indones J Electr Eng Comput Sci 27(1):52–61
23. Nugroho AS, Novitasari M, Pramono S, Wardihani E (2023) Performance of acidity nutrient automation using MQTT protocol in Dutch bucket system hydroponic. J Eng Sci Technol 18(4):214–228
24. Nurwarsito H, Pradana KD (2023) Implementation of Lora and CoAP protocol in hydroponic plant water pump control. J Inf Technol Comput Sci 8(3):263–272
25. Usman U, Achmad A, Yuyun Y, Satra R, Tribuana D, Konate S (2024) Monitoring and controlling humidity and pH use of LoRa in IoT-based hydroponic planting. Bullet Social Inform Theory Appli 8(1):107–123
26. Hanafi AM, Hussien SA, Elnahal DH, Ahmed SE, Salem MA, Zainhum AR, Elsayed AA, Ibrahim MA, Abdel Sattar YS (2025) Revolutionizing agriculture with IoT, Mobile apps, and computer vision in automated hydroponic greenhouses. Int J Eng Appl Sci 2(1):1–6
27. Aldhaheri L, Alshehhi N, Manzil II, Khalil RA, Javaid S, Saeed N, Alouini MS (2024) LoRa communication for agriculture 4.0: opportunities, challenges, and future directions. IEEE Internet Things J:1–27
28. Chairudin C, Fitria R, Harahap EJ, Afrillah M (2024) Application of various nutrition to the growth and production of melon (Cucumis melo L.) hydroponic DRFT (Dynamic Root Floating Technique). Jurnal Ilmu Pertanian Indonesia 29(3):372–376
29. Ramsari N, Hidayat T (2022) Monitoring system and hydroponic plant automation using microcontroller Internet of Things based (IoT). Compiler 11(2):59–74
30. Aurasopon A, Thongleam T, Kuankid S (2024) Integration of IoT Technology in Hydroponic Systems for enhanced efficiency and productivity in small-scale farming. Acta Technologica Agriculturae 27(4):203–211
31. Flores-Iwasaki M, Guadalupe GA, Pachas-Caycho M, Chapa-Gonza S, Mori-Zabarburú RC, Guerrero-Abad JC (2025) Internet of Things (IoT) sensors for water quality monitoring in aquaculture systems: a systematic review and bibliometric analysis. AgriEngineering 7(3):78
32. A ZAATOUT (2023) Intelligent Management of a Hydroponic Greenhouse Powered by a PV system using internet of things and artificial intelligence. Doctoral dissertation, Ibn Khaldoun University
33. Nizam MS, Abdullah E, Hidayat NM, Hashim NM, Hassan MA (2024) Real-time energy monitoring in renewable EV charging stations: an ESP32-based system integrating Modbus, MQTT, and ESP-NOW protocols. In: In2024 IEEE 22nd student conference on Research and Development (SCOReD), pp 339–344
34. Barus SP, Seo JI (2025) Temperature and humidity monitoring in hydroponic cultivation based on internet of things: dataset development for smart agriculture. Jurnal Sisfokom (Sistem Informasi dan Komputer) 14(1):63–68
35. Sidibe A, Ndeda R, Murimi E, Mutwiwa UN (2023) Implementation of a microcontroller-based neural network for prediction of pH and EC levels in hydroponics. J Sustain Res Eng 7(4):106–117
36. Sportelli M, Crivello A, La Rosa D, Bacco M, Incrocci L, Barsocchi P (2024) An IoT platform for smart hydroponics: building blocks and open challenges. In: In2024 IEEE international workshop on metrology for agriculture and forestry (MetroAgriFor), pp 112–117
37. Khalifeh A, Mazunga F, Nechibvute A, Nyambo BM (2022) Microcontroller unit-based wireless sensor network nodes: a review. Sensors 22(22):8937
38. Sulaiman H, Yusof AA, Mohamed Nor MK (2025) Automated hydroponic nutrient dosing system: a scoping review of pH and electrical conductivity dosing frameworks. AgriEngineering 7(2):43
39. Senathipathi K, Sathyapraba G, Sarnithasree RK, Swetha S (2022) Dashboard for alert monitoring system. In: 2022 8th International Conference on Advanced Computing and Communication Systems (ICACCS), vol 1. IEEE, pp 1229–1233

40. De Side GN, Putra GM, Setiawati DA (2025) IoT (internet of things) microclimate monitoring system using node-red platform at plant factory. InIOP Conf Ser: Earth Environ Sci, IOP Publishing 1441(1):12012
41. Venugopal V, Tanna P, Karnati R (2025) Handling sensor faults in hydroponics: a deep learning imputation technique for accurate tomato yield prediction
42. Soussi A, Zero E, Sacile R, Trinchero D, Fossa M (2024) Smart sensors and smart data for precision agriculture: a review. Sensors 24(8):2647
43. Ali MS, Nordin MK, Zaki MH, Saaid MF (2024) Optimizing plant growth in indoor NFT hydroponic systems: design, environmental monitoring, and analysis. In: In2024 IEEE international conference on applied electronics and engineering (ICAEE), pp 1–6
44. Sangeetha T, Periyathambi E (2024) Automatic nutrient estimator: distributing nutrient solution in hydroponic plants based on plant growth. PeerJ Comput Sci 10:e1871
45. Karimzadeh S, Li Z, Ahamed MS (2025) Machine learning-based fault detection and diagnosis of electrical conductivity and pH sensors in hydroponic systems. Comput Electron Agric 237:110544
46. Jamil A, Ting TS, Abidin ZZ, Othman M, Wahab MH, Abdullah MF, Homam MJ, Audah LH, Shah SM (2023) Polynomial regression calibration method of Total dissolved solids sensor for hydroponic systems. Pertanika J Sci Technol 31(6)
47. Bhargava YV, Chittoor PK, Bharatiraja C, Verma R, Sathiyasekar K (2022) Sensor fusion based intelligent hydroponic farming and nursing system. IEEE Sensors J 22(14):14584–14591
48. Ikrang EG, Ehiomogue PO, Udoumoh UI (2022) Hydroponics in precision agriculture–a review. Ann Facul Eng Hunedoara 20(2):143–148
49. Paul K, Chatterjee SS, Pai P, Varshney A, Juikar S, Prasad V, Bhadra B, Dasgupta S (2022) Viable smart sensors and their application in data driven agriculture. Comput Electron Agric 198:107096
50. Abu Bakar AA, Bakar ZA, Yusoff ZM, Mohamed Ibrahim MJ, Mokhtar NA, Zaiton SN (2025) IoT-based real-time water quality monitoring and sensor calibration for enhanced accuracy and reliability. Int J Interact Mobile Technol 19(1)
51. Hassani S, Dackermann U (2023) A systematic review of advanced sensor technologies for non-destructive testing and structural health monitoring. Sensors 23(4):2204
52. Bakirov K, Tussupov J, Tussupov A, Shayea I, Shoman A (2025) Application of a hybrid model for data analysis in hydroponic systems. Technologies 13(5):166
53. Basir MS, Zhang Y, Buckmaster D, Raturi A, Krogmeier JV (2025) Meta Ag: an automatic agricultural contextual metadata collection app. Smart Agric Technol 101073
54. Gecášek D, Solanik M, Genči J (2022) Which data format to store scientific data should I use? A performance analysis. Acta Electrotechnica et Informatica 22(3):32–40
55. Sherubha P, Bharathipriya C, Iqbal MM, Ganesh P, Sasirekha SP (2022) A real time IoT measurement system for hydroponics cultivation. In: In2022 2nd international conference on computing and information technology (ICCIT), IEEE, pp 432–435
56. Ramakrishnam Raju SV, Dappuri B, Ravi Kiran Varma P, Yachamaneni M, Verghese DM, Mishra MK (2022) Design and implementation of smart hydroponics farming using IoT-based AI controller with Mobile application system. J Nanomater 2022(1):4435591
57. Putra SD, Heriansyah H, Cahyadi EF, Anggriani K, Jaya MH (2024) Development of smart hydroponics system using AI-based sensing. Jurnal Infotel 16(3):474–485
58. Rofiansyah W, Zalianty FR, La Ito FA, Wijayanto I, Ryanu HH, Irawati ID (2025) IoT-based control and monitoring system for hydroponic plant growth using image processing and mobile applications. PeerJ Comput Sci 11:e2763
59. Ojo MO, Zahid A, Masabni JG (2024) Estimating hydroponic lettuce phenotypic parameters for efficient resource allocation. Comput Electron Agric 218:108642
60. Ding K, Sun S, Luo Y, Long C, Zhai J, Zhai Y, Wang G (2023) PlantCADB: a comprehensive plant chromatin accessibility database. Genomics Proteomics Bioinformatics 21(2):311–323

61. Gupta P, Naithani S, Preece J, Kim S, Cheng T, D'Eustachio P, Elser J, Bolton EE, Jaiswal P (2022) Plant reactome and PubChem: the plant pathway and (bio) chemical entity knowledgebases. Plant Bioinform: Meth Protocol 2443:511–525
62. Zhao K, Rhee SY (2022) Omics-guided metabolic pathway discovery in plants: resources, approaches, and opportunities. Curr Opin Plant Biol 67:102222
63. Zayed O, Hewedy OA, Abdelmoteleb A, Ali M, Youssef MS, Roumia AF, Seymour D, Yuan ZC (2023) Nitrogen journey in plants: from uptake to metabolism, stress response, and microbe interaction. Biomolecules 13(10):1443
64. Dale R, Oswald S, Jalihal A, LaPorte MF, Fletcher DM, Hubbard A, Shiu SH, Nelson AD, Bucksch A (2021) Overcoming the challenges to enhancing experimental plant biology with computational modeling. Front Plant Sci 12:687652
65. Majeed A, Mukhtar S (2023) Protein–protein interaction network exploration using cytoscape. In: Protein-protein interactions: methods and protocols. Springer, pp 419–427
66. Gong J, Peng Y, Yu J, Pei W, Zhang Z, Fan D, Liu L, Xiao X, Liu R, Lu Q, Li P (2022) Linkage and association analyses reveal that hub genes in energy-flow and lipid biosynthesis pathways form a cluster in upland cotton. Comput Struct Biotechnol J 20:1841–1859
67. Kok ZH, Shariff AR, Alfatni MS, Khairunniza-Bejo S (2021) Support vector machine in precision agriculture: a review. Comput Electron Agric 191:106546
68. Mokhtar A, El-Ssawy W, He H, Al-Anasari N, Sammen SS, Gyasi-Agyei Y, Abuarab M (2022) Using machine learning models to predict hydroponically grown lettuce yield. Front Plant Sci 13:706042
69. Ahmed AA, Reddy GH (2021) A mobile-based system for detecting plant leaf diseases using deep learning. AgriEngineering 3(3):478–493
70. Kushanov FN, Turaev OS, Ernazarova DK, Gapparov BM, Oripova BB, Kudratova MK, Rafieva FU, Khalikov KK, Erjigitov DS, Khidirov MT, Kholova MD (2021) Genetic diversity, QTL mapping, and marker-assisted selection technology in cotton (Gossypium spp.). Front Plant Sci 12:779386
71. Hakla HR, Sharma S, Urfan M, Mandlik R, Kumawat S, Rajput P, Khajuria B, Chowdhary R, Deshmukh R, Roychowdhury R, Pal S (2024) Genome-wide association study (GWAS) for identifying SNPs and genes related to phosphate-induced phenotypic traits in tomato (Solanum lycopersicum L.). Plants 13(3):457
72. Wade AR, Duruflé H, Sanchez L, Segura V (2022) eQTLs are key players in the integration of genomic and transcriptomic data for phenotype prediction. BMC Genomics 23(1):476
73. Gorai T, Yadav PK, Choudhary GL, Kumar A (2021) Site-specific crop nutrient management for precision agriculture—A review. Curr J Appl Sci Technol 40:37–52
74. Balafoutis AT, Kavroumatzi CK, Moraitis M, Vaiopoulos K, Mylonas N, Tsitsigiannis I, Ampatzidis Y, Fountas DS, Bochtis D (2021) Advanced crop protection techniques and technologies. In: Modeling for sustainable management in agriculture, food and the environment. CRC Press, pp 112–171
75. Chimmili SR, Kanneboina S, Hanjagi PS, PS B, Sakhare AS, P S, Kumar S, Kota S (2022) Integrating advanced molecular, genomic, and speed breeding methods for genetic improvement of stress tolerance in rice. InNext-generation plant breeding approaches for stress resilience in cereal crops, Springer Singapore pp. 263–283
76. Verma MS, Gawade SD (2021) A machine learning approach for prediction system and analysis of nutrients uptake for better crop growth in the hydroponics system. In: In2021 international conference on artificial intelligence and smart systems (ICAIS). IEEE, pp 150–156
77. Mu H, Wang B, Yuan F (2022) Bioinformatics in plant breeding and research on disease resistance. Plants 11(22):3118
78. Sadhukhan R, Kumar D, Sepat S, Ghosh A, Banerjee K, Shivay YS, Gawdiya S, Harish MN, Bhatia A, Kumawat A, Dutta S (2024) Precision nutrient management influences the productivity, nutrients use efficiency, N2O fluxes and soil enzymatic activity in zero-till wheat (Triticum aestivum L.). Field Crops Res 317:109526
79. Lyzenga WJ, Liu Z, Olukayode T, Zhao Y, Kochian LV, Ham BK (2023) Getting to the roots of N, P, and K uptake. J Exp Bot 74(6):1784–1805

80. Lv Z, Liu F, Zhang P, Li W, Song Q, Xu J (2023) Revealing the molecular mechanisms of zinc accumulation and zinc deficiency responses in quinoa, analyzed by high-throughput gene expression profiling under zinc depletion and resupply. Environ Exp Bot 210:105330
81. Daphal SD, Koli SM (2024) Enhanced deep learning technique for sugarcane leaf disease classification and mobile application integration. Heliyon 10(8):e29438
82. Zhou X, Yao X, He D, Sun H, Xie F (2024) Transcriptomic investigation unveils the role of energy metabolism under low phosphorus and salt combined stress in soybean (Glycine max). Physiol Plant 176(2):e14262
83. Kartseva T, Aleksandrov V, Alqudah AM, Schierenbeck M, Tasheva K, Börner A, Misheva S (2024) Exploring novel genomic loci and candidate genes associated with plant height in Bulgarian bread wheat via multi-model GWAS. Plants 13(19):2775
84. Toqeer A, Defourny P (2023) Developing wheat crop yield estimation method for Spain from remotely sensed metrics using artificial intelligence. Doctoral dissertation, University of Tartu
85. Roorkiwal M, Pandey S, Thavarajah D, Hemalatha R, Varshney RK (2021) Molecular mechanisms and biochemical pathways for micronutrient acquisition and storage in legumes to support biofortification for nutritional security. Front Plant Sci 12:682842
86. Kaur R, Shukla S, Sharma S, Dhillon BS (2023) Innovative nutrient management strategies for sustainable Rice-wheat cropping systems. Indian J Fertil 19(11):1134–1143
87. Stevens JD, Murray D, Diepeveen D, Toohey D (2023) Development and testing of an IoT spectroscopic nutrient monitoring system for use in micro indoor smart hydroponics. Horticulturae 9(2):185
88. Cho WJ, Kim DW, Kim HJ, Gang M, Jung D, Yang SH, Park SH (2024) Automated drift compensation system for electrical conductivity and pH probes in hydroponic systems. J ASABE 67(5):1203–1215
89. Lakshmanan R, Djama M, Selvaperumal SK, Abdulla R (2020) Automated smart hydroponics system using internet of things. Int J Elect Comput Eng (IJECE) 10(6):6389–6398
90. Rahman MA, Chakraborty NR, Sufiun A, Banshal SK, Tajnin FR (2024) An AIoT-based hydroponic system for crop recommendation and nutrient parameter monitorization. Smart Agric Technol 8:100472
91. Idoje G, Mouroutoglou C, Dagiuklas T, Kotsiras A, Muddesar I, Alefragkis P (2023) Comparative analysis of data using machine learning algorithms: a hydroponics system use case. Smart Agric Technol 4:100207

Chapter 15
Intellectual Property Rights in Smart Agriculture for Medicinal Plants

Hemant Sood, Shriyanshu Thakur, Neha Sharma, and Rolika Gupta

Abstract In the field of intellectual property rights (IPR), the incorporation of smart agricultural technology into the production of medicinal plants brings both new potential and difficulties. Medicinal plant farming is made more efficient, sustainable, and high-quality by smart agriculture, which uses biotechnology, blockchain, the Internet of Things, and artificial intelligence (AI). These developments, however, bring up moral and legal questions about fair benefit-sharing, ownership, and innovation protection. With an emphasis on patents, plant variety protection, and the preservation of traditional knowledge, this article examines the intellectual property rights (IPR) environment in smart agriculture for medicinal plants. Strong legal frameworks are necessary for the commercialization of AI-driven farming methods and genetically modified medicinal crops to protect inventors' rights and guarantee accessibility for farmers and indigenous people. The Convention on Biological Diversity (CBD) and the Trade-Related Aspects of Intellectual Property Rights (TRIPS) are two important international accords that define the legal limits of IPR in smart agriculture. Additionally, new technologies like blockchain can provide traceability of plant genetic resources and transparency in patent administration. This study emphasizes the necessity of well-balanced IPR laws that support innovation while preserving biodiversity and traditional knowledge. In the digital age, strengthening regulatory frameworks may promote equal access and technical breakthroughs, guaranteeing the long-term growth of medical plant cultivation.

Keywords Artificial intelligence · Biopiracy · Intellectual property rights · Smart agriculture · Traditional knowledge

H. Sood (✉)
Department of Biotechnology and Bioinformatics, Waknaghat, Himachal Pradesh, India
e-mail: hemant.sood@juitsolan.in

S. Thakur · R. Gupta
Division of Microbiology, Career Point University, Hamirpur, Himachal Pradesh, India

N. Sharma
Division of Crop Improvement and Seed Technology, ICAR-Central Potato Research Institute, Shimla, Himachal Pradesh, India

P. Kumar, A. R. Warghat (eds.), *Transforming Medicinal Plant Agriculture*, Smart Agriculture 16, https://doi.org/10.1007/978-981-95-3963-5_15

15.1 Introduction

Intellectual property rights (IPRs), which include patents, trademarks, copyrights, industrial designs, geographical indications (GI), and plant breeder's rights, are legal protections that regulate the use of works that are the result of human intelligence. IPR serves the dual purposes of recognizing innovative concepts and preventing third parties from utilizing protected subject matter without the owner's express approval for a predetermined amount of time. They foster innovation that accelerates social, economic, and scientific development by encouraging creators to benefit from their inventions and ensure fair trade. Initially, intellectual property laws were mainly centered on protecting inventions related to machinery, processes, and industrial applications and did not specifically cover agricultural innovations. However, as agricultural science and plant breeding advanced, this industry is seen as an important area to attract investment, resulting in the gradual extension of IPR in agriculture. IPR in agriculture is used to protect goods or services produced in the agricultural sector and mainly deals with patents, plant breeder's rights, trademarks, geographical indications, and trade secrets.

Previously, patent laws did not offer the protection for inventions related to plant products or methods; however, in 1930, a special type of Plant Patent Act was introduced for vegetatively propagated materials in the USA [1]. Before 1970, plant patents were typically denied due to plant inventions not meeting the required legal criteria for patentability. A key argument against patenting plant breeder's products, including modified ones, was that they lacked creativity, a crucial requirement for patents, and were seen as products of nature with minimal human intervention. Additionally, IPR protection for agricultural products was delayed due to the belief that they lacked industrial applicability. Novel varieties of plants with improved yields, better quality and enhanced resistance to various stresses are developed due to continued breeding endeavors and selective inheritance, which is expensive, time-consuming, and labor-intensive. Therefore, legal protection for plant breeding and innovation in the agricultural sector creates an opportunity for incentive for this investment. In India, the Patent Act 1970 and its subsequent amendments provided patents for agricultural tools and machinery or the processes of development of agricultural chemicals. The Protection of Plant Varieties and Farmers' Rights (PPV&FR) Act, which was passed by the Indian government in 2001, was the first intellectual property rights law focused on plant varieties that recognized and protected the rights of farmers and breeders who maintain traditional landraces (The Gazette of India Extraordinary, 2001, and The Gazette of India, 2003). The Act gives farmers the freedom to save, use, sow, re-sow, exchange, share, or sell their produce, including seeds from a protected variety, as long as they do not brand or package the variety for commercial use. At the same time, breeders have the sole right to commercially produce, sell, market, distribute, and export their protected varieties. Furthermore, certain agricultural products grown in a particular region explicit the importance of GI and guarantee more locally produced, authentic, and superior food items in a competitive agricultural food market. Therefore, it allows farmers to sell their goods, increase their revenues and boost the local economy.

Medicinal plants have always been used in traditional medicine practices to treat a condition or disease since prehistoric times [2]. This generates interest among individuals and companies, encouraging them to make use of these natural resources. Unfortunately, the widespread exploitation of medicinal plants has led to a significant decline in their availability along with the threats of biopiracy and breaches of customary laws. The cultivation of medicinal plants has undergone significant transformations in the recent decades. As a result, there is a clear need for legal protection and rights in this area. IPR helps protect medicinal plants, as well as indigenous communities, individuals, and companies, from losses due to overexploitation. In this chapter, we have explained the IPR scopes in agriculture and discussed the different challenges for smart agriculture of medicinal plants.

15.2 IPR Scopes in Agriculture

The use of IPRs (plant breeders rights, plant patents, and utility patents) in agriculture has increased in recent decades for several reasons. (i) changes in the quantity and quality of demand for agricultural products, which resulted in changes in their production; (ii) technological advancements, such as the expansion of biotechnology applied to agriculture, which has increased private investments and changed innovation activities; (iii) the TRIPS agreement, which mandated patent protection for plant varieties or a sui generis system, as well as patent protection for other related products like microorganisms. Therefore, using an IP protection index for the agricultural sector recently developed by [3], this study looks at the effects of strengthening IPR systems in the agricultural sector for the post-TRIPS period (1995–2011) on traded volumes, bilateral trade flows, and trade margins for a set of 60 countries, comprising 28 developed and 32 developing nations.

15.2.1 Patents: Protecting New Plant Varieties and Agricultural Innovations

Farmers frequently create new plant varieties in response to market demands to produce larger or more attractive fruit, stronger, disease-resistant plants, or larger and faster harvests. In the past, it was believed that their whole effort in agricultural research and development was put forth for the good of society rather than private profit. The demand to extend intellectual property protection to agriculture in developing nations has often been met with opposition from those who think these innovations are lagging and therefore do not deserve private protection. Investors have countered with claims for the protection of plant varieties, breeders' rights, and farmers' rights. The most challenging endeavor for a breeder is to create a superior plant variety. To impact different characteristics or traits, a vast number of genes interact in intricate and unknown ways. To create such an enhanced plant variety, it is necessary to study several plants in various growth conditions and seasons. Establishing a plant variety is very difficult, expensive, and takes several years [4].

15.2.1.1 Need for Plant Variety Production

The global loss of agricultural harvests in the wake of World War II led to a need for increased agricultural production. As a result, agricultural technology improved. The goal of the inventors and those with traditional or indigenous knowledge was to protect their interests at every stage. Legal action was thus taken to protect the rights of indigenous knowledge holders, breeders, and inventors. Patent protection was granted to plant varieties to improve agricultural quality, honor the work of researchers, and produce products like crops resistant to pests. The goal of an IPR policy is to promote scientific advancement by temporarily preserving inventors' exclusive rights to their unique discoveries. Promoting innovation and creativity is the goal.

Plant varieties are developed by years of selected inheritance of traits that offer better resistance to such plant variations, higher quality, and superior yields. Acquiring high-performing cultivars requires the development of newer plant production techniques. These improved high-performing plant varieties have contributed significantly to the growth in agricultural productivity in many regions of the world. They are also essential for raising agricultural revenue and promoting overall economic development. Because it takes a lot of time and money to produce plants, breeding efforts cannot be sustained unless there is a chance of high returns. To promote the development of novel plant varieties for the good of society, an efficient mechanism for protecting plant varieties must be provided. Plant breeders should be protected against unapproved exploitation and given the chance to get just compensation for their labor, investment, and hard work in breeding. The incompatibility of the plant types with the legal requirements to fulfill the patent law standards resulted in technical issues. This led to the creation of a system that was unique to the traits of different plant varieties and the needs of farmers, merchants, and breeders. It was predicted that the use of IPRs in agriculture would boost private sector development and transfer of relevant technology, leading to agricultural development and increased food production [5]. The goal of IPRs is to encourage ideas and developments that will make them useful. For at least a crucial time, they are supposed to provide the public and the rest of the industry with information about the breakthrough that would otherwise be kept secret [6].

15.2.2 Patents Grant Inventors Exclusive Rights to Their Inventions and Provide the Right

Patents provide inventors the exclusive right to their creations, guaranteeing them legal protection against illicit use, manufacture, or sale for a certain amount of time, usually 20 years from the date of filing. This exclusive right is a vital motivator for companies and inventors because it enables them to profit monetarily from their R&D expenditures and innovative endeavors. Through the protection of these rights, patents promote technical development and a market that is competitive and constantly innovating. Inventors may be deterred from devoting time and money to creating new technologies if they lack patent protection because they worry that their ideas may be used without their consent [7].

Promoting public access to innovations while simultaneously fostering creativity is one of the core tenets of the patent system. In return for exclusive rights, inventors are required to make their innovations publicly available so that others can build upon and learn from them. This prerequisite makes sure that industries can develop together and that society gains from new knowledge. The inventions eventually become public domain after patents expire, enabling unrestricted usage by everyone and fostering future advancements in technology and the economy. In industries like information technology, engineering, and medicine, this cycle of innovation and knowledge exchange is crucial to advancement. Furthermore, by encouraging investment and entrepreneurship, patents significantly contribute to economic growth. Patents are essential for businesses and individual inventors to draw in investors, establish alliances, and get capital for more research and development. Patent protection is especially advantageous for startups as it gives them a competitive advantage against larger, more well-funded companies. Furthermore, patents help create jobs by incentivizing the industry to spend money on R&D, which results in the launch of new companies and the growth of already-existing ones [7].

15.2.2.1 The Three Main Categories of Patent

Utility, design, and plant patents are the three main categories into which patents often fall. Each has a specific function in protecting various innovations.

(i) **Utility Patent**

The most prevalent kind of patents are utility patents, which include novel and practical discoveries or innovations about machinery, processes, manufactured goods, or material compositions [8]. These patents are crucial in sectors like technology, engineering, and medicine because they safeguard an invention's functional elements. Many pharmaceutical firms, for instance, get utility patents for recently created medications, which enables them to recoup their R&D expenses before the production of generic copies [9].

Three essential requirements must be fulfilled by an innovation to receive a utility patent includes, 'novelty' showing the invention must be brand-new and unreported before; 'usefulness' which means that the product must have a real-world application and lastly 'non-obviousness' that is it shouldn't be a clear advancement over current technologies [10].

(ii) **Design Patent**

Patents of this kind are granted for unique, inventive, and elaborate designs applied to manufactured goods. The owner of this kind of patent prevents others from making, using, or commercializing the design. This kind of patent is granted if the design is inseparable from its goal.

Instead of protecting an invention's utilitarian qualities, design patents cover its decorative or artistic elements. In sectors like fashion, automobiles, and consumer electronics, where a product's aesthetic appeal has a big influence on its marketability, these patents are essential [11]. For instance, Apple Inc. has protected the appearance of its iPhones with many design patents, making it impossible for rivals to imitate their distinctive style [12]. Design patents only cover an innovation's outward design and usually endure for 15 years from the date of issuance, in contrast to utility patents, which protect how an invention function [13].

Despite their advantages, design patents have several challenges and drawbacks. Firstly, limited Scope, i.e., a design patent does not cover a product's operation; it only covers its aesthetic features [14]. Secondly, Patent Infringement Disputes, for example, expensive legal disputes over design patents in Apple's case against Samsung over identical smartphone designs [15]. Finally, International Variations which is complicated by the disparate design of protection laws in different nations [16].

(iii) **Plant Patent**

New plant varieties that have undergone asexual reproduction are covered under plant patents. These patents provide the creators of novel plant species like crops resistant to disease or unusually attractive flowers with control over how they are propagated and marketed. Agricultural firms, for example, frequently patent genetically engineered plants that are more resilient to environmental factors. Improvements in food production and sustainability have been made possible by the growth of plant patents, which has had a major impact on the agricultural sector [17].

15.2.3 Plant Breeder's Rights: Ensuring Breeders' Rights Over New Plant Varieties

Plant Breeder's Rights (PBRs), often referred to as Plant Variety Protection (PVP), are a type of intellectual property that gives breeders sole authority over the sale and propagation of newly created plant varieties. These rights are intended to encourage plant breeding innovation by enabling breeders to profit from their innovations and recover their investments. Plant varieties need to be unique, stable, uniform, and distinct to be eligible for PBRs. Stability ensures that the variety's key traits will not change throughout successive generations; uniformity ensures consistency across specimens; distinctiveness necessitates a clear differentiation from existing varieties; and novelty ensures that the variety has not been previously exploited [18].

15.2.3.1 Essentially Derived Varieties and Breeder's Exemption

Another important aspect of PBRs is the breeder exemption, which permits other breeders to use protected varieties to produce new ones without asking for the original breeder's consent. In plant breeding, this exception encourages continued creativity and genetic variety. The concept of "essentially derived varieties" (EDVs), however, was introduced by the 1991 UPOV Act and limits the breeder's exemption by prohibiting the commercialization of new varieties that are mostly derived from an original protected variety without significant modification [19].

15.2.3.2 Economic Effects

Studies have examined the economic effects of PBRs and found that robust PBR systems can encourage more funding for plant breeding initiatives. For example, research examining the economic impact of PBR discovered that nations with strong PBR frameworks saw greater levels of private sector investment in plant breeding, which resulted in the creation of better plant varieties and increased agricultural output [20].

15.2.3.3 Criticisms and Challenges

PBRs have been criticized for their efficacy and applicability in modern agriculture, despite their advantages. PBRs may not offer breeders enough security, according to some, particularly in nations with lax enforcement. PBRs' suitability for safeguarding contemporary plant advances has also been questioned in light of the development of biotechnology and the patenting of genetically modified organisms (GMOs). The existing PBR system may not be up to date in addressing the complexity of contemporary plant breeding, according to a descriptive and empirical evaluation of PBRs in Australia. It also raised concerns about how well it encourages innovation [21].

15.2.4 GI: Protecting Products Specific to Certain Regions

GIs are distinguishing markers that designate a product as coming from a particular place where its reputation, quality, or other attributes are inextricably tied to its place of origin. GIs are frequently linked to agricultural products, although they can also apply to industrial and handicraft goods. French champagne, Indian Darjeeling tea, and Italian Parmigiano Reggiano cheese are a few examples [22].

15.2.4.1 Key components of Geographical Indications:

Product Specification A thorough explanation of the features, manufacturing processes, and place of origin of the product. Link to Origin: Proof of the relationship between a product's attributes or reputation and its geographic setting, including both natural (such as soil and climate) and man-made (such as conventional manufacturing methods) aspects.

Control Mechanism Developed protocols, frequently overseen by regulatory agencies or producer groups, which guarantee constant quality and conformity to specified manufacturing standards.

Legal Protection Framework National and international regulations that protect geographical indications (GIs) against infringement or misuse, guaranteeing that the protected term may only be used by authorized producers in the defined area [23].

15.2.5 Trade Secrets: Safeguarding Proprietary Agricultural Practices

In agriculture, trade secrets are an essential part of intellectual property rights (IPR) because they allow organizations to safeguard exclusive knowledge, methods, and procedures that provide them a competitive advantage. Trade secrets, as opposed to patents, are exempted from public disclosure requirements, which make them ideal for protecting exclusive farming techniques [24].

15.3 Important Features of Trade Secrets in Agriculture

1. **Definition and scope:** Formulas, procedures, designs, instruments, and compilations of information that are not widely known or easily ascertainable are all examples of the wide range of proprietary information that is included in the category of trade secrets. In the agricultural industry, this might entail customized

equipment designs, exclusive crop management strategies, or exclusive breeding methods.

2. **Law Enforcement Framework:**
   The Uniform Trade Secrets Act (UTSA), which defines a trade secret as information that has economic value due to its lack of public knowledge and is subject to reasonable attempts to keep it secret, offers a legal foundation for trade secret protection in the USA. Federal civil lawsuits for trade secret misuse are also permitted by the Defend Trade Secrets Act (DTSA) of 2016 [25].
3. **Safety precautions:**
   - Agricultural organizations frequently use strategies like non-disclosure.
   - Agreements (NDAs) with partners, contractors, and workers to protect trade secrets.
   - Restricted availability of private data.
   - Routine information flow monitoring and audits.
   - Instruction on confidentiality procedures for employees.
4. **Economic Espionage Concerns:**
   There have been cases of economic espionage in the agriculture industry, when foreign entities target confidential agricultural data, such as plant genomes and breeding plans. Such actions have increased, according to the FBI, highlighting the necessity of strong security measures.
5. **Challenges in the Digital Era:**
   The preservation of trade secrets faces additional difficulties with the rise of big data and digital agriculture. Farm data, including production figures, soil health indicators, and input use, is very valuable. Carefully navigating the current trade secret regulations is necessary to ascertain ownership and guarantee the secrecy of such material.
6. **Balancing Collaboration and Protection:**
   Cooperation among farmers may drive productivity and creativity, but it also requires careful supervision to avoid inadvertent trade secret exposure. In addition to highlighting the advantages of sharing information, historical examples like the Grange movement also emphasize how crucial it is to protect unique processes [26].
7. **Global Considerations:**
   Trade secret protection differs among countries, with domestic laws being influenced by accords such as the Trans-Pacific Partnership (TPP). For agricultural organizations that operate internationally, it is essential to comprehend these international frameworks [27].

15.4 Smart Agriculture Technologies

15.4.1 Overview of Smart Agriculture and Its Impact on Medicinal Plant Cultivation

Precision agriculture, another name for smart agriculture, incorporates cutting-edge technologies into farming methods to improve sustainability, resource efficiency, and crop productivity. By increasing growing conditions, enhancing the quality of active compounds, and guaranteeing a steady supply chain, smart agriculture provides substantial advantages in the field of medical plant production.

15.4.2 Important Technologies in Medicinal Plant Smart Agriculture

15.4.2.1 Controlled-Environment Agriculture (CEA)

Temperature, humidity, and nutrient supply may all be precisely controlled with methods like hydroponics and aeroponics. Higher yields and better quality medicinal plants are the results of this management. Hydroponic systems, for example, have been demonstrated to improve the biochemical makeup and development of medicinal plants, providing long-term answers to agricultural problems made worse by climate change [28].

15.4.2.2 Vertical Farming

Vertical farming reduces the requirement for huge land areas and transportation expenses by allowing the production of medicinal plants in urban areas through the use of vertical space and controlled surroundings. Additionally, year-round cultivation is made possible by this technique, guaranteeing a consistent supply of therapeutic plants [29].

15.4.2.3 Sensors with the Internet of Things (IoT)

By using sensors to track soil moisture, nutrient levels, and plant health in real time, watering and fertilization may be done precisely, saving resources and promoting plant development. Throughout their growing cycle, medicinal plants are given the best care possible thanks to this real-time monitoring [30].

15.4.2.4 Machine Learning and Artificial Intelligence (AI)

AI systems may evaluate data from several sources to forecast plant diseases, schedule harvests, and enhance supply chain operations. Convolutional Neural Networks (CNNs), for instance, have been used in agriculture to estimate yield and identify plant diseases, both of which are critical for preserving the quality of medicinal plants [30].

15.4.3 Impact on Medicinal Plant Cultivation

15.4.3.1 Improved Yield and Quality

Smart agriculture techniques increase yields and enhance concentrations of active chemicals in medicinal plants by optimizing resource utilization and environmental conditions. It has been demonstrated that controlled-environment systems, such as vertical farms, increase the production of medicinal plants by suggesting environmental management techniques that boost phytochemical content and growth.

15.4.3.2 Sustainable Practices

By using less water, fewer chemicals, and a smaller carbon impact than conventional agricultural techniques, smart agriculture encourages sustainable farming. For example, aeroponics, a technique that consumes less water and enables precise fertilizer management by suspending plant roots in the air and misting them with nutrient solutions, results in better plants and a smaller environmental effect [31].

15.4.3.3 Supply Chain Efficiency

Including technology in the growing process guarantees a steady and traceable supply of medicinal plants, which is essential for satisfying customer demands and legal requirements. A research study that examined the Lucanian experience made clear how crucial it is to integrate technology at every stage of production in order to improve the supply chain for medicinal plants [31].

15.5 Examples of Smart Technologies: Precision Farming, IoT, and AI in Agriculture

One of the technologies that is expanding the quickest in the modern world is artificial intelligence (AI). AI is being widely used in many different fields, from environmental monitoring to robotics development. Unquestionably, artificial intelligence (AI) has applications in healthcare systems, self-driving cars, satellite imaging, landscape mapping, and climate change monitoring [32]. AI has accelerated the global development of robots, which has improved people's quality of life and made living easier [33]. AI-mediated automation in several industries, including agriculture, will provide high-quality goods with fewer adverse environmental effects. AI uses a range of intelligent sensors to process data more accurately in real time [34]. AI uses a range of intelligent sensors to process data more accurately in real time. As a result, AI has increased any device's performance multiple times while producing results that are more sensitive and accurate. The Internet of Things (IoT) is a cutting-edge technology that was created by combining AI with smart sensors. IoT serves as a hub for wireless systems, and AI processes data in real time to produce relevant results. This technique is now used in nearly every field of study and science. Digital technologies based on the Internet of Things have the potential to produce novel models for improved consumer goods manufacturing, including agricultural products [35].

Furthermore, AI and IoT provide great opportunities for real-time monitoring of several parameters, including temperature, humidity, radiation, water content, pollution, and soil quality. Better farm job management and quick decision-making based on data are two ways that smart farming helps with accuracy [36]. Together with a variety of smart sensors, the Internet of Things helps farmers manage their crops and vegetables more effectively over a wider area in less time. Furthermore, IoT might help in understanding when and how pesticides and fertilizers should be applied in the field. Using pesticides and fertilizers excessively can contaminate the environment and pose health risks [37]. The Internet of Things has the potential to preserve farmers' incomes and cut down on the waste of different resources [38–42]. Additionally, by enabling precise crop disease diagnostics, IoT, and smart sensors may be integrated with a range of current technologies to further improve agricultural output [43].

15.6 IoT and Agriculture

In order to monitor soil health, water contents, crop quality, and crop output on a site, real-time data collection and analysis is made possible by the Internet of Things (IoT), smart sensors, and artificial intelligence (AI) [34]. Smart farming, which is defined by higher production, is taking the place of conventional agricultural practices thanks to the Internet of Things and smart sensors. In agriculture, IoT-enabled

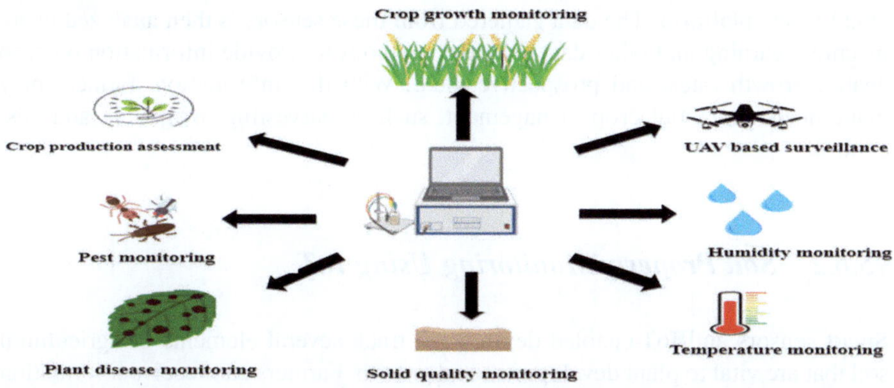

Fig. 15.1 Uses of smart sensors and integrated IoT for precision farming. Smart sensors based on the Internet of Things can precisely track environmental variables, including humidity, wetness, and temperature. Certain sensors use water content and nitrate levels to evaluate the condition of the soil. With the use of a high-resolution camera and a GPRS system, plant diseases and insect pests may be identified. Crop development and agricultural land morphology may be monitored with the use of UAV-based monitoring. Mass flow sensors that are automated can be used to estimate crop productivity

technical methods also help assess crop quality, fertilizer requirements, soil fertility, soil health, and soil erosion [32]. At the moment, most of the equipment developed and utilized is for gathering agricultural and crop data, tracking environmental data, and tracking animal movements. In agricultural settings, physical property sensors, biosensors, and electromechanical sensors are all very important. While physical property sensors use devices sensitive to changes in the physical environment, biological sensors can detect biologically sensitive elements from the outside world Fig. 15.1 [32].

15.6.1 IoT-Based Crop Monitoring

Crop monitoring has experienced a paradigm shift from an experience-based qualitative undertaking to a quantitative and data-driven one with the introduction of IoT in soil, microclimate, and crop sensing. IoT-based solutions allow farmers to precisely monitor the growth and condition of their crops [44]. It also aids field workers in the real-time assessment of insect assaults and plant diseases. The real-time data gathering and processing offered by IoT-tagged sensors for intelligent management of agricultural cultivation, fertilizer application, irrigation, and plant environments may be extremely advantageous to farmers and researchers [45].

Sensors placed in the field to gather information on a range of environmental parameters, such as temperature, humidity, soil moisture, and nutrient levels, are the main components of Internet of Things-based smart crop monitoring systems. The frequent networking of these sensors enables real-time data transfer to a centralized cloud-based platform. The data gathered from these sensors is then analyzed using machine learning and other data analytics methods to provide information on crop health, growth rates, and prospective yield. With this information, farmers may make decisions about crop management, such as harvesting, irrigation, and pest control.

15.6.2 Soil Property Monitoring Using IoT

Smart sensors and IoT-enabled devices can track several elements of agricultural soil that are vital to plant development and growth. Farmers can receive information from IoT about soil moisture, temperature, and fertilizer levels [46]. Farmers may remotely access and calibrate this data to take adequate precautions against crop illnesses and insect attacks. Consequently, smart farming may be beneficial to farmers [47]. Farmers can also take preventative action by using IoT to monitor soil pH and rhizosphere zone multiparameters from a distance [48]. It is also possible to track the growth and development of plants in different types of soil [48].

15.6.3 Smart Farming Sensors

In smart farming, the usage of several sensors is becoming increasingly important. These sensors might be useful for automated agricultural harvesting, environmental condition monitoring, and crop yield assessment. Sensors that have been integrated with electronics are known as smart sensors. Smart sensors can more reliably and automatically gather a range of environmental data and other pertinent information in agricultural settings. These sensors may then store this data on drives. Microprocessors are used to process this data to comprehend and analyze them. The internet is used to transfer data, and smart sensors are essential parts of the Internet of Things. It consists of an actuator and wireless network system with a few to hundreds of nodes linked to sensor hubs Fig. 15.2 [49].

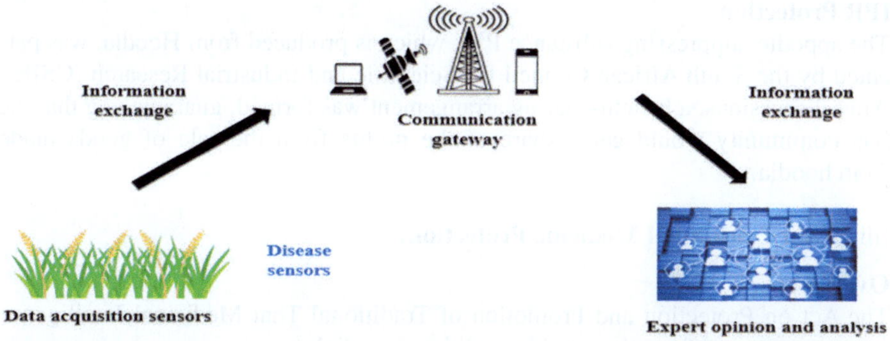

Fig. 15.2 Combining human networks, agriculture, and Internet of Things devices. IoT-based sensors used in agricultural areas detect several important variables, including temperature, humidity, nutrient levels, etc., and transmit the information to communication gateways, including satellites, cloud servers, and computer systems. Human networks then employ these IoT-enabled devices for analysis, decision-making, and other necessary agricultural tasks

15.7 Case Studies of Medicinal Plant

15.7.1 Examples of Successful IPR Protection in Medicinal Plants

In order to protect inventions pertaining to medicinal plants and to guarantee that inventors and communities are acknowledged for their contributions, intellectual property rights, or IPR, are essential. Here are some noteworthy instances of efficient IPR defense in the medicinal plant industry:

(i) **Traditional Knowledge Digital Library (TKDL):**

Overview
In order to stop the exploitation of its traditional medical knowledge, India founded the TKDL. Patent examiners from all around the world can access the comprehensive database that records conventional compositions and methods.

Impact
The TKDL has played a crucial role in halting the issuance of patents on already-existing traditional knowledge by presenting proof of prior art. It has, for example, successfully contested patent applications that sought to assert ownership of conventional Ayurvedic formulas [50].

(ii) **Hoodia Plant (Hoodia Gordonii): Background:**

The San people have long utilized the Hoodia plant, which is indigenous to Southern Africa, to quench their thirst and hunger.

IPR Protection
The appetite-suppressing substance P57, which is produced from Hoodia, was patented by the South African Council for Scientific and Industrial Research (CSIR). After discussions, a benefit-sharing arrangement was formed, guaranteeing that the San community would get a share of the profits from the sale of goods made from hoodia.

(iii) **Thai Traditional Medicine Protection:**

Overview
The Act on Protection and Promotion of Traditional Thai Medicinal Intelligence protects texts and formulae used in traditional medicine.

Impact
By limiting access to research, development, and manufacturing of medications utilizing traditional Thai medical knowledge to registered persons or businesses, this law protects indigenous customs and fosters fair benefit-sharing.

(iv) **Cannabis Plant Varieties: Background:**

The creation of novel cannabis plant kinds has exploded as a result of the changing legal status of cannabis in different nations.

IPR Protection
To safeguard new cannabis strains, breeders have made use of their rights as plant breeders under international frameworks like the International Union for the Protection of New Varieties of Plants (UPOV). This encourages innovation in the production of medicinal cannabis by guaranteeing breeders exclusive rights to their new cultivars [51].

15.8 Challenges Faced in Protecting Traditional Knowledge and Biodiversity

There are many obstacles to overcome in order to preserve biodiversity and traditional knowledge, especially when it comes to intellectual property rights and the maintenance of indigenous customs. Important difficulties include:

(i) **Inadequate frameworks and legal complexity:**

Traditional knowledge is susceptible to appropriation due to the lack of comprehensive international legal frameworks, which makes its protection more difficult [52].

(ii) **Issues with Documentation and Preservation:**

Traditional methods deteriorate due to a lack of systematic recording, which makes it challenging to enforce intellectual property rights [53].

(iii) **Unauthorized Use and Biopiracy:**

Commercial enterprises' unauthorized use of biological resources and related traditional knowledge is still a major problem [54].

(iv) **Bureaucratic Hurdles:**

Excessively intricate administrative procedures can impede scientific investigation and the successful preservation of biodiversity [54].

(v) **Climate Change Impacts:**

Traditional knowledge systems are upset by environmental changes, which reduces their dependability and jeopardizes their ability to be passed on to future generations [55].

(vi) **Cultural Sensitivity and Intellectual Property Conflicts:**

Intellectual property laws' exploitation of revered traditional knowledge can result in moral quandaries and cultural insensitivity [56].

(vii) **Future Perspectives:**

The patent system is nevertheless essential in the context of research and development, especially in industries like biopharmaceuticals that need large upfront investments. By providing transitory property rights, it promotes innovation and the creation of new technology and solutions. The system must, however, change to meet the particular difficulties presented by new technologies and guarantee that it keeps promoting innovation in an efficient manner.

15.9 Emerging Trends in IPR and Smart Agriculture

Rapid technical breakthroughs and changing legal frameworks are driving the confluence of smart agriculture and intellectual property rights (IPR), which is changing the agricultural landscape. The extension of IPR rights in agricultural seeds, especially in the US, is one noteworthy development. New crop varieties have been produced as a result of higher research and development expenditures prompted by improved IPR regulations. But this has also led to increased market concentration, with large companies controlling seed markets, which raises questions about accessibility and competition [57].

Another significant advancement in smart agriculture is the emergence of digital technology, particularly artificial intelligence (AI). The agricultural industry has seen a sharp rise in patent applications pertaining to AI, which is indicative of the increasing use of AI in farming methods. Data analytics, Internet of Things (IoT) devices, and precision farming equipment are all part of this digitalization, which brings with it new IPR issues, especially when it comes to safeguarding software inventions and data-driven approaches [58].

15.10 Potential Solutions for Enhancing IPR Protection in Medicinal Plants

Improving medicinal plant Intellectual Property Rights (IPR) protection is essential for preserving traditional knowledge, encouraging innovation, and stopping biopiracy. Building thorough databases that record conventional medical information is one such tactic. This strategy is demonstrated by India's Traditional Knowledge Digital Library (TKDL), which diligently documents traditional pharmaceutical formulas to guard against illegal patents and guarantee that such information stays in the public domain.

It is also crucial to modify current IPR frameworks in order to properly incorporate conventional medical knowledge. Legal changes that acknowledge and safeguard indigenous knowledge systems are necessary, as seen by attempts to incorporate traditional medicine into the patent system in China, for example. Additionally, international cooperation is essential to bolstering medicinal plant IPR protection. Countries may handle biopiracy issues together and guarantee fair benefit-sharing with indigenous groups by coordinating legislative frameworks and exchanging best practices. This cooperative approach promotes worldwide respect for biodiversity and cultural heritage while also strengthening the preservation of information about therapeutic plants [59].

To successfully safeguard the intellectual property linked to medicinal plants, putting these techniques into practice calls for a multipronged strategy that incorporates technical advancements, legislative reforms, and international collaboration.

15.11 Conclusion

Smart agriculture's incorporation of intellectual property rights (IPR), especially about medicinal plants, offers both substantial potential and difficult obstacles. Intellectual property rights (IPR) in agriculture have been essential in promoting investment and innovation in the sector, from its initial focus on industrial uses to its recent extension to encompass plant varieties. Legal frameworks such as India's Protection of Plant Varieties and Farmers' Rights Act, 2001, promote the development of novel plant varieties while conserving traditional knowledge to balance the interests of farmers and breeders. The advent of smart agriculture technologies like precision farming, the Internet of Things, and artificial intelligence (AI) has revolutionized the production of medicinal plants. These innovations enhance efficiency, yield, and quality, but also introduce new IPR considerations. It becomes crucial to protect proprietary technologies and data, which calls for a sophisticated approach to IPR that takes into account the rapid pace at which technology is developing.

However, there are still problems, particularly with finding a balance between innovation and the preservation of old knowledge. The instances of biopiracy, in which indigenous knowledge and biological resources are used without just compensation, highlight the necessity of strong legal safeguards and moral research procedures. It is still crucial to make sure that Indigenous populations receive fair benefit sharing. In the future, it will be crucial to create IPR frameworks that can be modified to accommodate new developments in smart agriculture. This involves addressing challenges connected to digital sequence information and ensuring that the benefits of innovation are shared equally among all stakeholders. By fostering collaboration between policymakers, researchers, industry players, and indigenous communities, it is possible to create a sustainable and equitable future for medicinal plant cultivation within the realm of smart agriculture. In conclusion, IPR stimulates innovation in smart agriculture, especially about medicinal plants, but it is crucial to carefully consider the difficulties that come with it. By maintaining traditional knowledge while protecting new discoveries and guaranteeing equitable benefit-sharing, we can fully utilize smart agriculture to support economic growth and societal well-being.

References

1. Bugos GE, Kevles DJ (1992) Plants as intellectual property: American practice, law, and policy in world context. Osiris 7:75–104
2. Prasathkumar M, Anisha S, Dhrisya C, Becky R, Sadhasivam S (2021) Therapeutic and pharmacological efficacy of selective Indian medicinal plants—a review. Phytomedicine Plus 1(2):100029
3. Campi M, Nuvolari A (2015) Intellectual property protection in plant varieties. A new worldwide index (1961–2011). Res Policy 4(44):951–964
4. Sethi MR, Singh V (2023) Navigating the complexities of ip in agriculture: milestones and challenges. Russ Law J 11(3):1619–1632
5. Singh KK (2023) Patent and pandemic: exploring duties, obligations and responsibilities. In: Relevance of duties in the contemporary world: with special emphasis on Gandhian thought Singapore. Springer, Singapore, pp 367–389
6. William S (2024) The protection of plant varieties and farmers rights act, 2001 and its impact on farmers' rights in India: a study with special reference to Kerala
7. Ebrahim TY (2024) Corpus linguistics at the US Patent and Trademark Office. BYU L Rev 50:49
8. Pareek V, Bist R, Pareek H (2019) Intellectual property rights of inventions in biological science. Hospice Palliative Med Int J 3:178–190
9. Dash DK, Vaiswade R, Gupta G (2023) A review on the Indian patent system and its implication on the pharmaceutical industry. J Health Sci Med Res 41(3):2023926

10. Organisation for Economic Co-operation and Development (OECD) (2022) Patent statistics and their economic impact
11. Kim Y (2023) Design patents and market competition. Harv Bus Rev
12. Lee J, Cohen M (2023) Apple's patent strategy: a case study in design protection. Stanford Law Rev
13. United States Patent and Trademark Office (USPTO) (2023) Types of patents and their legal framework
14. European Patent Office (EPO) (2023) Design patents and their global impact. Retrieved from www.epo.org
15. Graham S, Mowery D (2023) Patent law and market competition in technology
16. World Intellectual Property Organization (WIPO) (2023) Understanding design patents and their applications. Retrieved from www.wipo.int
17. Bagade OM, Pujari RR, Vanave MD, Shete AM, Kharat PP (2024) Evolving pace of patent in India and its corollary in past, present and future. Adv Drug Deliv Rev 2:503–521
18. Bostyn SJ (2021) Towards a fair scope of protection for plant breeders' rights in an era of new breeding techniques: proposals for a modernization of the essentially derived variety concept. Agronomy 11(8):1511
19. Lawson C (2023) Propagating materials and harvested materials: clarifying the scope of plant variety or breeder's rights. J Intellect Property Law Practice 18(9):655–672
20. Hervouet A, Langinier C (2018) Plant breeders' rights, patents, and incentives to innovate. J Agric Resour Econ 43:118–150
21. Hervouet A, Langinier C (2015) Plant breeders' rights, patents and incentives to innovate. J Agric Resour Econ 49
22. Török Á, Jantyik L, Maró ZM, Moir HV (2020) Understanding the real-world impact of geographical indications: a critical review of the empirical economic literature. Sustainability 12(22):9434
23. Cardoso VA, Lourenzani AEBS, Caldas MM, Bernardo CHC, Bernardo R (2022) The benefits and barriers of geographical indications to producers: a review. Renew Agric Food Syst 37(6):707–719
24. Mödlhamer C (2021) Intellectual property rights in preferential trade agreements. Doctoral dissertation, Doctoral Dissertation, University of Salzburg. https://eplus.unisalzburg.at/obvusbhs/content/titleinfo/7332917/full.pdf
25. Leopold B (2018) Forecasting change: examining the future of agricultural data processors and ownership rights. J Corp Law 44:403
26. Linton K (2016) The importance of trade secrets: new directions in international trade policy making and empirical research. J Int Econ 1
27. Garg D, Alam M (2023) Smart agriculture: a literature review. J Manag Anal 10(2):359–415
28. Dsouza A, Dixon M, Shukla M, Graham T (2025) Harnessing controlled-environment systems for enhanced production of medicinal plants. J Exp Bot 76(1):76–93
29. Dsouza A, Newman L, Graham T, Fraser ED (2023) Exploring the landscape of controlled environment agriculture research: a systematic scoping review of trends and topics. Agric Syst 209:103673
30. El Sakka M, Mothe J, Ivanovici M (2024) Images and CNN applications in smart agriculture. Eur J Remote Sens 57(1):2352386
31. Goel RK, Yadav CS, Vishnoi S, Rastogi R (2021) Smart agriculture–urgent need of the day in developing countries. Sustain Comput Inform Syst 30:100512
32. Sishodia RP, Ray RL, Singh SK (2020) Applications of remote sensing in precision agriculture: a review. Remote Sens 12(19):3136
33. Shepherd M, Turner JA, Small B, Wheeler D (2020) Priorities for science to overcome hurdles thwarting the full promise of the 'digital agriculture' revolution. J Sci Food Agric 100(14):5083–5092
34. Ayaz M, Ammad-Uddin M, Sharif Z, Mansour A, Aggoune EHM (2019) Internet-of-Things (IoT)-based smart agriculture: toward making the fields talk. IEEE Access 7:129551–129583

35. Lajoie-O'Malley A, Bronson K, van der Burg S, Klerkx L (2020) The future(s) of digital agriculture and sustainable food systems: an analysis of high-level policy documents. Ecosyst Serv 45:101183
36. Wolfert S, Goense D, Sørensen CAG (2014) A future internet collaboration platform for safe and healthy food from farm to fork. In: 2014 annual SRII global conference, pp 266–273
37. Wang YS, Blackwell PG, Merkle JA, Potts JR (2019) Continuous time resource selection analysis for moving animals. Methods Ecol Evol 10(10):1664–1678
38. Sarkar S, Rajak P, Roy S (2018) Toxicological evaluation of a new lepidopteran insecticide, flubendiamide, in non-target drosophila melanogaster Meigen (Diptera: Drosophilidae). Iran J Toxicol 12(3):45–50
39. Ghanty S, Mandi M, Ganguly A, Das K, Dutta A, Nanda S, Biswas G, Rajak P (2022) Lung surfactant proteins as potential targets of prallethrin: an insilico approach. Toxicology and Environmental Health Sciences 14(1):89–100
40. Rajak P, Roy S, Pal AK, Paramanik M, Dutta M, Podder S, Sarkar S, Ganguly A, Mandi M, Dutta A, Das K (2021) In silico study reveals binding potential of rotenone at multiple sites of pulmonary surfactant proteins: a matter of concern. Curr Res Toxicol 2:411–423
41. Rajak P, Roy S, Podder S, Dutta M, Sarkar S, Ganguly A, Mandi M, Dutta A, Nanda S, Khatun S (2022) Synergistic action of organophosphates and COVID-19 on inflammation, oxidative stress, and renin-angiotensin system can amplify the risk of cardiovascular maladies. Toxicol Appl Pharmacol 456:116267
42. Rajak P, Roy S, Ganguly A, Mandi M, Dutta A, Das K, Nanda S, Ghanty S, Biswas G (2023) Agricultural pesticides–friends or foes to biosphere. J Hazard Mater Adv 10:100264
43. Navarro E, Costa N, Pereira A (2020) A systematic review of IoT solutions for smart farming. Sensors 20(15):4231
44. Kaloxylos A, Wolfert J, Verwaart T, Terol CM, Brewster C, Robbemond R, Sundmaker H (2013) The use of future internet technologies in the agriculture and food sectors: integrating the supply chain. Procedia Technol 8:51–60
45. FuBing (2012) Research on the agriculture intelligent system based on IOT. In: 2012 international conference on image analysis and signal processing, pp 1–4
46. Kiani F, Seyyedabbasi A (2018) Wireless sensor network and internet of things in precision agriculture. Int J Adv Comput Sci Appl 9
47. Navulur S, Prasad MG (2017) Agricultural management through wireless sensors and internet of things. Int J Elect Comput Eng 7(6):3492
48. Zhang LF (2014) Design of agricultural environmental parameters monitoring system based on internet of things. Appl Mech Mater 608:1115–1119
49. La Rosa R, Dehollain C, Costanza M, Speciale A, Viola F, Livreri P (2022) A battery-free wireless smart sensor platform with bluetooth low energy connectivity for smart agriculture. In: 2022 IEEE 21st Mediterranean Electrotechnical Conference (MELECON). IEEE, pp 554–558
50. Chaudhary A, Singh N (2012) Intellectual property rights and patents in perspective of Ayurveda. AYU (An International Quarterly Journal of Research in Ayurveda) 33(1):20–26
51. Willcox M, Diallo D, Sanogo R, Giani S, Graz B, Falquet J, Bodeker G (2015) Intellectual property rights, benefit-sharing and development of "improved traditional medicines": a new approach. J Ethnopharmacol 176:281–285
52. Hossain K, Ballardini RM (2021) Protecting indigenous traditional knowledge through a holistic principle-based approach. Nord J Hum Rights 39(1):51–72
53. Parween R, Marchant R (2022) Traditional knowledge and practices, sacred spaces and protected areas, technological progress: their success in conserving biodiversity. Conserv Sci Pract 4(5):12643
54. Reddy P, Lakshmikumaran M (2015) Protecting traditional knowledge related to biological resources: is scientific research going to become more bureaucratized? Cold Spring Harb Perspect Med 5(10):020974

55. Malhotra N (2024) A systematic literature review study on the social inclusion of the indigenous people. In: Sustainable pathways: the role of indigenous tribes and native practices in India's economic model. Emerald Publishing, pp 41–69
56. Hossain K, Ballardini RM (2021) Protecting indigenous traditional knowledge through a holistic principle-based approach. Nordic J Human Rights 39(1):51–72
57. Amani B, Ncube CB, Rimmer M (2024) Conclusion: 'blueprints for a better world': the future of intellectual property and sustainable development. In: The Elgar companion to intellectual property and the sustainable development goals. Edward Elgar Publishing, pp 582–610
58. Packin NG (2024). Emerging compliance in the generative decentralized era. Available at SSRN
59. Xia N (2023) Intellectual property protection for traditional medical knowledge in China's context: a round peg in a square hole. Med Law Rev 31(3):358–390

Chapter 16
Marketing Medicinal Plants in the Digital Age: Strategies for Soilless Cultivation Systems

Avika Subba and Meenakshi Thakur

Abstract The convergence of traditional medicinal plant knowledge with modern controlled environment agriculture represents a paradigm shift in how botanical therapeutics are cultivated, processed, and marketed globally. This chapter examines comprehensive marketing strategies specifically developed for medicinal plants grown through hydroponic and aeroponic systems, addressing the critical transition from conventional soil-based agricultural marketing to technology-driven approaches. While the global medicinal plant market demonstrates robust growth, traditional wild harvesting practices face sustainability challenges due to overexploitation and habitat degradation. Controlled environment agriculture systems offer compelling advantages including consistent quality, contamination-free production, year-round availability, and optimized phytochemical profiles. However, successful commercialization requires sophisticated marketing strategies that address unique challenges including consumer perception, premium pricing justification, and supply chain optimization. This chapter systematically analyzes market dynamics, consumer segmentation, and positioning strategies through established frameworks including SWOT analysis and the 7Ps marketing mix. The discussion encompasses digital marketing innovations, certification-based branding, traceability systems, and emerging business models that capitalize on the growing demand for clean-label, sustainably produced medicinal plants. Through detailed case studies and comparative analysis, this work provides actionable insights for entrepreneurs, policymakers, and investors navigating the intersection of agricultural innovation and market success.

Keywords Medicinal plants · Controlled environment agriculture · Marketing strategies · Soilless cultivation · Value chain · Sustainable agriculture · Market positioning

A. Subba (✉) · M. Thakur
Regional Cum Facilitation Centre Northern Region-I, NMPB, Ministry of Ayush GOI 1st Floor, CoEDG Building, Research Institute in Indian Systems of medicine (RIISM), Mandi, HP, India

National Medicinal Plants Board, New Delhi, India

P. Kumar, A. R. Warghat (eds.), *Transforming Medicinal Plant Agriculture*, Smart Agriculture 16, https://doi.org/10.1007/978-981-95-3963-5_16

16.1 Introduction

The global healthcare landscape has witnessed a remarkable renaissance in botanical therapeutics, driven by increasing consumer awareness of natural wellness solutions and growing scepticism toward synthetic pharmaceuticals. This transformation has been particularly pronounced following the COVID-19 pandemic, which catalyzed a fundamental shift in consumer behavior toward preventive healthcare and immune-supporting natural products (IBEF, 2024) [1]. Traditional medicinal systems including Ayurveda, Traditional Chinese Medicine, Unani, and indigenous healing practices worldwide have experienced unprecedented commercial validation, creating substantial market opportunities for innovative cultivation and marketing approaches.

India's position as a global leader in medicinal plant biodiversity and traditional knowledge systems presents unique advantages in this evolving market. The country's AYUSH sector has demonstrated exceptional growth trajectories, with manufacturing values increasing from USD 18 billion in 2020 to USD 24 billion by 2024, while the broader industry valuation exceeded USD 50 billion [1]. Export performance reflects this domestic growth, with AYUSH and herbal products generating USD 651.17 million in FY 2023–24, representing a 3.6% year-on-year increase. Market projections indicate continued expansion, with the sector expected to reach USD 200 billion by 2030, representing a compound annual growth rate of approximately 17% [2].

However, this growth trajectory faces significant sustainability challenges. The global medicinal plant trade has historically depended on wild harvesting, leading to overexploitation of natural resources and threatening biodiversity conservation [3, 4]. Climate change, habitat fragmentation, and increasing urbanization further compound these challenges, creating supply chain vulnerabilities that threaten long-term market stability. Approximately 880 medicinal plant species participate in India's domestic trade; yet sustainable cultivation practices remain limited, with wild collection continuing to dominate many high-value species [5].

Controlled environment agriculture, encompassing hydroponic and aeroponic cultivation systems, emerges as a transformative solution addressing both sustainability imperatives and market quality demands. These soilless cultivation methods enable precise control over growing conditions, resulting in consistent phytochemical profiles, elimination of soil-borne contaminants, and dramatic improvements in resource efficiency. Research demonstrates that controlled environment systems can achieve 80%–90% water savings compared to conventional cultivation while producing higher concentrations of bioactive compounds through environmental optimization [6, 7].

The strategic importance of controlled environment agriculture extends beyond production efficiency to encompass quality assurance critical for pharmaceutical and nutraceutical applications. Traditional soil-based cultivation often results in variable phytochemical content, pesticide residues, heavy metal contamination, and microbial adulteration—issues that compromise product safety and efficacy [8].

Soilless systems eliminate these concerns while enabling standardization requirements essential for modern medicinal plant commerce.

Despite these compelling advantages, successful commercialization of soilless-grown medicinal plants requires sophisticated marketing strategies that differ fundamentally from conventional agricultural approaches. Consumer acceptance must be cultivated through education about production benefits, premium pricing must be justified through demonstrated value propositions, and distribution channels must be optimized for specialized market segments. This chapter addresses these challenges by providing comprehensive marketing frameworks specifically adapted for the unique requirements of controlled environment medicinal plant production, offering practical guidance for stakeholders seeking to capitalize on the convergence of agricultural innovation and market opportunity.

16.2 Global Market Dynamics and Trade Patterns

16.2.1 Market Size and Growth Trajectories

The global medicinal plant market demonstrates exceptional dynamism, reflecting increasing consumer preference for natural health solutions and growing integration of traditional medicine with modern healthcare systems. International trade data reveals remarkable expansion, with export and import values increasing by 97.8% and 98.1%, respectively, from 2010 to 2023, reaching USD 4.18 billion and USD 4.25 billion [9]. This growth pattern indicates robust global demand and expanding supply chain networks.

India has emerged as a dominant exporter in the global medicinal plant sector, achieving a 240% increase in export value over recent years, second only to China [10, 11]. This growth is supported by India's biodiversity, traditional medicinal systems, and expanding processing infrastructure [12, 13].

16.2.2 Trade Classification and Product Categories

International trade in medicinal plants operates under specific classification systems, primarily HS Code 1211, covering plants and plant parts used in pharmacy, perfumery, insecticides, and similar applications. Comprehensive analysis of trade databases including COMTRADE and Trade Map reveals that global trade encompassed approximately 702,813 tons valued at USD 3.60 billion in 2014, representing significant volume and value growth over the analyzed period [14].

The industry demonstrates consistent expansion with average annual growth rates of 2.4% in volume and 9.2% in value, reflecting product upgrading and a shift toward value-added systems that emphasize traceability and quality control [14].

This divergence between volume and value growth suggests opportunities for value-added production systems, particularly those emphasizing quality, consistency, and traceability characteristics inherent in controlled environment agriculture.

16.2.3 Emerging Market Opportunities

Market analysis reveals significant opportunities in specialized segments including organic certification, pharmaceutical-grade raw materials, and standardized extracts. The World Health Organization estimates the global herbal products market at USD 6.2 billion currently, with projections reaching USD 5 trillion by 2050 [5]. These projections reflect increasing mainstream acceptance and integration of botanical therapeutics into conventional healthcare systems.

Controlled environment agriculture (CEA) systems are particularly well-positioned to capture premium market segments through demonstrated advantages in quality consistency, contamination control, and environmental sustainability attributes that align with evolving consumer preferences and regulatory requirements in developed markets, creating substantial competitive advantages [15–17]. These emerging market opportunities, particularly in premium pharmaceutical-grade raw materials and standardized extracts, create an imperative for production systems that can deliver consistent quality, contamination-free products, and verifiable traceability. Controlled environment agriculture systems emerge as the technological solution to bridge this market demand-supply quality gap, offering the precision and control necessary to meet these evolving market requirements.

16.3 Controlled Environment Agriculture: Technology and Market Integration

16.3.1 Technological Foundations and Advantages

Controlled environment agriculture encompasses sophisticated production systems that precisely manage all growth variables including nutrients, water, light, temperature, humidity, and atmospheric composition. Hydroponic systems utilize nutrient-rich water solutions delivered to plant roots through various media including expanded clay, coir, perlite, and vermiculite, enabling optimized nutrient uptake and accelerated growth cycles [18].

Aeroponic systems represent a sophisticated advancement in soilless cultivation, where plant roots are suspended in air and nourished via nutrient-rich mist. This environment enhances root oxygenation, resulting in faster growth, increased metabolic activity, and greater biosynthesis of secondary metabolites the bioactive compounds critical to medicinal efficacy [19–21]. Studies on *Withania somnifera*,

Valeriana jatamansi, and *Artemisia annua* reveal significantly higher levels of compounds such as withanolides and valerenic acid under aeroponic conditions compared to soil or hydroponic cultivation.

Research demonstrates that controlled environment cultivation of medicinal plants results in superior quality profiles compared to field-grown alternatives. Studies by [6, 8] established that greenhouse hydroponic cultivation under controlled lighting produces more consistent bioactive compound concentrations year-round, addressing critical quality variability issues that compromise traditional cultivation approaches.

16.3.2 Market Size and Growth Projections

India's controlled environment agriculture sector demonstrates remarkable expansion, with the hydroponics market valued at USD 506.7 million in 2024 and projected to reach USD 2292.7 million by 2033, representing a compound annual growth rate of 16.91% [2]. Alternative projections suggest even more aggressive growth, with estimates ranging from USD 1.4 billion in 2022 to USD 5.3 billion by 2031 [22].

Aeroponic systems making up approximately 25% of vertical farming are increasingly used for high-value medicinal crops due to their ability to deliver controlled, high-density phytochemical production suitable for pharmaceutical and nutraceutical standards [23, 24].

16.3.3 Species Selection and Cultivation Success

Several high-value medicinal species have demonstrated exceptional performance in controlled environment systems. Notable examples include *Bacopa monnieri* (Brahmi) for cognitive enhancement applications, *Withania somnifera* (Ashwagandha) for adaptogenic products, and *Valeriana officinalis* (Valeriana) for natural sedative preparations. These species respond particularly well to optimized environmental conditions, producing enhanced biomass and elevated secondary metabolite concentrations [25].

The controlled nature of these systems enables manipulation of environmental stresses that trigger secondary metabolite production, allowing producers to optimize bioactive compound profiles for specific market applications. While Controlled Environment Agriculture (CEA) systems offer significant agronomic and environmental benefits such as improved biomass yield, enhanced phytochemical content, and reduced contamination the successful commercialization of CEA-grown medicinal plants requires strategically aligned marketing efforts. These must move beyond traditional agricultural promotion and focus on articulating clear value propositions that emphasize quality consistency, traceability, and sustainability. Effective marketing strategies are essential not only to justify premium pricing but also to foster consumer trust and market differentiation in increasingly competitive and regulated herbal product markets [15, 16, 23].

16.4 Strategic Marketing Framework

16.4.1 Marketing Conceptual Foundation

Marketing in the controlled environment medicinal plant sector requires sophisticated understanding of value creation, consumer behavior, and market positioning strategies. Philip As per Kotler marketing is defined as "*discovering, developing, and delivering value to meet the needs of specific market segments profitably*" provides foundational guidance for developing effective approaches in this specialized sector.

The marketing process encompasses comprehensive activities including market research, product development, pricing strategies, distribution channel optimization, and promotional campaigns. Success requires continuous adaptation to evolving consumer preferences, regulatory changes, and technological innovations that characterize the dynamic medicinal plant marketplace.

16.4.2 Concept and Definition

> Marketing is both a science and an art focused on discovering, developing, and delivering value to meet the needs of specific market segments profitably. It involves identifying unmet customer needs, evaluating the size and potential of those market opportunities, and determining which segments the business can serve most effectively. Marketing then involves designing and promoting suitable products or services to cater to those target customers (Philip Kotler).

Marketing is a structured and ongoing process that involves multiple coordinated activities such as market research, planning, budgeting, product development, and promotion (Fig. 16.1). These functions are implemented in a strategic sequence to address consumer needs and market dynamics, which collectively form what, is known as a marketing strategy. This strategy serves to connect a product or service with its target audience, guiding the transition from potential interest to active and loyal consumption. As long as a business operates, marketing remains a continuous effort aimed at maximizing value exchange between the firm and its customers [26].

In today's business environment, marketing has expanded beyond satisfying customers to delighting them. Companies now prioritize identifying valuable customer segments, nurturing loyalty, and building long-term relationships. This relationship-driven approach is central to Customer Relationship Management (CRM), which

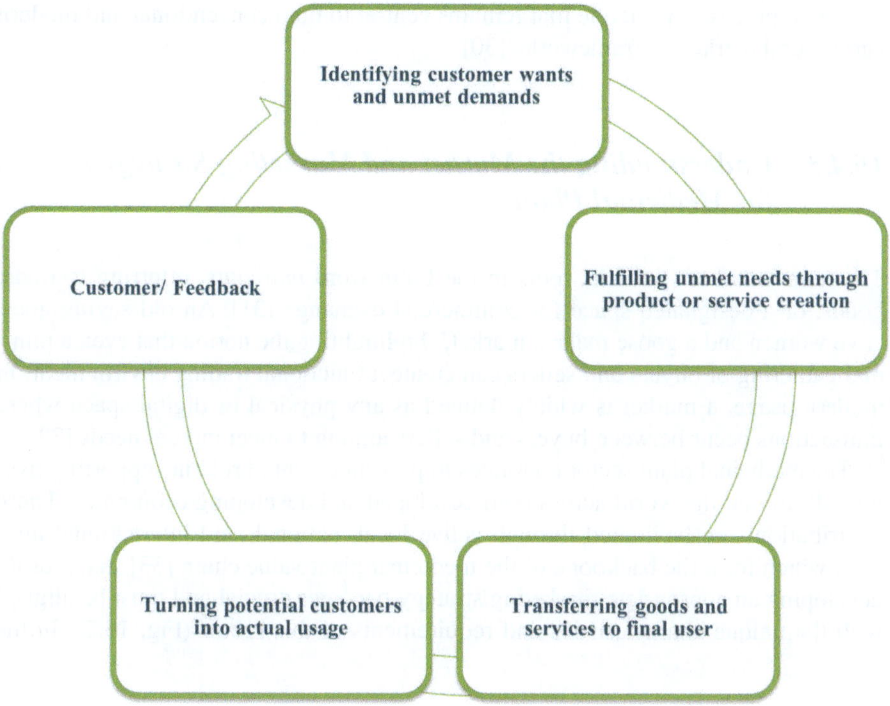

Fig. 16.1 Marketing as an ongoing-dynamic process

integrates data and communication to enhance customer experiences, ensure retention, and boost business growth [27].

Marketing has consistently served as a cornerstone of agricultural enterprises, functioning with significance comparable to core agronomic practices. While the use of quality planting material, scientifically validated cultivation techniques, and robust post-harvest measures are pivotal for enhancing the quality and quantity of farm outputs, the implementation of a well-planned marketing strategy is equally indispensable. The long-term sustainability and profitability of farming ventures depend not only on what is grown but also on how effectively products are positioned, promoted, and distributed within the market ecosystem [26].

Farms that adopt strategic marketing approaches such as direct-to-consumer models, intermediated sales, and regional market channels tend to exhibit greater economic resilience and financial viability, even though growth trajectories may vary by operational scale [28, 29]. In this context, agricultural marketing a term derived from agriculture and marketing refers to the full suite of services that facilitate the movement of farm products from producers to end consumers. This includes modernized value chains that incorporate controlled environment agriculture (CEA) systems, where produce flows from protected cultivation units to companies, processors, and ultimately, the consumer. As famously articulated by Philip Kotler, marketing is a human activity directed at satisfying needs and wants through an exchange process a principle that remains central to both conventional and modern agricultural marketing frameworks [30].

16.4.3 Understanding the Market and Marketing Strategies for Medicinal Plant

The term "market" finds its roots in the Latin word *marcatus*, referring to trade, goods, or a designated space for commercial exchange [31]. An old saying goes, "two women and a goose make a market," highlighting the notion that even a minimal gathering of buyers and sellers can create a functional trading environment. In modern usage, a market is widely defined as any physical or digital space where transactions occur between buyers and sellers aiming to meet mutual needs [32].

The medicinal plant sector continues to play an essential role in supporting livelihoods around the world across both developed and developing economies. These contributions are facilitated through active local, national, and international markets, which form the backbone of the medicinal plant value chain [33]. As a result, developing an appropriate marketing strategy becomes crucial and must be aligned with the unique characteristics and requirements of this sector (Fig. 16.2). In the

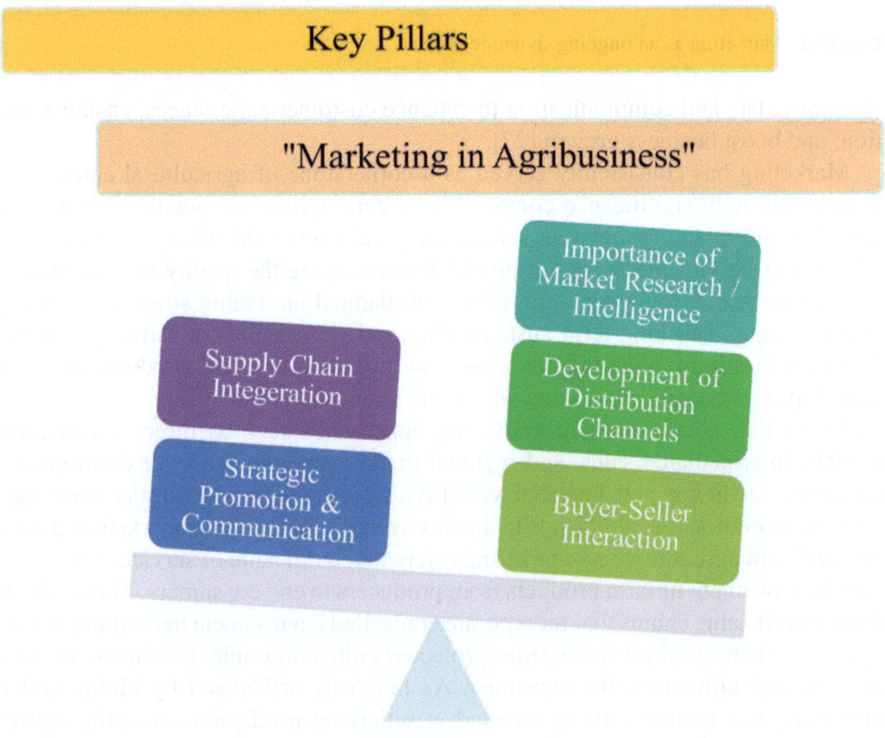

Fig. 16.2 Key pillars of Marketing in Agribusiness

context of strategic management, Moutinho [34] emphasizes that strategy formulation is about setting collective long-term goals that define an organization's future direction. These goals guide major business decisions, such as product development, market entry, regional expansion, diversification, acquisitions, and forming alliances like joint ventures. Such strategic initiatives typically demand high-level decision-making and substantial resource commitment, underscoring their significance to sustainable business growth.

Market strategies (Fig. 16.3) are essential tools for business expansion and competitiveness, especially in dynamic sectors such as medicinal plant trade and agri-based enterprises. These strategies are broadly classified into four categories: market penetration, market development, product development, and differentiation each serving a specific business objective based on market conditions and product life cycles [35].

1. **Market penetration strategy:** It focuses on increasing the sales volume of existing products within current markets. This approach aims to strengthen market share and enhance customer loyalty, particularly where growth potential still

Fig. 16.3 Marketing strategies

exists and the company's current presence is limited. It is commonly measured through the volume or value of goods exported to established markets.

2. **Market development strategy:** This strategy employed when the existing market becomes saturated or growth stagnates. Companies adopt this strategy to enter new geographic regions or target different customer segments with their existing product line. Its effectiveness is often evaluated by the number of new export markets accessed or expanded into.
3. **Product development strategy**: This is suited for businesses with the capacity to innovate. This involves designing and offering new products to the existing customer base, especially when demand for current products has fluctuated. Metrics such as the number of newly introduced product variants or brands are used to assess the success of this approach.
4. **Differentiation strategy:** It entails the development of entirely new products aimed at new markets. Though this strategy involves higher levels of risk and investment particularly in research, development, and marketing it enables firms to distinguish themselves from competitors and create new demand. Performance in this area is often gaged by the variety of packaging formats or product innovations introduced.

16.4.4 Market Segmentation, Targeting, and Positioning (STP Framework)

Segmenting, Positioning, and Targeting Theory: Market segmentation is an evolving process that shifts over time in response to changing consumer behavior, preferences, and market conditions. For an effective marketing strategy, the implementation of segmentation methods is crucial. In today's dynamic marketplace, market segmentation is a fundamental process that helps businesses stay responsive to evolving consumer needs, preferences, and behaviors. It involves dividing a broad consumer or business market into sub-groups of consumers based on shared characteristics. This process enables marketers to gain a structured understanding of the market and identify distinct groups that may respond differently to marketing strategies [36, 37].

After segmentation, companies engage in targeting, this involves evaluating each segment's attractiveness and selecting one or more to serve. Targeting decisions are influenced by factors such as market potential, competition, and the organization's strategic priorities. Since different analysts may approach segmentation differently, the resulting "market maps" often vary, reflecting alternate strategic perspectives [38, 39].

The final stage is positioning, where businesses aim to create a specific image of their product or service in the minds of the target audience. This is achieved by tailoring the marketing mix product features, pricing strategies, distribution methods, and promotional tools to align with the preferences of the targeted segment. A well-defined positioning strategy allows firms to clearly differentiate their offerings in competitive markets, improving customer perception and engagement [26, 34].

Together, Segmentation, Targeting, and Positioning (STP) (Fig. 16.4) provide a cohesive framework that guides marketers in designing effective strategies for identifying opportunities, reaching the right customers, and delivering relevant value propositions. When the principles of segmentation, targeting, and positioning are effectively applied, they form the foundation for marketing mix strategies aimed at maximizing product performance. The ultimate goal is to achieve customer satisfaction, which also supports a more accurate and demand-oriented sales forecast [40].

Inaccurate market segmentation can significantly undermine the effectiveness of decision-making, particularly when identifying target markets and establishing product positioning in the minds of consumers [41, 42]. Therefore, successful segmentation demands both skill and market knowledge, serving as a strategic foundation for any business. When applied effectively, segmentation enables a company/entrepreneur to identify and focus on specific market groups that are more likely to respond positively to its marketing efforts, thereby offering a competitive advantage.

To effectively reach and serve a chosen target market, a marketer must adopt both strategic thinking and creative approaches [40, 43]. Researches outline four essential criteria for identifying an optimal target market:

1. Responsiveness defines the selected market segment should respond positively to the company's product offerings and marketing strategies. If there is little or

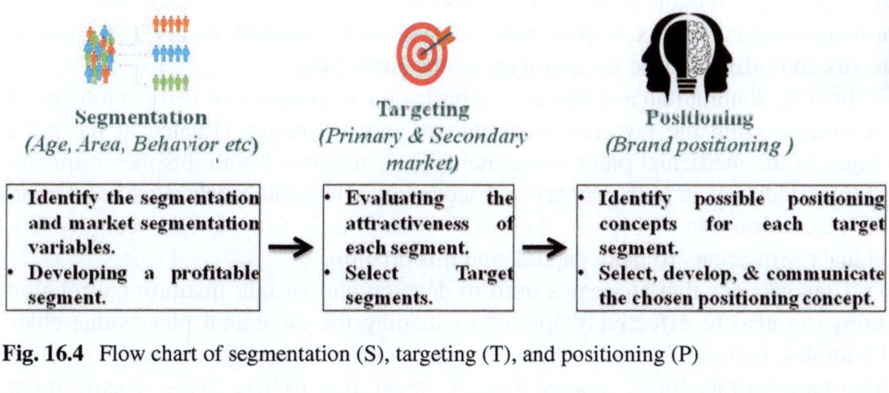

Fig. 16.4 Flow chart of segmentation (S), targeting (T), and positioning (P)

no response, it becomes essential to investigate the reasons behind this lack of engagement. Without a well-defined segmentation and a clearly identified target market, businesses risk directing their efforts inefficiently potentially leading to significant losses and missed opportunities.

2. Sales Potential refers to the overall capacity of a target market to generate revenue. A larger market often implies greater value, but this isn't based solely on population size. True potential also depends on the consumers' purchasing power and their willingness or desire to buy the product.
3. Adequate growth implies that market expansion is typically gradual at first. Over time, as awareness and demand increase, the growth rate accelerates before eventually stabilizing as the market reaches maturity.
4. Media reach refers to the ability to effectively connect with the target audience through appropriate communication channels. Selecting the right media ensures that marketing messages are delivered efficiently, maximizing visibility and engagement with the intended market segment.

16.4.5 Marketing Avenues in the Medicinal Plant Sector: (Traditional Approach of Marketing)

The supply chain (distribution chain) (Fig. 16.5) for medicinal plants is often extensive, involving multiple stages ranging from primary collectors and cultivators to local contractors, regional and large wholesale markets, and specialized traders. Each stage in this chain has unique requirements for cultivation, wild resource management, harvesting, processing, and especially marketing. In rural areas, medicinal plant enterprises often function with minimal intermediaries. Here, producers themselves take responsibility for multiple stages (harvesting, drying, and selling) often through direct markets or local fairs. However, in extended value chains catering to urban or export markets, responsibilities are more fragmented, involving formalized actors and infrastructure at each stage of the chain [44].

In [45], Shahidullah and Haque emphasize the importance of vertical integration in strengthening the position of producers and processors situated at the initial stages of the medicinal plant value chain. Their research highlights the dominance of intermediaries in both primary and secondary wholesale markets and challenges the perception that medicinal plant cultivation is only feasible for wealthier individuals with access to land, capital, and information.

They propose that strategies used to develop and sustain institutional relationships can also be effectively applied to shaping the medicinal plant value chain. Examples include formal contracts, quasi-vertical integration (involving close, long-term partnerships), tapered vertical integration (where firms procure inputs both externally from independent suppliers and internally), cost-plus agreements (where a contractor is compensated with a predetermined amount irrespective of actual costs), as well as joint ventures and strategic alliances [45]. In order to sustain

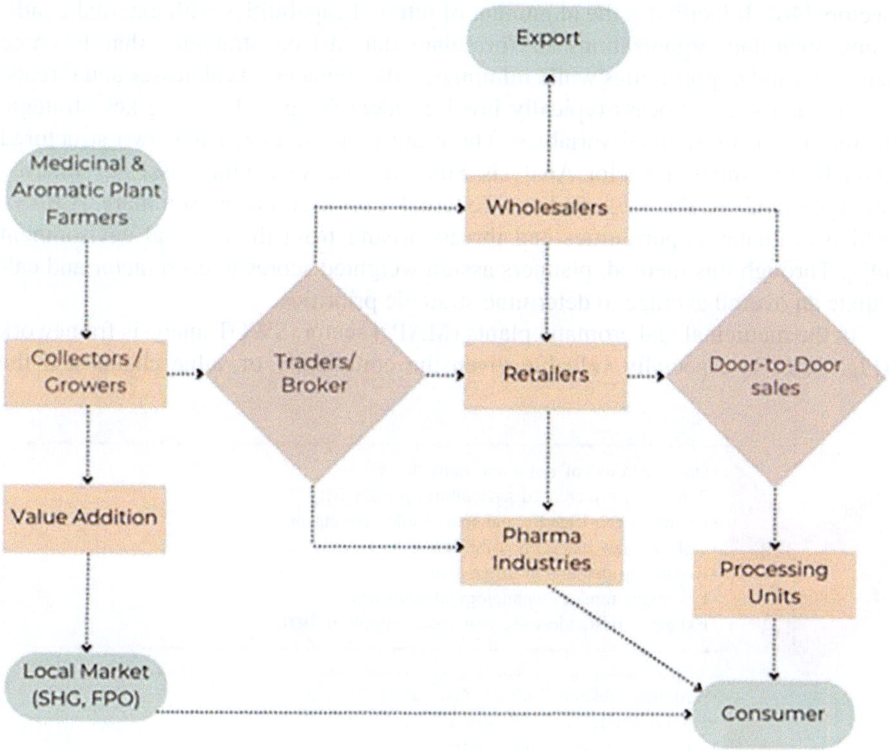

Fig. 16.5 Flow diagram of medicinal plant supply chain

growth in medicinal plant production, fair distribution of gross margins to primary producers is essential. The current value chain shows that most profits go to intermediaries due to structural weaknesses. A vertically integrated model limited to producers, processors, and supported by NGOs could create a more equitable and economically resilient system. Having established the theoretical foundations of marketing strategy and the STP framework, the practical application of these concepts to medicinal plant enterprises requires systematic analysis of internal capabilities and external market conditions. The SWOT analysis framework provides a structured approach to evaluate the strategic position of controlled environment medicinal plant operations.

16.4.6 SWOT Analysis Framework

Initially developed in the 1970s as a tool for business strategy, SWOT analysis an acronym for Strengths, Weaknesses, Opportunities, and Threats has since evolved into a versatile framework for strategic assessment and planning across diverse sectors [46]. It facilitates the alignment of internal capabilities with external conditions, enabling organizations to formulate data-driven strategies that leverage strengths and opportunities while minimizing the impact of weaknesses and threats.

The analytical process typically involves identifying and scoring key strategic factors based on defined variables. These are then organized into two structured models: the Internal Factor Analysis Summary (IFAS), which assesses internal strengths and weaknesses, and the External Factor Analysis Summary (EFAS), which evaluates opportunities and threats arising from the external environment [47]. Through this method, planners assign weighted scores to each factor and calculate an overall average to determine strategic priorities.

In the medicinal and aromatic plants (MAPs) sector, SWOT analysis framework (Fig. 16.6) is especially valuable given the complexity of value chains and the

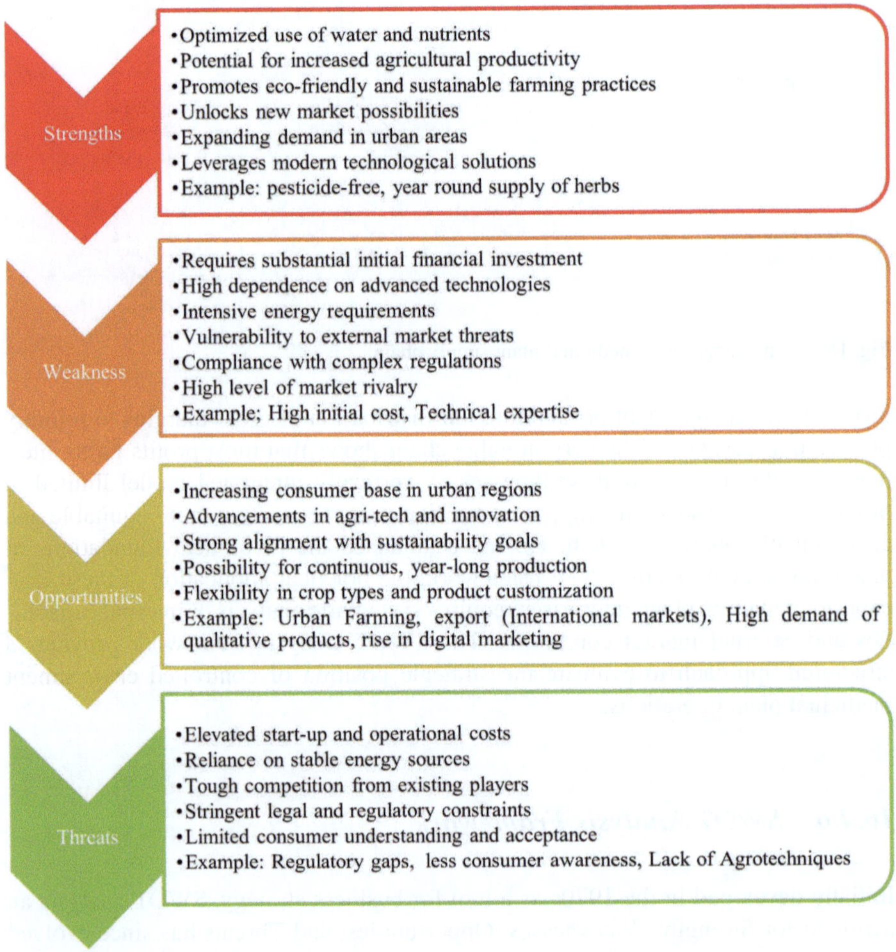

Fig. 16.6 SWOT analysis of soil-less cultivation farm business

diversity of stakeholders involved. A strategic SWOT assessment may reveal the following:

- **Strengths**: Agro-climatic advantages, biodiversity richness, indigenous knowledge of herbal medicine, and supportive government policies as exemplified in regions like the Western Ghats of Karnataka, India [48].
- **Weaknesses**: Overdependence on wild harvesting, lack of value-added processing infrastructure, inadequate cold storage facilities, and limited access to market intelligence [49].
- **Opportunities**: Global demand for herbal and natural remedies is growing at an estimated 7% annually, driven by increasing consumer preferences for organic, traceable, and eco-certified botanical products [12].
- **Threats**: Unsustainable harvesting, biodiversity degradation, market price instability, and vulnerability to climate-related risks such as floods and droughts [50].

The strategic application of SWOT supports policy formulation, enterprise development, and resource allocation. It enables stakeholders to develop models such as vertical integration, certification schemes, and value chain optimization tailored to specific crop systems. International examples further validate its relevance Iran and Germany's Baden-Württemberg region have successfully employed SWOT frameworks to guide species selection, regulatory planning, and investment in sustainable MAP production [35].

16.5 Marketing Mix Strategy

Marketers understand that it is impractical and unprofitable to serve every customer in a diverse marketplace equally. Therefore, they divide the market into distinct segments, target the most promising ones, and develop positioning strategies that attract new customers through perceived superior value while enhancing satisfaction and loyalty among existing clients [36] The marketing mix represents a fundamental framework in the field of marketing, encompassing a set of controllable tools that organizations use to influence consumer behavior and achieve desired outcomes in target markets. The term "marketing mix" was first introduced during a 1953 presidential address to the American Marketing Association. It was later conceptualized and popularized by [51, 52], who identified four key elements Product, Price, Place (distribution), and Promotion collectively known as the Four P's of marketing. These components serve as the cornerstone of marketing strategy, providing a practical structure for translating marketing plans into actionable outcomes. The main 4 P's Strategy is as follows:

- **Product Strategy:** A product is anything offered to the market to get the attention of buyers, users or consumers that can fulfill the user's desires or needs.

- **Price Strategy:** The price of a good or service is a determinant of market demand; price can influence a company's competitive position and also influence its sales market.
- **Place (Distribution Strategy):** Place which is located not far from consumers and hotels, restaurants and eateries so that distribution is easier from producers to consumers to meet their needs.
- **Promotion Strategy:** Promotional activities involves in businesses are very important activities which influence the marketing of the products.

Over time, the original Four Ps model was expanded to address the growing complexity of service-oriented and experiential markets. Researchers, extended the framework by adding three additional elements: People (Personnel), Process, and Physical Evidence. This broader model, often referred to as the 7Ps of Marketing, offers a more comprehensive perspective (Table 16.1), particularly relevant for sectors such as healthcare, education, hospitality, and herbal product marketing [53].

The 7Ps marketing mix framework provides the tactical foundation for implementing marketing strategies in controlled environment medicinal plant enterprises. However, understanding the broader market forces that drive adoption and acceptance of these innovative cultivation methods requires examination of the push and pull dynamics that influence both producer decisions and consumer behavior in the medicinal plant sector.

Table 16.1 7 P's of marketing mix–framework

7 P's framework		
P's	Definition	Medicinal plant Marketing Example (CEA system)
Product	Any offering that satisfies customer needs	Raw herbs, powders, oils, certified extracts (organic, GAP, Ayurvedic)
Price	Value charged, influencing demand and competitive positioning	Premium pricing for traceable, high potency extracts
Place	Distribution channels enabling product access	D2C platforms, B2B to pharma/wellness, online retailers, FPCs, export channels
Promotion	Communication and branding activities	Digital campaigns, "clean and green" branding, e-commerce strategies, traceability messaging
People	Individuals delivering value to customers	Trained staff, agronomists, harvest teams ensuring quality control
Process	Systems and procedures ensuring efficient service delivery	SOP-based harvesting, lab testing, logistics, CRM systems
Physical evidence	Tangible elements that reinforce product quality and brand trust	Certifications, packaging with traceability labels, facility hygiene and branding aesthetics

16.6 Push and Pull Marketing Dynamics Driving Market Adoption

The expansion of the medicinal and aromatic plants (MAPs) sector is largely driven by two forces: pull and push factors (Fig. 16.7) of marketing [54].

- **Pull Factors:** refer to incentives that attract farmers toward MAP cultivation over conventional crops. These include higher market prices, pre-established supply chains, price assurance by intermediaries, and strong market demand for specific species.
- **Push Factors:** arise from declining profitability and unpredictability in traditional farming due to fluctuating yields, poor market infrastructure, and income instability, pushing farmers to seek alternative crops such as medicinal plants.

In this context, direct marketing (aligned with push strategies) as defined [55] involve farmers selling directly to consumers via farmers' markets, herbal fairs, or farm shops. This approach helps them bypass intermediaries, secure better margins, and maintain control over pricing. Conversely, indirect marketing (aligned with pull strategies) as explained by Shehzadi et al. [56] includes promotional tactics such as awareness campaigns, digital content about plant benefits, trade exhibitions, and social media. These activities help build long-term demand and trust for niche medicinal crops, effectively pulling both consumers and farmers into the MAP market

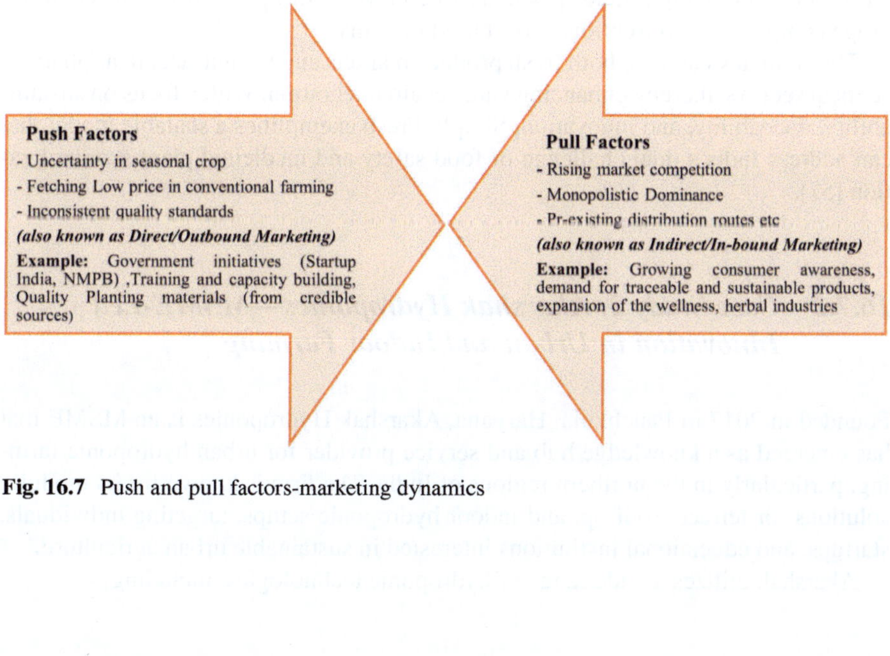

Fig. 16.7 Push and pull factors-marketing dynamics

16.7 Case Studies

These case studies demonstrate the current state of successful implementation in controlled environment medicinal plant cultivation and marketing. The strategies employed by Simply Fresh, Akarshak Hydroponics, and Saffron Tech provide valuable insights into effective approaches, while also revealing emerging trends that will shape the future landscape of this rapidly evolving sector. Building on these foundational successes, the next phase of development will be characterized by technological advancement, market sophistication, and strategic innovation.

16.7.1 Case Study 1: Simply Fresh—Revolutionizing Herbal Cultivation with Hydroponics

Simply Fresh is a state-of-the-art hydroponic agritech enterprise headquartered in Telangana, India, and represents a benchmark in the integration of controlled environment agriculture (CEA) and medicinal plant cultivation. Established with the mission of producing nutrient-rich, chemical-free, and traceable herbs, the company has developed a "Farm-in-a-Box" model an AI-driven, sensor-integrated hydroponic solution that controls every variable from humidity to nutrient delivery. Operating across a 150-acre campus with 22 acres under protected cultivation**,** Simply Fresh cultivates a range of MAPs including thyme, lavender, oregano, rosemary, sage, cinnamon, and Ashwagandha**.** Its precision agriculture approach enables the production of high-yield, pesticide-free crops with up to 90% reduced water usage compared to conventional soil-based systems.

The company caters to both fresh produce markets and the nutraceutical/pharmaceutical **sectors**, thereby enhancing value chain integration. With a focus on sustainability, traceability, and innovation**,** Simply Fresh exemplifies a scalable model that can address India's dual challenge of food safety and medicinal plant standardization [57].

16.7.2 Case Study 2: Akarshak Hydroponics—MSME-Led Innovation in Urban and Indoor Farming

Founded in 2017 in Panchkula, Haryana**,** Akarshak Hydroponics is an MSME that has emerged as a knowledge hub and service provider for urban hydroponic farming, particularly in the northern regions of India. The firm is known for its modular solutions for terrace, rooftop, and indoor hydroponic setups**,** targeting individuals, startups, and educational institutions interested in sustainable urban agriculture**.**

Akarshak utilizes a wide range of hydroponic technologies, including:

- **NFT (Nutrient Film Technique)**—for leafy greens.
- **DWC (Deep Water Culture)** and **Floating Rafts**—for root crops.
- **Dutch Buckets and Grow Bags**—for vine crops.
- **Aeroponics**—for high-value medicinal crops.

The company has applied these systems to the cultivation of medicinal plants like saffron, turmeric, and ginger, showcasing the potential of hydroponics for high-value, climate-sensitive crops. Their model demonstrates up to 90% water savings, minimal land use, and year-round production, positioning hydroponics as a viable solution for urban food security and medicinal herb cultivation in space-constrained settings [58].

16.7.3 Case Study 3: Saffron Tech: Pioneering Aeroponic Innovation in Medicinal Plant Cultivation

Saffron Tech, established in 2020, is an innovative agritech enterprise leading the transformation of saffron production in India. Traditionally grown under specific climatic conditions, saffron has long been a challenging and high-value medicinal crop. Saffron Tech has overcome these limitations by adopting advanced aeroponic and Controlled Environment Agriculture (CEA) systems to facilitate multi-cycle, year-round cultivation of pharmaceutical-grade saffron.

Based in India, the company has developed a proprietary soil-free aeroponic model that delivers nutrient-rich mist directly to plant roots. This system is integrated with AI-enabled **automation** to precisely regulate environmental variables such as temperature, humidity, light, and nutrient flow. The result is a dramatic increase in efficiency with up to 44 time's higher yield and more than 90% water savings compared to traditional methods.

Saffron Tech's diverse product portfolio includes:

- High-quality saffron spice for culinary and retail markets.
- Standardized saffron extracts for nutraceuticals and cosmetics.
- Proprietary branded wellness products.
- Technology licensing for vertical farms and greenhouse operators.

Beyond product innovation, the company is addressing critical challenges in India's Medicinal and Aromatic Plants (MAPs) value chain particularly traceability, adulteration, and resource constraints. By promoting a clean-label, tech-enabled cultivation model, Saffron Tech is setting new benchmarks in sustainability, scalability, and supply chain integrity.

With ongoing expansion through international collaborations, Saffron Tech aims to replicate its aeroponic model for other high-value herbs, offering a blueprint for the future of soilless medicinal plant farming in both domestic and global markets [59].

16.8 Future Prospects Innovations: Marketing Strategies

With rising global demand for herbal and plant-based wellness products, marketing strategies for medicinal plants cultivated via soilless methods (e.g., hydroponics, aeroponics) are evolving rapidly. Future-focused marketing will rely heavily on technology, sustainability branding, consumer trust, and value chain integration.

(a) **Digital-Driven Marketing:** The future of marketing in this niche lies in digital transformation including the use of social media, e-commerce, and content-driven inbound strategies to reach niche markets.

- Inbound marketing (blogs, storytelling, SEO, traceability content) is more effective than traditional push marketing in promoting eco-friendly herbal products [60]. Platforms like Instagram, YouTube, and health blogs help build consumer trust in "clean-label" medicinal herbs grown via soilless methods).

(b) **Value Chain Integration and B2B Models:** Marketing will increasingly involve backward and forward integration:

- Contract farming models with wellness brands or pharma firms ensure steady demand and quality compliance [61]. Business-to-business (B2B) strategies, including white-label supply and cluster-based production, allow scale-up while maintaining product traceability and identity.

(c) **Consumer-Centric Customization:** Growing awareness of personalized herbal medicine will shape marketing:

- Soilless cultivation enables customized growth of plants with specific active compounds [e.g., high-bacoside (Brahmi)]. Marketing tailored to niche segments (e.g., immunity boosters, adaptogens, women's health) will enhance positioning.

(d) **AI, Data & Predictive Marketing**

- AI-driven customer segmentation, predictive analysis of demand trends, and real-time market intelligence will play a central role in decision-making [41]. Platforms like Google Trends and AI dashboards can help farmers and marketers forecast shifts in demand for specific botanicals.

16.9 Conclusion

The commercialization of medicinal plants has progressed far beyond traditional supply chains and informal marketplaces. Driven by rising global demand for natural health products, Ayurvedic remedies, and herbal supplements, marketing strategies in this sector must now evolve to reflect the growing complexities of consumer expectations, regulatory frameworks, and technological advancements.

This chapter highlights that both traditional and controlled environment agriculture (CEA), including hydroponics and aeroponics have critical roles in the medicinal plant value chain, but they demand distinctly different marketing approaches. Traditional models often depend on mandis, middlemen, and informal trade routes, which can reduce producer margins and compromise quality. On the other hand, soilless cultivation systems integrated with traceability tools, quality certifications, and digital platforms offer higher levels of control, product consistency, and consumer trust.

Effective marketing in this evolving landscape requires a hybrid and strategic approach, including:

- Segmented targeting through Segmentation, Targeting, Positioning (STP) strategies;
- A strong value proposition built on purity, efficacy, and sustainability;
- Tailored implementation of the 7Ps of marketing, especially within niche B2B and B2C herbal segments;
- Robust supply chain management from farm to consumer, ensuring product integrity and delivery efficiency.

Additionally, marketing strategies for medicinal plants like other agricultural commodities must account for the unique characteristics of each crop, including its shelf life, perish ability, seasonality, and target market. Unlike durable industrial goods, medicinal and perishable crops are vulnerable to quality degradation and market loss if not sold in time. Therefore, market research must be crop-specific and consider consumer demographics, usage patterns, and channel preferences.

In rural settings, traditional sales channels such as local markets, regional mandis, and **cooperatives** remain prevalent. In contrast, urban consumers are increasingly accessing herbal and agricultural products via modern retail outlets like Reliance Fresh, BigBasket**,** JioMart, and Nature's Basket. These platforms provide enhanced visibility, quality assurance, and packaging innovation. The emergence of e-commerce tools, digital marketplaces, and social media platforms is also transforming the agrimarketing landscape, enabling farmers and producers to connect directly with end consumers, bypassing traditional intermediaries and enhancing margins.

Furthermore, the future growth of this sector depends on several non-marketing enablers:

- Consumer education to raise awareness about medicinal plant benefits and usage;
- Policy support to streamline regulations, certifications, and exports;
- Infrastructure development such as cold chains and rural connectivity; and.
- Economic feasibility studies to ensure the sustainability of cultivation and distribution systems.

In essence, marketing medicinal plants today demands a hybrid approach one that harmonizes traditional agricultural wisdom with modern cultivation technologies, digital engagement, and intelligent market research. The path forward lies in creating transparent, traceable, high-quality, and sustainable value chains that meet the expectations of both domestic and global consumers. With coordinated efforts from farmers, marketers, researchers, and policymakers, biodiversity-rich nations like India can position themselves as global leaders in the herbal products market driving not only economic growth but also contributing meaningfully to public health and global sustainability goals.

References

1. India Brand Equity Foundation (n.d.) AYUSH industry overview. Retrieved June 27, 2025, from https://www.ibef.org/industry/ayush
2. IMARC Group (2024) India hydroponics market: industry trends, share, size, growth, opportunity and forecast 2024–2033. https://www.imarcgroup.com/india-hydroponics-market
3. Olsen CS (1998) The trade in medicinal and aromatic plants from central Nepal to northern India. *Econ Bot*:279–292
4. Kuniyal CP, Kuniyal PC, Butola JS, Sundriyal RC (2013) Trends in the marketing of some important medicinal plants in Uttarakhand, India. Int J Biodivers Sci Ecosyst Serv Manage 9(4):324–329
5. Kumar MR, Janagam D (2011) Export and import pattern of medicinal plants in India. Indian J Sci Technol 4(3):245–248
6. Dorais M, Papadopoulos AP, Leonhart S, Gosselin A, Gaudreau L (2000) Soilless greenhouse production of medicinal plants in north eastern Canada. Acta Hort 554:297–303
7. Léonhart S, Pedneault K, Gosselin A, Angers P, Papadopoulos AP, Dorais M (2000, November) Diversification of greenhouse crop production under supplemental lighting by the use of new cultures with high economic potential. In: IV international ISHS symposium on artificial lighting 580, pp 249–254
8. Pedneault K, Léonhart S, Gosselin A, Papadopoulos AP, Dorais M, Angers P (2002) Variations in concentration of active compounds in four hydroponically and field grown medicinal plant species. Acta Hort (ISHS) 580:255–262
9. Zamani S, Fathi M, Ebadi MT, Máthé Á (2025) Global trade of medicinal and aromatic plants. A review. J Agric Food Res 101910
10. Export-Import Bank of India (EXIM Bank) (2021) Indian Medicinal Plant Sector: potential and challenges. EXIM Bank Occasional Paper No. 203. Retrieved from https://www.eximbankindia.in
11. NMPB (National Medicinal Plants Board) (2023) Annual report 2022–23. Ministry of AYUSH, Government of India. Retrieved from https://nmpb.nic.in
12. FAO (Food and Agriculture Organization of the United Nations) (2019) Medicinal and aromatic plants: trade and sustainable use. FAO, Rome. Retrieved from https://www.fao.org
13. Kumar R, Tewari AK. Isolation of medicinally important constituents from rare and exotic medicinal plants. InSynthesis of Medicinal Agents from Plants 2018 Jan 1 (pp. 229–256). Elsevier.
14. Vasisht K, Sharma N, Karan M (2016) Current perspective in the international trade of medicinal plants material: an update. Curr Pharm Des 22(27):4288–4336
15. Graamans L, Baeza E, Van Den Dobbelsteen A, Tsafaras I, Stanghellini C (2018) Plant factories versus greenhouses: comparison of resource use efficiency. Agr Syst 160:31–43
16. Gómez C, Currey CJ, Dickson RW, Kim HJ, Hernández R, Sabeh NC et al (2019) Controlled environment food production for urban agriculture. HortScience 54(9):1448–1458

17. Shamshiri R, Kalantari F, Ting KC, Thorp KR, Hameed IA, Weltzien C et al (2018) Advances in greenhouse automation and controlled environment agriculture: a transition to plant factories and urban agriculture. Int J Agric Biolog Eng 11(1):1–22
18. Shrestha A, Dunn B (2010) Hydroponics production systems. Oklahoma Cooperative Extension Service
19. Kumar S, Singh A, Pandey S, Kumari N (2020) Aeroponic systems for sustainable crop production: a review. Int J Agric Environ Biotechnol 13(4):567–578
20. Partap M, Kumar P, Kumar A, Joshi R, Kumar D, Warghat AR (2020) Effect of elicitors on morpho-physiological performance and metabolites enrichment in Valeriana jatamansi cultivated under aeroponic conditions. Front Plant Sci 11:01263
21. Paponov M, Ziegler J, Paponov IA (2023) Light exposure of roots in aeroponics enhances the accumulation of phytochemicals in aboveground parts of the medicinal plants Artemisia annua and Hypericum perforatum. Front Plant Sci 14:1079656
22. DataM Intelligence (2023) India hydroponics market—size, share, trends, and forecast (2022–2031). https://www.datamintelligence.com/research-report/india-hydroponics-market
23. Sharma N, Acharya S, Kumar K, Singh N, Chaurasia OP (2018) Hydroponics as an advanced technique for vegetable production: An overview. J Soil Water Conserv 17(4):364–371
24. Mehandru P, Shekhawat NS, Rai MK, Kataria V, Gehlot HS (2014) Evaluation of aeroponics for clonal propagation of Caralluma edulis, Leptadenia reticulata and Tylophora indica–three threatened medicinal Asclepiads. Physiol Mol Biol Plants 20:365–373
25. Tabatabaie SJ, Azarmi F, Dadpour MR, Nazemiyeh H, Nazari J (2006, February) Evaluation of valerian and lemon verbena production using different soilless and soil cultivations under glasshouse. In: VIII international symposium on protected cultivation in mild winter climates: advances in soil and soilless cultivation under 747, pp 465–470
26. Kotler P, Keller KL (2016) Marketing management, 15th edn. Pearson Education
27. Buttle F, Maklan S (2019) Customer relationship management: Concepts and technologies, 4th edn. Routledge
28. Bauman et al. (2018) Agricultural and Resource Economics Review — on how direct and intermediated marketing affect farm income
29. National Research Council (2010) Toward sustainable agricultural systems in the 21st century. The National Academies Press. https://doi.org/10.17226/12832
30. Kotler P (1973) The major tasks of marketing management. J Market 37(4):42–49. https://doi.org/10.1177/002224297303700407
31. Oxford English Dictionary (2024) Market. Retrieved from https://www.oed.com/
32. Kotler P, Armstrong G (2020) Principles of marketing, 18th edn. Pearson Education
33. Olsen CS, Helles F (2009) Market efficiency and benefit distribution in medicinal plant markets: empirical evidence from South Asia. Int J Biodiv Sci Manage 5(2):53–62
34. Moutinho L (2000) Segmentation, targeting, positioning and strategic marketing. Strategic management in tourism, pp 121–166. Sharma U, Barupal M, Shekhawat NS, Kataria V (2018) Aeroponics for propagation of horticultural plants: an approach for vertical farming. Hortic Int J 2: 443–444
35. Karim MH, Karbasi A, Mohammadzadeh SH (2020) Marketing strategies and export of Iranian medicinal plants. J Med Plants By-Prod 9(1)
36. Armstrong PK (2006) Principle of marketing, 11th edn. Prentice-Hall, Upper Saddle River
37. Dibb S (1998) Market segmentation: strategies for Success. Mark Intell Plan 16(7):394–406. https://doi.org/10.1108/02634509810244390
38. Zimmerman A, Blythe J (2018) Business-to-business marketing management: strategies, cases, and solutions. Routledge
39. Bowen JT (1998) Market segmentation in hospitality research: no longer a sequential process. Int J Contemp Hospit Manage 10(7):289–296. https://doi.org/10.1108/09596119810240859
40. Kasali R (2005) Membidik pasar Indonesia: Segmentasi, targeting, positioning. PT Gramedika Pustaka Utama, Jakarta

41. Roy A (2025) Marketing 5.0: technology for humanity: Philip Kotler, Hermawan Kartajaya, and Iwan Setiawan. Wiley, Hoboken. 2021. 206 pp. ISBN 978-1119668510. List price: $28.00. Hardcover
42. Lasmini. (2008) Segmenting, targeting, dan positioning kopi bubuk Bali Banyuatis di Kota Denpasar (Master's thesis). Program Studi Agribisnis, Program Pascasarjana, Universitas Udayana, Denpasar
43. Clancy KJ, Shulman RS (1991) The marketing revolution: a radical manifesto for dominating the marketplace. HarperBusiness, New York
44. Hishe M, Asfaw Z, Giday M (2016) Review on value chain analysis of medicinal plants and the associated challenges. J Med Plants Stud 4(3):45–55
45. Shahidullah AKM, Haque CE (2010) Linking medicinal plant production with livelihood enhancement in Bangladesh: implications of a vertically integrated value chain. J Transdiscip Environ Stud 9(2):1
46. David FR (2017) Strategic management: a competitive advantage approach, 16th edn. Pearson
47. Lantarsih C, Nugroho A, Pramono R (2023) SWOT analysis in sustainable agroforestry planning. IOP Conf Ser: Earth Environ Sci 1143:012070
48. FAO & IUCN (2003) Biodiversity and the ecosystem approach in agriculture, forestry and fisheries—SWOT of Karnataka's MAP sector.
49. TRAFFIC & FAO (2016) Conservation and sustainable use of medicinal plants: problems, progress, and prospects. Chin Med 11(37). https://doi.org/10.1186/s13020-016-0108-7
50. FAO (2016) Medicinal plants: current status and future perspectives. Food and Agriculture Organization of the United Nations
51. McCarthy EJ, Perreault WD Jr (1960) Basic marketing: a managerial approach. Irwin, Homewood
52. Donnelly J, George W (1981) Marketing strategies and organization structures for service firms. Technical report, American Marketing Association
53. Awan WA, Salam A, Inam SG (2015) The impact of marketing mix (5 Ps) elements on sales of UPS: a case of Karachi Market–Buyer's perspective. VFAST Trans Educ Social Sci 3(1):50–56
54. Ramana Rao DGS (2023) Challenges and strategies of marketing of medicinal plants. Int J Horticul Agric Food Sci 7(5):06–09
55. Murrow M, Hyman MR (1994) The ethics of database marketing. J Business Ethics 13(11):919–926. https://doi.org/10.1007/BF00871970
56. Shehzadi H, Ahmad S, Iqbal A, Nawaz MM (2010) Impact of advertisement on brand awareness and consumer preference: a case study of soft drinks industry in Southern Punjab–Pakistan. J Business Manage 3(2):1–10
57. Simply Fresh (n.d.) Simply Fresh: pesticide-free, traceable produce grown in high-tech farms. Retrieved June 27, 2025, from https://www.simplyfresh.co.in/
58. Akarshak Hydroponics (2024) Hydroponics [Web page]. Retrieved June 27, 2025, from https://akarshakfarms.com/hydroponics/
59. Saffron Tech. (2024). Where tradition meets innovation. Retrieved June 27, 2025, from https://saffron-tech.ag/
60. Chaffey D, Ellis-Chadwick F (2019) Digital marketing. Pearson uk
61. Acharya SS, Agarwal NL (2004) Agricultural marketing in India, 4th edn. Oxford & IBH Publishing

MIX
Papier aus verantwortungsvollen Quellen
Paper from responsible sources
FSC® C105338

If you have any concerns about our products, you can contact us on
ProductSafety@springernature.com

In case Publisher is established outside the EU, the EU authorized representative is:
Springer Nature Customer Service Center GmbH
Europaplatz 3, 69115 Heidelberg, Germany

Printed by Libri Plureos GmbH
in Hamburg, Germany